Anatomy and Physiology Laboratory Textbook

Anatomy and Physiology Laboratory Textbook

Third Edition   Short Version

Harold J. Benson
Pasadena City College

Stanley E. Gunstream
Pasadena City College

Arthur Talaro
Pasadena City College

Kathleen P. Talaro
Pasadena City College

wcb

Wm. C. Brown Company Publishers

Dubuque, Iowa

Book Team

Edward G. Jaffe, Editor
Barbara J. Grantham, Designer
Maxine Kollasch, Production Editor
Shirley M. Charley, Visual Assistant
Mavis M. Oeth, Permissions Editor

wcb
group

Wm. C. Brown, Chairman of the Board
Mark C. Falb, Executive Vice President

wcb

Wm. C. Brown Company Publishers, College Division

Lawrence E. Cremer, President
David Wm. Smith, Vice-President, Marketing
E. F. Jogerst, Vice-President, Cost Analyst
David A. Corona, Assistant Vice-President/Production Development and Design
James L. Romig, Executive Editor
Marcia H. Stout, Marketing Manager
Marilyn A. Phelps, Manager of Design
William A. Moss, Production Editorial Manager
Mary M. Heller, Visual Research Manager

The cover photograph, a product of computer graphics at Lawrence Livermore, shows a transparent DNA molecule with an ethidium ion between the center two base pairs. The print has Newton's rings, produced by the glass mount used in printing from the half-frame 35mm original. The photograph was produced by Nelson L. Max, John Blunden, and George Watchmaker of Lawrence Livermore National Laboratory, Livermore, Calif.

2-04739-02

Contents

Contents

Preface

The basic format of this third edition is essentially the same as the first and second editions; its content, however, has been considerably broadened to accommodate the needs of students in a wider variety of disciplines. Since instructors in nursing, medical assisting, X-ray technology, dental assisting, dental hygiene, and several other paramedical programs have found the previous editions of considerable value to their students, we have responded to their varied suggestions for changes in content.

Originally, this shortened version satisfied the requirements of a very simple course in anatomy and physiology with primary emphasis on anatomy. While this pleased a large group of instructors, a much larger group decried the paucity of physiological experimentation. In the second edition the physiology portion was expanded somewhat, but the clamor for more physiology has persisted, and with this edition we have succumbed to further expansion of the physiology content.

In addition to broadening the physiology offering, we have added several new exercises that pertain to cytology, anatomy, and instrumentation. New concepts in endocrinology, neurology, and immunology have been incorporated where necessary. Some terminology has been changed to conform to broader usage and many illustrations have been replaced for greater clarity of understanding.

The principal application of this manual is for full-year courses in which students meet three hours per week in the laboratory. Since all cat anatomy of the complete version has been deleted it is assumed that dissections will be confined to the rat and certain animal parts (sheep and cattle). If one or more cadavers are available in the laboratory the instructor may elect to omit the animal parts dissections, relying entirely on cadaver studies.

Some instructors may wish to use this book in single-semester courses in which the primary emphasis in lecture is placed on physiology. Where the time span is so short and the subject matter so extensive, there is a procedure that works quite well. To effectively utilize the limited amount of laboratory time, students are expected to label all anatomical illustrations prior to entering the laboratory. This maneuver reserves all laboratory time for dissections and physiological experimentation. Since dissections are considerably outnumbered by physiological experiments, the majority of laboratory time will, thus, be devoted to physiology. Another ploy that works well in such courses is to use a lecture textbook that is devoted entirely to physiological principles; the laboratory manual, then, becomes the anatomy text. With this scenario the student will learn most of the anatomy by labeling illustrations, doing dissections, and answering the questions on the Laboratory Report sheets. A course taught in this manner requires considerable effort on the part of the student. An expectation of this sort is not unusual in programs where a great deal of information must be assimilated in a short period of time.

This shorter version of the Anatomy and Physiology Laboratory Textbook differs from the complete version in several respects other than size and the absence of cat anatomy. Not only are there fewer exercises (57 vs. 80), but there are also some basic differences in the content of some of the exercises. For one thing, some anatomical minutiae have been deleted in certain sections. This applies primarily to exercises on the skeleton and muscles; result: fewer muscles and skeletal structures are included in this book. Another topic that is less comprehensive in this book is instrumentation. In the complete version a whole section of five exercises is devoted to instrumentation; only one instrumentation exercise of eight pages was needed for this version.

The two senior authors of this book feel greatly strengthened by the support and contributions of Arthur and Kathleen Talaro. In addition, the four of us have received considerable assistance from our colleague, Dr. Terry Pavlovitch, here at Pasadena City College. Our combined teaching experiences, plus the assistance of many physiologists and anatomists in other institutions, have done much to mold this book into its present form. As with previous editions, we welcome your contributions and criticisms.

Critical Readers

Richard Steadman, Clark County Community College
Kathy Vogel, University of New Mexico
William Kleinelp, Middlesex County College
Dr. Terry Pavlovich, Pasadena City College

Introduction

These laboratory exercises have been developed to provide you with a basic understanding of anatomical and physiological principles that underlie medicine, nursing, dentistry, and other related health professions. Laboratory procedures that reflect actual clinical practices are included wherever feasible. In each exercise you will find essential terminology that will become part of your working vocabulary. Mastery of all concepts, vocabulary, and techniques will provide you with a core of knowledge crucial to success in your chosen profession.

During the first week of this course your instructor will provide you with a schedule of laboratory exercises in the order of their performance. There is an implied expectation that you will have familiarized yourself with the content of each experiment prior to that week's session, thus ensuring that you will be properly prepared so as to minimize disorganization and mistakes.

The *Laboratory Reports* coinciding with each exercise are located at the back of the book. They are perforated for easy removal; be sure to remove each sheet as necessary. This will facilitate data collection, completion of answers, and grading. Your instructor may give further procedural details on the handling of these reports.

The exercises in this laboratory guide consist essentially of four kinds of activities: (1) illustration labeling; (2) anatomical dissections; (3) physiological experiments; and (4) microscopic studies. The following suggestions should be helpful in performing these assignments.

Labeling The activity of labeling illustrations is essentially a determination of your understanding of the written text. Since all labeled structures are explicitly described in the manual, all that is necessary is to read the manual very carefully. Incorrectly labeled illustrations usually indicate a lack of comprehension.

Once the illustrations are labeled they can be useful to you in other ways. First, the illustrations may be used for reference purposes in dissections or examinations of anatomical specimens. This is particularly true in the skeletal and nervous systems. Secondly, the illustrations can be used for review purposes. If the number legend of the labels is covered over as you mentally attempt to name the structures on the illustration you can easily determine your level of understanding. Periodic review of this type during the semester will be very helpful.

In general, the labeling of illustrations will usually be performed *prior* to entering the laboratory. In this way the laboratory time will be used primarily for dissections, experimentation, or microscopic examinations.

Dissections Rats, frogs, and mammalian animal parts (sheep and cattle) will be used for various dissections. Cadavers or parts of cadavers may also be available. If cadavers are conservatively used and not dissected, handle them carefully to avoid unnecessary mutilation.

When using live animals in experimental procedures it is imperative that they be handled with great care. Consideration must be exercised to minimize pain in all experiments on vertebrate animals. Inconsiderate or haphazard treatment of any animal will not be tolerated.

Physiological Experiments Before performing any physiological experiments be sure that you understand the overall procedure. Reading over the experiment prior to entering the laboratory will help a great deal.

Handle all instruments carefully. Most pieces of equipment are expensive, may be easily damaged, and are often irreplaceable. The best insurance against breakage or damage is to thoroughly understand how the equipment is expected to function.

Maintain astuteness in observation and record keeping. Record data immediately; postponement detracts from precision. Insightful data interpretation will also be expected.

Microscopic Studies Cytological and histological studies will be made to lend meaning to text descriptions. An illustration in a textbook of a specific type of tissue loses its abstractness with the study of the actual material under the microscope. If drawings are required, execute them with care. Label those structures that are significant.

Laboratory Efficiency Success in any science laboratory situation requires a few additional disciplines:

1. Always follow the instructor's verbal comments at the beginning of each laboratory session. It is at this time that difficulties will be pointed out, group assignments will be made, and procedural changes will be announced. Take careful notes on substitutions or changes in methods or materials.

2. Don't be late to class. Since so much takes place during the first ten minutes, tardiness can cause confusion. If you are tardy don't expect the instructor to be very helpful.

3. Keep your work area tidy at all times. Books, bags, purses, and extraneous supplies should be located away from the work area. Tidiness should also extend to assembly of all apparati.

4. Abstain from eating, drinking, or smoking within the confines of the laboratory.

5. Report immediately to the instructor any injuries that occur.

6. Be serious-minded and methodical. Horseplay, silliness, or flippancy will not be tolerated during experimental procedures.

7. Work independently, but cooperatively, in team experiments. Attend to your assigned responsibility, but be willing to lend a hand where necessary. Participation and development of laboratory techniques are an integral part of the course.

Laboratory Reports When seeking answers to the questions and problems on the Laboratory Reports, work independently. The effort you expend to complete these reports is as essential as doing the experiment. The easier route of letting someone else solve the problems for you will handicap you at examination time. You are taking this course to learn anatomy and physiology. No one else can learn it for you.

Part 1 Some Fundamentals

The student who has had a previous course in biological science has gained much that will be helpful in the study of anatomy and physiology. Basic terminology, microscopy, and the study of cells and tissues are common to all of the introductory courses in the life sciences. It is with these units that this first portion of the manual is concerned.

When you start Exercise 1 scan the entire exercise before attempting to label any of the drawings. Note that the first two pages consist of descriptive terminology, followed by an Assignment which describes what you are to label. This, in turn, is followed by additional descriptive text and another Assignment on page 7. After all the drawings have been labeled the Laboratory Report sheet at the back of the manual should be completed. This will be the procedure to follow on all laboratory exercises.

Exercise 2 is an introduction to all the organ systems of the body. A freshly killed rat will be dissected to study the organ systems. This dissection will provide an opportunity for you to examine organs and tissues in a condition that closely resembles actual life.

Exercise 3 presents the fundamental principles of microscopy. This exercise should be studied prior to coming to class so that you have a good understanding of the operation of this instrument prior to using it.

Exercises 4 and 5 relate to cellular anatomy and physiology. Exercise 6 ("The Integument") is a logical extension of the preceding experiments since the skin contains some of the tissues studied in Exercise 5.

Exercise 7 pertains to some of the mechanics of the transmission of substances across the cell membrane.

Anatomical Terminology, Body Cavities and Membranes

Anatomical description would be extremely difficult without specific terminology. A consensus prevails among many students that anatomists synthesize multisyllabic words in a determined conspiracy to harass the beginner's already overburdened mind. Naturally, nothing could be further from the truth.

Scientific terminology is created out of necessity. It functions as a precise tool which allows people to say a great deal with a minimum of words. Conciseness in scientific discussion not only saves time, but it usually promotes clarity of understanding as well.

Most of the exercises in this laboratory manual employ the terms defined in this exercise. They are used liberally to help you to locate structures that are to be identified on the illustrations. If you do not know the exact meanings of these words, obviously you will be unable to complete the required assignments. First of all, read over all of the material carefully; then read the specific assignment for this exercise.

Relative Positions

Descriptive positioning of one structure with respect to another is accomplished with the following pairs of words. Their Latin derivations are provided to help you understand their meanings.

Superior and Inferior These two words are used to denote vertical levels of position. The Latin word *super* means *above;* thus, a structure that is located above another one is said to be superior. Example: The nose is *superior* to the mouth.

The Latin word *inferus* means *below* or *low;* thus, an inferior structure is one that is below or under some other structure. Example: The mouth is *inferior* to the nose.

Anterior and Posterior Fore and aft positioning of structures are described with these two terms. The word anterior is derived from the Latin *ante,*

meaning *before.* A structure that is anterior to another one is in front of it. Example: Bicuspids are *anterior* to molars.

Anterior surfaces are the most forward surfaces of the body. The front portions of the face, chest, and abdomen are anterior surfaces.

Posterior is derived from the Latin *posterus,* which means *following.* The term is the opposite of anterior. Example: The molars are *posterior* to the bicuspids.

Cephalad and Caudad These terms are sometimes used in place of anterior and posterior. The word cephalad is derived from the Latin *cephalicus,* meaning head. The word caudad is derived from the Latin *cauda,* meaning tail. These words are frequently used in describing the position of structures in the cat, dog, or other four-legged animals.

Dorsal and Ventral These terms, as used in comparative anatomy of animals, assume all animals, including man, to be walking on all fours. The dorsal surfaces are thought of as *upper* surfaces, and the ventral surfaces as *underneath* surfaces.

The word *dorsal* (Latin: *dorsum,* back) not only applies to the back of the trunk of the body, but may also be used in speaking of the back of the head and the back of the hand.

Standing in a normal posture, man's dorsal surfaces become posterior. A four-legged animal's back, on the other hand, occupies a superior position.

The word *ventral* (Latin: *venter,* belly) generally pertains to the abdominal and chest surfaces. However, the underneath surfaces of the head and feet of four-legged animals are also often referred to an ventral surfaces. Likewise, the palm of the hand may also be referred to as being ventral.

Proximal and Distal These terms are used to describe parts of a limb with respect to the point of reference such as the attachment of the appendage to the trunk of the body. *Proximal,* (Latin:

proximus, nearest) refers to that part of the limb nearest to the point of attachment. Example: The upper arm is the *proximal* portion of the arm.

Distal (Latin: *distare,* to stand apart) means just the opposite of proximal. Anatomically, the distal portion of a limb or other part of the body is that portion of the structure that is most remote from the point of reference (attachment). Example: The hand is *distal* to the arm.

Medial and Lateral These two terms are used to describe surface relationships with respect to the median line of the body. The *median line* is an imaginary line on a plane which divides the body into right and left halves.

The term *medial* (Latin: *medius,* middle) is applied to those surfaces of structures that are closest to the median line. The medial surface of the arm, for example, is the surface next to the body because it is closest to the median line.

As applied to the appendages, the term *lateral* is the opposite of medial. The Latin derivation of this word is *lateralis,* which pertains to *side.* The lateral surface of the arm is the outer surface, or that surface furthest away from the median line. The sides of the head are said to be lateral surfaces.

Body Sections

To observe the structure and relative positions of internal organs it is necessary to view them in sections that have been cut through the body. Considering the body as a whole, there are only three planes to identify. Figure 1.1 shows these three sections.

Sagittal Sections A section parallel to the long axis of the body (longitudinal section) which divides the body into right and left sides is a *sagittal section.* If such a section divides the body into equal halves, as in figure 1.1, it is said to be a *midsagittal section.*

Frontal Sections A longitudinal section which divides the body into front and back portions is a *frontal* or *coronal section.* The other longitudinal section seen in figure 1.1 is of this type.

Transverse Sections Any section which cuts through the body in a direction which is perpendicular to the long axis is a *transverse* or *cross section.*

tion. This is the third section which is shown in figure 1.1. It is parallel to the ground.

Although these sections have been described here only in relationship to the body as a whole, they can be used on individual organs such as the arm, finger, or tooth.

Assignment:

To test your understanding of the above descriptive terminology, identify the labels in figures 1.1 and 1.2 by placing the correct numbers in front of the terms to the right of each illustration. Also, record these numbers on the Laboratory Report.

Regional Terminology

Various terms such as flank, groin, brachium, and hypochondriac have been applied to specific regions of the body to facilitate localization. Figures 1.3 and 1.4 pertain to some of the more predominantly used terminology.

Trunk

The anterior surface of the trunk may be subdivided into two pectoral, two groin, and the abdominal regions. The upper chest region may be designated as **pectoral** or **mammary** regions. The anterior trunk region not covered by the ribs is the **abdominal** region. The depressed area where the thigh of the leg meets the abdomen is the **groin.**

The posterior surface or dorsum of the trunk can be differentiated into the costal, lumbar, and buttocks regions. The **costal** (Latin: *costa,* rib) portion is that part of the dorsum which lies over the rib cage. The lower back region between the ribs and hips is the **lumbar** or **loin** region. The **buttocks** are the rounded eminences of the rump formed by the gluteal, muscles; also called the **gluteal** region.

The side of the trunk which adjoins the lumbar region is called the **flank.** The armpit region which is between the trunk and arm is the **axilla.**

Arm

To differentiate the parts of the arm the term **brachium** is used for the upper arm and **antebrachium** for the forearm (between the elbow and wrist). The elbow area which is on the posterior surface of the

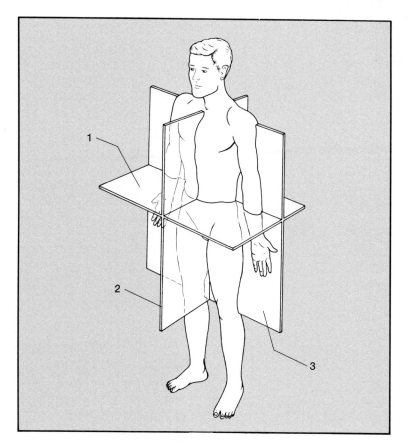

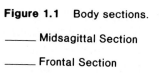

Figure 1.1 Body sections.

_____ Midsagittal Section

_____ Frontal Section

_____ Transverse Section

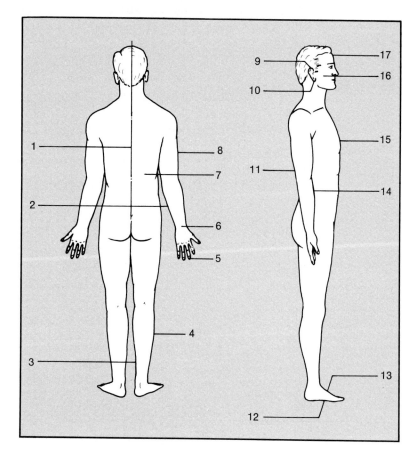

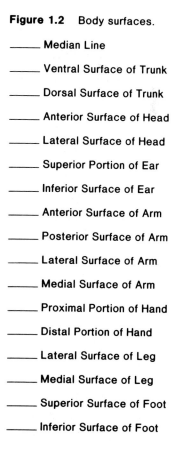

Figure 1.2 Body surfaces.

_____ Median Line

_____ Ventral Surface of Trunk

_____ Dorsal Surface of Trunk

_____ Anterior Surface of Head

_____ Lateral Surface of Head

_____ Superior Portion of Ear

_____ Inferior Surface of Ear

_____ Anterior Surface of Arm

_____ Posterior Surface of Arm

_____ Lateral Surface of Arm

_____ Medial Surface of Arm

_____ Proximal Portion of Hand

_____ Distal Portion of Hand

_____ Lateral Surface of Leg

_____ Medial Surface of Leg

_____ Superior Surface of Foot

_____ Inferior Surface of Foot

arm is the **cubital** area. That area on the opposite side of the elbow is the **antecubital** area. It is also correct to refer to the entire anterior surface of the antebrachium as being antecubital.

Leg and Foot

While the upper portion of the leg is designated as the **thigh,** the lower fleshy posterior portion is called the **calf.** Between the thigh and calf on the posterior surface, opposite to the knee, is a depression called the **ham** or **popliteal** region. The sole of the foot is the **plantar** surface.

Abdominal Divisions

The abdominal surface may be divided into quadrants or into nine distinct areas. To divide the abdomen into nine regions one must establish four imaginary planes: two that are horizontal and two that are vertical. These planes and areas are shown in figure 1.4. The **transpyloric plane** is the upper horizontal plane which would pass through the lower portion of the stomach (pyloric portion). The **transtubercular plane** is the other horizontal plane which touches the top surfaces of the hipbones (iliac crests). The two vertical planes, or **right** and

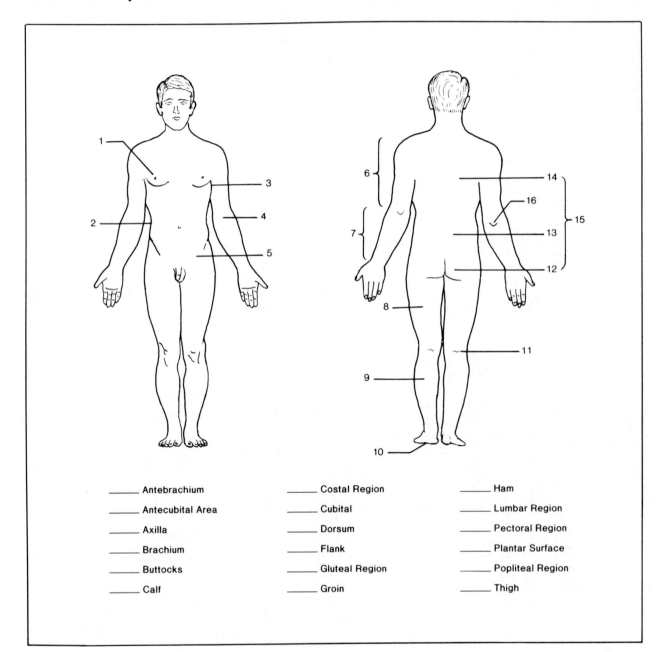

_____ Antebrachium	_____ Costal Region	_____ Ham
_____ Antecubital Area	_____ Cubital	_____ Lumbar Region
_____ Axilla	_____ Dorsum	_____ Pectoral Region
_____ Brachium	_____ Flank	_____ Plantar Surface
_____ Buttocks	_____ Gluteal Region	_____ Popliteal Region
_____ Calf	_____ Groin	_____ Thigh

Figure 1.3   Specific regional terminology.

left lateral planes, are approximately halfway between the midsagittal line and the crests of the hips.

The above planes describe the following nine areas. The **umbilical** region lies in the center, includes the navel, and is bordered by the two horizontal and two vertical planes. Immediately above the umbilical area is the **epigastric,** which covers much of the stomach. Below the umbilical zone is the **hypogastric** or *pubic area.* On each side of the epigastric is a right and left **hypochondriac.** Beneath the hypochondriac areas are the right and left **lumbar** areas. On each side of the hypogastric area are the right and left **iliac** areas.

Assignment:
Label figures 1.3 and 1.4 and transfer these numbers to the Laboratory Report.

Body Cavities
Figure 1.5 illustrates seven principal cavities of the body. The two major cavities are the dorsal and the ventral cavities. The **dorsal cavity,** which is nearest to the dorsal surface, includes the cranial and spinal cavities. The **cranial cavity** is the hollow portion of the skull that contains the brain. The **spinal cavity** is a long tubular canal within the vertebrae which contains the spinal cord. The **ventral cavity** is the large cavity which encompasses the chest and abdominal regions.

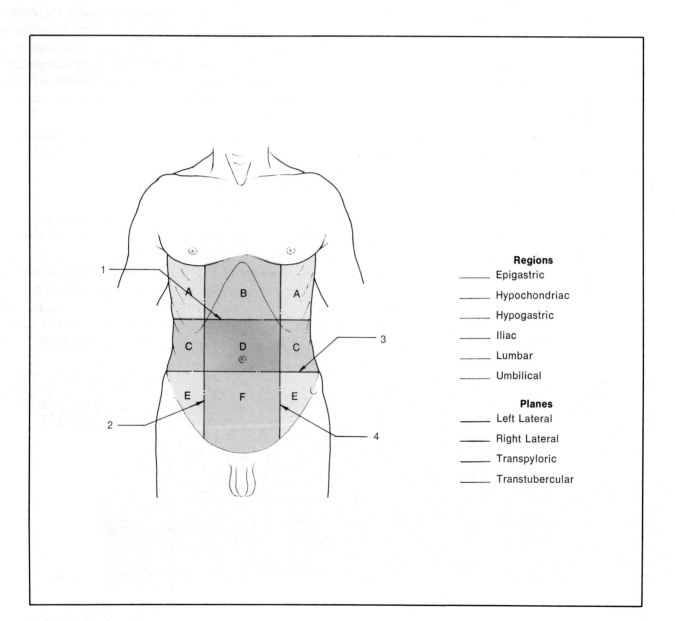

Regions
_____ Epigastric
_____ Hypochondriac
_____ Hypogastric
_____ Iliac
_____ Lumbar
_____ Umbilical

Planes
_____ Left Lateral
_____ Right Lateral
_____ Transpyloric
_____ Transtubercular

Figure 1.4 Abdominal regions.

The upper and lower portions of the ventral cavity are separated by a dome-shaped thin muscle, the **diaphragm.** The thoracic cavity which is that part of the ventral cavity superior to the diaphragm is separated into right and left compartments by a membranous partition or septum called the **mediastinum.** The lungs are contained in these right and left compartments. The heart, trachea, esophagus, and thymus gland are enclosed within the mediastinum. Figure 1.6 reveals the relationship of the lungs to the structures within the mediastinum. Note that within the thoracic cavity there exist right and left **pleural cavities** which contain the lungs, and a **pericardial cavity** which contains the heart.

The **abdominopelvic cavity** is that portion of the ventral cavity which is inferior to the diaphragm. It consists of two portions: the abdominal and pelvic cavities. The **abdominal** cavity contains the stomach, liver, gall bladder, pancreas, spleen, kidneys, and intestines. The **pelvic cavity** is the most inferior portion of the abdominopelvic cavity which contains the urinary bladder, sigmoid colon, rectum, uterus, and ovaries.

Body Cavity Membranes

The body cavities are lined with *serous membranes* which provide a smooth surface for the enclosed internal organs. Although these membranes are quite thin they are strong and elastic. Their surfaces are moistened by a self-secreted *serous fluid* which facilitates ease of movement of the viscera against the cavity walls.

Thoracic Cavity Membranes

The membranes that line the walls of the right and left thoracic compartments are called **parietal**

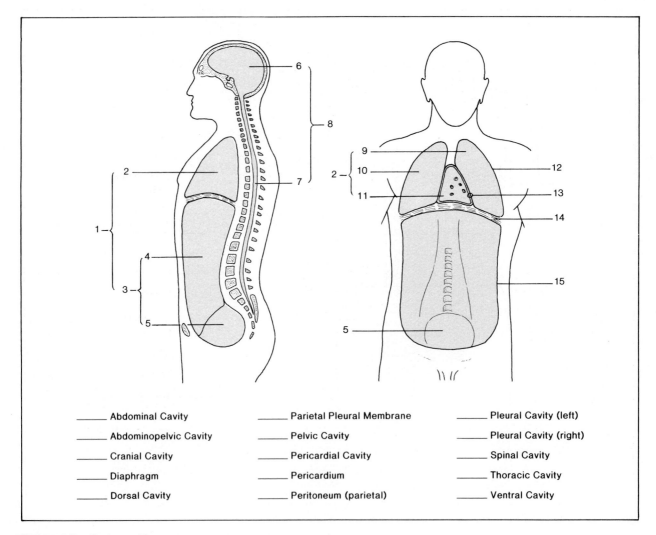

_____ Abdominal Cavity	_____ Parietal Pleural Membrane	_____ Pleural Cavity (left)
_____ Abdominopelvic Cavity	_____ Pelvic Cavity	_____ Pleural Cavity (right)
_____ Cranial Cavity	_____ Pericardial Cavity	_____ Spinal Cavity
_____ Diaphragm	_____ Pericardium	_____ Thoracic Cavity
_____ Dorsal Cavity	_____ Peritoneum (parietal)	_____ Ventral Cavity

Figure 1.5 Body cavities.

pleurae. The lungs, in turn, are covered with **visceral (pulmonary) pleurae.** Note in figure 1.6 that these pleurae are continuous with each other and the mediastinum. The potential cavity between the parietal and visceral pleurae is the **pleural cavity.** Inflammation of the pleural membranes results in a condition called *pleurisy.*

Within the broadest portion of the mediastinum lies the heart. It, like the lungs, is covered by a thin serous membrane, the **visceral pericardium,** or **epicardium.** Surrounding the heart is a double-layered fibroserous sac, the **parietal pericardium.** The inner layer of this sac is a serous membrane that is continuous with the epicardium of the heart. Its outer layer is fibrous, which lends considerable strength to the structure. A small amount of serous fluid produced by the two serous membranes lubricates the surface of the heart to minimize friction as it moves within the parietal pericardium. The potential space between the visceral and parietal pericardia is called the **pericardial cavity.**

Abdominal Cavity Membranes

The serous membrane of the abdominal cavity is the **peritoneum.** It does not extend deep down into the pelvic cavity, however; instead its most inferior boundary extends across the abdominal cavity at a level which is just superior to the pelvic cavity. The top portion of the urinary bladder is covered with peritoneum. In addition to lining the abdominal cavity, the peritoneum has double-layered folds called **mesenteries,** which extend from the body wall to the viscera, holding these organs in place. These mesenteries contain blood vessels

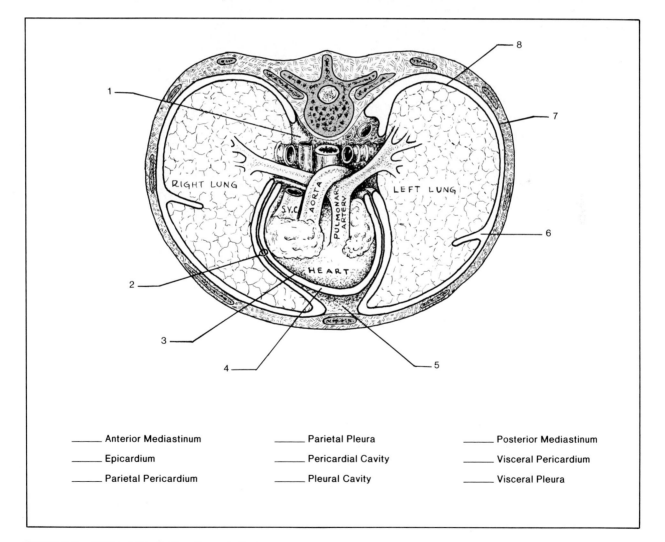

RIGHT LUNG

LEFT LUNG

S.V.C.

AORTA

PULMONARY ARTERY

HEART

_____ Anterior Mediastinum	_____ Parietal Pleura	_____ Posterior Mediastinum
_____ Epicardium	_____ Pericardial Cavity	_____ Visceral Pericardium
_____ Parietal Pericardium	_____ Pleural Cavity	_____ Visceral Pleura

Figure 1.6 Transverse section through thorax.

and nerves that supply the viscera enclosed by the peritoneum. That part of the peritoneum attached to the body wall is the **parietal peritoneum.** The peritoneum that covers the visceral surfaces is **visceral peritoneum.** The potential cavity between the parietal and visceral peritoneums is called the **peritoneal cavity.** Organs such as the kidneys, which lie posterior to the parietal peritoneum, are said to be **retroperitoneal.**

Figure 1.7 reveals how the various abdominal organs and the body wall are covered by peritoneal membranes. Note that like the pleurae of the thoracic cavity, the parietal peritoneum, visceral peritoneum, and mesenteries are continuous with each other.

Assignment:

Label figures 1.5, 1.6, and 1.7.

Laboratory Report

After transferring all the labels from figures 1.1 through 1.7 to the proper columns on the Laboratory Report, answer all the questions.

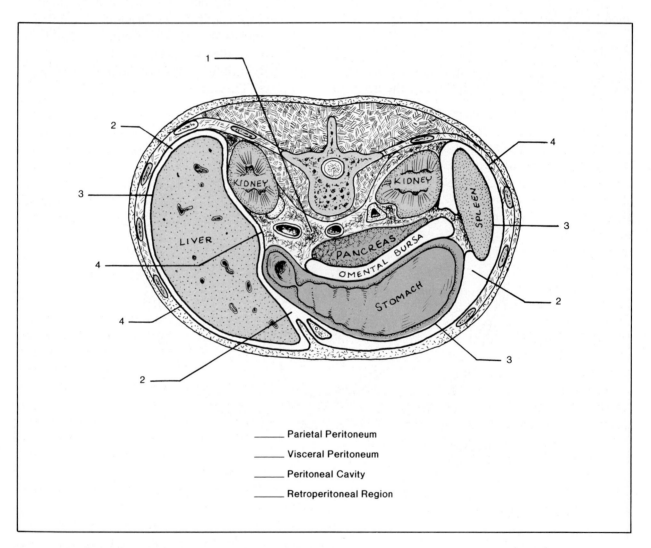

_____ Parietal Peritoneum

_____ Visceral Peritoneum

_____ Peritoneal Cavity

_____ Retroperitoneal Region

Figure 1.7   Transverse section through the abdominal cavity.

Gross Structures: Organ Systems

Various physiological activities, such as breathing, digestion, and removal of waste materials, are performed by specific organs of different systems of the body. Each organ is unique in design to accomplish its task. By definition, an **organ** is described as *a structure composed of two or more tissues which performs an essential physiological function.* The heart, for example, consists of muscle, epithelial, and other tissues. It is an organ which acts as a pump to move blood throughout the blood vessels of the body. A **system,** on the other hand, is a *group of organs that are directly related to each other functionally.* The circulatory system, which includes the heart, also includes the arteries, veins, capillaries, and spleen. Its primary concern is with the overall distribution of food, gases, hormones, and other materials to all parts of the body. There are eleven such groups of organs (organ systems) in the body.

In this exercise we will dissect a freshly killed rat to study the organs of most of these systems. This will be a cursory type of observation intended just to familiarize you with major structures and body cavities. No attempt will be made to seek anatomical minutiae at this time.

In preparation for this dissection you should read over the introductory material concerning the eleven systems and complete the questions on the Laboratory Report. This report should be turned in to the instructor at the beginning of the laboratory period.

The Integumentary System

The Latin word *integumentum* means covering. The surface of the body which consists of skin, hair, and nails comprises the integumentary system.

The skin consists of two layers: an inner layer, called the **dermis,** and an outer layer, the **epidermis.** These layers are also present in the lining of the mouth, indicating that the mucous membrane of the oral region is a part of this sytem.

The integumentary system performs many functions. One of its prime responsibilities is to prevent bodily invasion by harmful miscroorganisms. Chemical substances in perspiration and sebum (oil from sebaceous glands) are antibacterial.

The skin also aids in temperature regulation and excretion. The evaporation of perpsiration cools the body. The fact that perspiration contains the same excretory products found in urine indicates that the kidneys are aided by the skin in the elimination of water, salts, and some nitrogenous wastes from the blood.

As long as the skin is intact, the internal environment is protected: serious damage to the skin by burns, for example, may result in serous fluid and electrolyte imbalances.

The Skeletal System

The skeletal system forms a solid framework around which the body is constructed. It consists of bones, cartilage, and ligaments. This system provides support and protection for the softer parts of the body. Delicate organs such as the lungs, heart, brain and spinal cord are protected by the bony enclosure of the skeletal system.

In addition to protection, the bones provide points of attachment for muscles which act as levers when the muscles contract. This makes movement possible.

Another very important function of the skeletal system is the production of various types of blood cells. The central hollow portion of the bones contains yellow marrow. Red marrow, which is present in the porous ends of the bone, gives rise to red blood cells and certain types of white blood cells.

The Muscular System

Attached to the skeletal framework of the body are muscles that make up nearly half the weight of the body. These skeletal muscles consist primarily of long multinucleated cells. The ability of these cells to shorten, when stimulated by nerve impulses, enables the muscles to move parts of the body in walking, eating, breathing, and other activities.

The Nervous System

The adjustment to external and internal environmental changes is the function of the nervous system. It is the most highly organized system of the body, accomplishing its function by means of sensory receptors, conduction pathways, and interpretation centers.

The **sensory receptors,** which are activated by various kinds of stimuli, may be internal or external. They may be affected by pressure, heat, chemicals, light, sound, and other stimuli. The ears, eyes, nose, tongue, and skin contain external receptors *(exteroceptors).* Receptors in the deeper tissues and organs are *interoceptors.*

The **conduction pathways,** which carry the nerve impulses, may be located in nerves, the spinal cord, or brain. Nerve cell fibers, acting much like wires in a telephone system, carry the messages from the receptors through these pathways to centers or switchboards of the nervous system.

Interpretation centers are located in the brain. They receive these nerve cell messages and make us aware of environmental conditions. Correct responses, which may be muscular, are then achieved by the nervous system via outgoing conduction pathways.

The Circulatory System

The circulatory system provides transportation of various materials from one part of the body to another. Digested food that is absorbed through the intestinal wall must be carried to various parts of the body for utilization. Oxygen in the lungs must be delivered to all cells and carbon dioxide, produced by those same cells, must be carried back to the lungs. Hormones produced in various endocrine glands can reach their destinations only via the circulatory system. Even the transportation of heat from the muscles to the surface of the

body for cooling is achieved by the circulatory system.

In addition to the transportation of food, gases, and hormones, the blood is the body's primary defense against microbial invasion. The presence of phagocytic (cell-eating) white blood cells, antibodies and special enzymes in the blood prevents invading organisms from destroying the body.

The circulatory system includes the heart, arteries, veins, capillaries, spleen, and blood: **Arteries** are thick-walled vessels that carry blood *from the heart* to microscopic **capillaries** of all organs of the body. **Tissue fluid,** containing nutrients and oxygen, leaves the blood through the capillary walls and passes into the spaces between the cells. Some substances, such as carbon dioxide, diffuse back into the capillaries from the tissue cells. The tissue fluid is picked up by the lymphatic system and carried back to the circulatory system. **Veins** are large blood vessels that carry the blood *from the organs back to the heart.* Veins differ from arteries in that they are thinner walled and have valves to prevent reversal of blood flow.

The **spleen** is an oval structure on the left side of the abdominal cavity that acts to some extent as a blood reservoir. It also plays an important role in the removal of fragile red blood cells.

The Lymphatic System

The lymphatic system is, essentially, a part of the circulatory system in that its prime function is to return tissue fluid to the blood. It is also involved in the absorption of fats from the intestines. Whereas carbohydrates and proteins pass directly into the blood from the intestines, fat reaches the blood indirectly by being absorbed first into the lymphatic system. This system consists of a network of lymphatic vessels in the legs, arms, and head that empty into a pair of large veins, the subclavians, in the neck region.

Before **lymph** (name for tissue fluid after it enters the lymphatic vessels) is returned to the blood, it passes through nodules of lymphoid tissue called **lymph nodes.** This lymphoid tissue contains phagocytic cells of the reticuloendothelial system that remove bacteria and other foreign materials from the lymph, thus purifying it before it is returned to the blood. The **reticuloendothelial system (RES)** consists of fixed phagocytic cells that are located throughout the body in lymphoid tissue of structures such as the liver, spleen, thymus, adenoids, appendix, Peyer's patches, and bone marrow.

The spleen is as much a part of the lymphatic system as of the circulatory system, since it functions in both systems. Histologically, it resembles lymph nodes in that it contains lymphoidal tissue; however, instead of filtering lymph it filters blood, removing worn out blood cells.

The Respiratory System

The respiratory system consists of two portions: (1) the air passageways and (2) the respiratory portion. The actual exchange of gases between the blood and the air in the lungs occurs in the respiratory portion. The lungs contain many tiny sacs called **alveoli,** which greatly increase the surface area for the transfer of oxygen and carbon dioxide in breathing. The passageways consist of the **nasal cavity, nasopharynx, larynx, trachea,** and **bronchi.**

The Digestive System

The preparation of food for absorption into the blood or lymph is accomplished by the digestive system. Mechanical breakdown of the food begins in the mouth with the chewing action of the teeth. As the food is mixed with saliva, chemical breakdown, or digestion, starts. *Enzymes* in gastric, pancreatic, and intestinal juices eventually complete this process, reducing the food to molecular sizes small enough to be absorbed into the blood and lymph. The parts of the digestive system are the **mouth, salivary glands, esophagus, stomach, small intestine, large intestine, rectum, pancreas,** and **liver.** An understanding of this system includes food chemistry, mechanisms of food movement, kinds and actions of enzymes, control of enzyme production, mechanisms of absorption, and elimination.

The Urinary System

Cellular metabolism produces waste materials such as carbon dioxide, excess water, nitrogenous wastes, and excess metabolites. Nitrogenous wastes include such materials as urea, uric acid, and creatinine. The removal of these substances from the body is a function of various systems.

There are four channels for the elimination of these unneeded materials from the body: the kidneys, the skin, the lungs, and the large intestine. The skin, lungs, and intestine are parts of other systems which function only secondarily in excretion. The **kidneys** are the essential excretory organs of the urinary system; consequently, they are concerned with the removal of the bulk of these wastes. In addition to the kidneys, this system includes the two **ureters** which drain the kidneys, the **urinary bladder** which receives and stores urine from the kidneys, and the **urethra** which drains the bladder.

The Endocrine System

Small masses of glandular tissue which secrete hormones directly into the circulatory system make up the endocrine system. These clumps of cells that are thoroughly permeated by blood capillaries are called **endocrine glands.** Their secretions, the *hormones,* perform various duties such as (1) integrating various physiological activities; (2) regulating the growth of the skeleton, muscles, viscera, etc.; (3) directing the differentiation and maturation of the ovaries, testes, and secondary sex characteristics; and (4) regulating specific enzymatic reactions.

The glands of the endocrine system do not form a physical integrated system in continuity, as do the organs of other systems. The individual glands are widely separated in different regions of the body, to form a system only from a functional standpoint.

The endocrine system includes the following glands: pituitary, thyroid, parathyroid, thymus, adrenal, pancreas, pineal, ovaries, and testes. It even includes certain tissue of the placenta and the digestive tract lining.

The Reproductive System

Continuity of the species is the function of the reproductive system. Spermatozoa are produced in the testes of the male and ova are produced by the ovaries of the female. In addition to the testes the male reproductive system is comprised of the **scrotum, penis, accessory glands,** and **ducts.** The female reproductive system includes the **vagina, uterus, Fallopian tubes,** and **accessory glands.**

Laboratory Report

Complete the Laboratory Report for this exercise. It is due at the beginning of the laboratory period in which the rat dissection is scheduled.

Rat Dissection

You will work in pairs to perform this part of the exercise. Your principal objective in the dissection is to *expose the organs for study, not to simply cut up the animal.* Most cutting will be performed with scissors. Whereas the scalpel blade will be used only occasionally, the flat blunt end of the handle will be used frequently for separating tissues.

Materials:

> freshly killed rat
> dissecting pan (with wax bottom)
> dissecting kit
> dissecting pins

Skinning the Ventral Surface

1. Pin the four feet to the bottom of the dissecting pan as illustrated in figure 2.1. Before making any incision examine the oral cavity. Note the large **incisors** in the front of the mouth that are used for biting off food particles. Force the mouth open sufficiently to examine the flattened **molars** at the back of the mouth. These teeth are used for grinding food into small particles. Note that the **tongue** is attached at its posterior end. Lightly scrape the surface of the tongue with a scalpel to determine its texture. The roof of the mouth consists of an anterior **hard palate** and a posterior **soft palate.** The throat is the **pharynx,** which is a component of both the digestive and respiratory systems.

2. Lift the skin along the mid-ventral line with your forceps and make a small incision with scissors as shown in figure 2.1. Cut the skin upward to the lower jaw, turn the pan around, and complete this incision to the anus, cutting around both sides of the genital openings. The completed incision should appear as in figure 2.3.

3. With the handle of the scalpel, separate the skin from the musculature as shown in figure 2.4. The fibrous connective tissue which lies

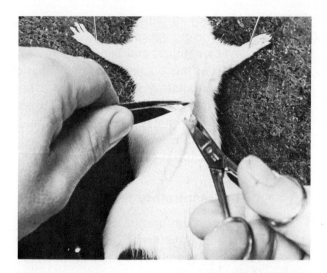

Figure 2.1 Incision is started on the median line with a pair of scissors.

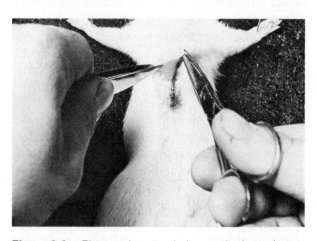

Figure 2.2 First cut is extended up to the lower jaw.

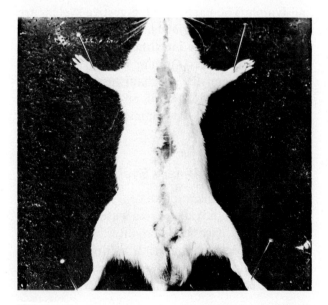

Figure 2.3 Completed incision from the lower jaw to the anus.

between the skin and musculature is the **superficial fascia** (Latin: *fascia,* band).

4. Skin the legs down to the "knees" or "elbows" and pin the stretched out skin to the wax. Examine the surfaces of the **muscles** and note that **tendons** which consist of tough fibrous connective tissue, attach the muscles to the skeleton. Covering the surface of each muscle is another thin gray felt-like layer, the **deep fascia.** Fibers of the deep fascia are continuous with fibers of the superficial fascia, so that considerable force with the scalpel handle

dle is necessary to separate the two membranes.

5. At this stage your specimen should appear as in figure 2.5. If your specimen is a female, the mammary glands will probably remain attached to the skin.

Opening the Abdominal Wall

1. As shown in figure 2.5, make an incision through the abdominal wall with a pair of scissors. To make the cut it is necessary to hold

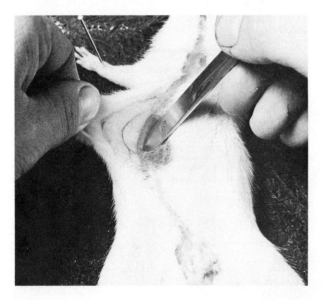

Figure 2.4 Skin is separated from musculature with scalpel handle.

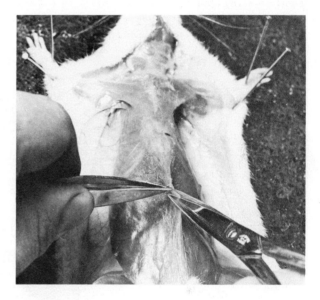

Figure 2.5 Incision of musculature is begun on the median line.

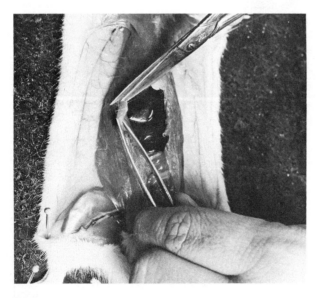

Figure 2.6 Lateral cuts at base of rib cage are made in both directions.

Figure 2.7 Flaps of abdominal wall are pinned back to expose viscera.

the muscle tissue with a pair of forceps. **Caution:** Avoid damaging the underlying viscera as you cut.

2. Cut upward along the midline to the rib cage and downward along the midline to the genitalia.

3. To completely expose the abdominal organs make two lateral cuts near the base of the rib cage—one to the left and the other to the

right. See figure 2.6. The cuts should extend all the way to the pinned back skin.

4. Fold out the flaps of the body wall and pin them to the wax as shown in figure 2.7. The abdominal organs are now well exposed.

5. Using figure 2.8 as a reference, identify all the labeled viscera without moving the organs out of place. Note in particular the position and structure of the **diaphragm.**

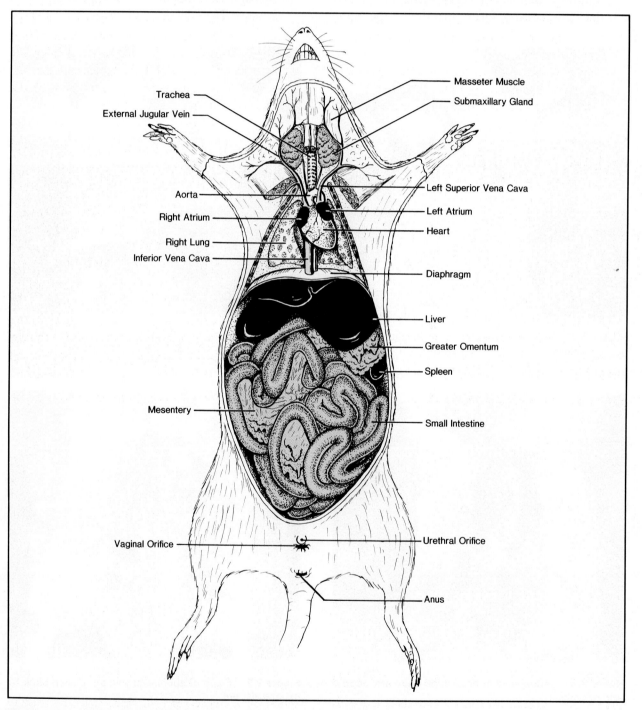

Figure 2.8   Viscera of a female rat.

Examination of Thoracic Cavity

1. Using your scissors, cut along the left side of the rib cage as shown in figure 2.9. Cut through all of the ribs and connective tissue. Then, cut along the right side of the rib cage in a similar manner.

2. Grasp the xiphoid cartilage of the sternum with forceps as shown in figure 2.10 and cut the diaphragm away from the rib cage with your scissors. Now you can lift up the rib cage and look into the thoracic cavity.

3. With your scissors, complete the removal of the rib cage by cutting off any remaining attachment tissue.

4. Now, examine the structures that are exposed in the thoracic cavity. Refer to figure 2.8 and identify all the structures that are labeled.

5. Note the pale-colored **thymus gland,** which is located just above the heart. Remove this gland.

6. Carefully remove the thin **pericardial membrane** that encloses the **heart.**

7. Remove the heart by cutting through the major blood vessels attached to it. Gently sponge away pools of blood with *Kimwipes* or other soft tissues.

8. Locate the **trachea** in the throat region. Can you see the **larynx** (voice box) which is located at the anterior end of the trachea? Trace the trachea posteriorly to where it divides into two **bronchi** that enter the **lungs.** Squeeze the lungs with your fingers, noting how elastic they are. Remove the lungs.

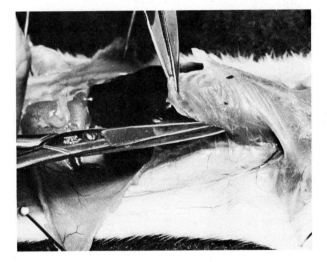

Figure 2.9 Rib cage is severed on each side with scissors.

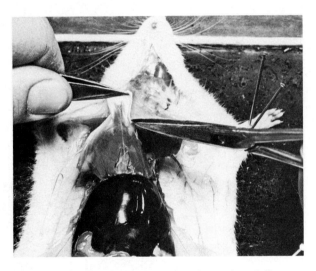

Figure 2.10 Diaphragm is cut free from edge of rib cage.

Figure 2.11 Thoracic organs are exposed as rib cage is lifted off.

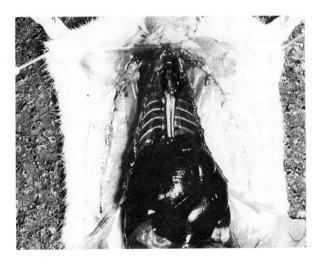

Figure 2.12 Specimen with heart, lungs, and thymus gland removed.

9. Probe under the trachea to locate the soft tubular **esophagus** that runs from the oral cavity to the stomach. Excise a section of the trachea to reveal the esophagus as illustrated in figure 2.13.

Deeper Examination of Abdominal Organs

1. Lift up the lobes of the reddish-brown liver and examine them. Note that rats lack a **gall** **bladder.** *Carefully excise the liver* and wash out the adominal cavity. The stomach and intestines are now clearly visible.

2. Lift out a portion of the intestines and identify the membranous **mesentery** which holds the intestines in place. It contains blood vessels and nerves that supply the digestive tract. If your specimen is a mature healthy animal the mesenteries will contain considerable fat.

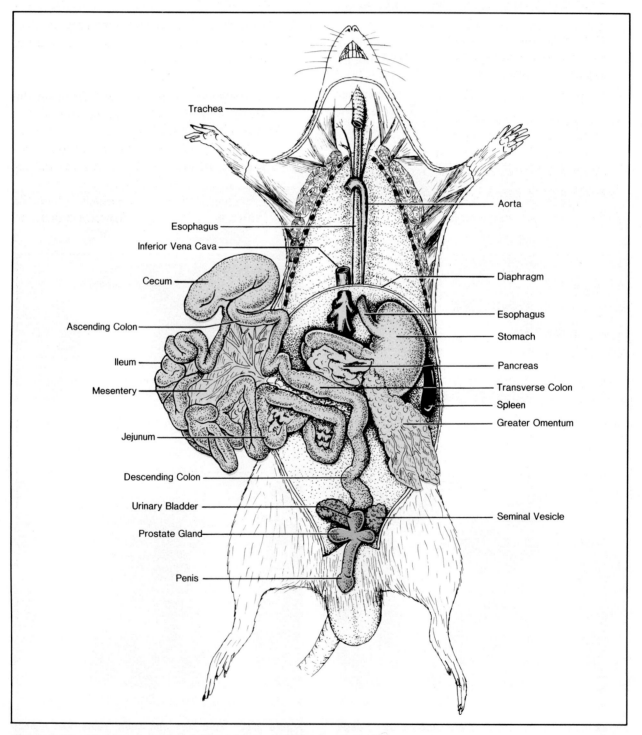

Figure 2.13 Viscera of a male rat (heart, lungs, and thymus removed).

3. Now, lift the intestines out of the abdominal cavity, cutting the mesenteries, as necessary, for a better view of the organs. Note the great length of the small intestine. Its name refers to its diameter, not its length. The first portion of the small intestine, which is connected to the stomach, is called the **duodenum.** At its distal end the small intestine is connected to a large sac-like structure, the **cecum.** The *appendix* in a man is a *vestigal* portion of the cecum. The cecum communicates with the **large intestine.** This latter structure consists of the **ascending, transverse, descending,** and **sigmoid** divisions. The last of these portions empties into the **rectum.**

4. Try to locate the **pancreas** which is embedded in the mesentery alongside the duodenum. It is often difficult to see. Pancreatic enzymes enter the duodenum via the **pancreatic duct.** See if you can locate this minute tube.

5. Locate the **spleen** which is situated on the left side of the abdomen near the stomach. It is reddish brown and held in place with mesentery. Do you recall the functions of this organ?

6. Remove the digestive tract by cutting through the esophagus next to the stomach and through the sigmoid colon. You can now see the descending **aorta** and the **inferior vena cava.** The aorta carries blood posteriorly to the body tissues. The **inferior vena cava** vein returns blood from the posterior regions to the heart.

7. Peel away the peritoneum and fat from the posterior wall of the abdominal cavity. *Removal of the fat will require special care to avoid damaging important structures.* This will make the kidneys, blood vessels, and reproductive structures more visible. Locate the two **kidneys** and **urinary bladder.** Trace the two **ureters** which extend from the kidneys to

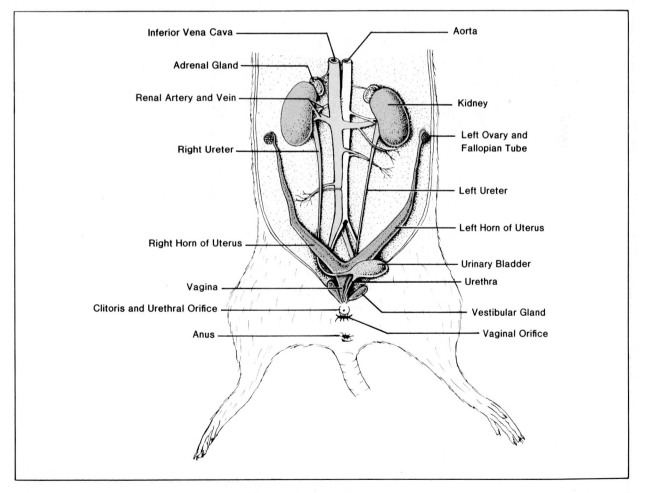

Figure 2.14 Abdominal cavity of female rat (intestines and liver removed).

the bladder. Examine the anterior surfaces of the kidneys and locate the **adrenal glands,** which are important components of the endocrine gland system.

8. **Female.** If your specimen is a female compare it with figure 2.14. Locate the two **ovaries,** which lie lateral to the kidneys. From each ovary a **fallopian tube** leads posteriorly to join the **uterus.** Note that the uterus is a Y-shaped structure joined to the **vagina.** If your specimen appears to be pregnant, open up the uterus and examine the developing embryos. Note how they are attached to the uterine wall.

9. **Male.** If your specimen is a male compare it with figure 2.15. The **urethra** is located in the **penis.**

   Apply pressure to one of the **testis** through the wall of the **scrotum** to see if it can be forced up into the **inguinal canal.**

   Carefully dissect out the testis, **epididy-mis,** and **vas deferens** from one side of the scrotum and, if possible, trace the vas deferens over the urinary bladder to where it penetrates the **prostate gland** to join the urethra.

In this cursory dissection you have become acquainted with the respiratory, circulatory, digestive, urinary, and reproductive systems. Portions of the endocrine system have also been observed. Five systems (integumentary, skeletal, muscular, lymphatic, and nervous) have been omitted at this time. These will be studied later. If you have done a careful and thoughtful rat dissection, you should have a good general understanding of the basic structural organization of the human body. Much that we see in rat anatomy has its human counterpart.

Clean-up Dispose of the specimen as directed by your instructor. Scrub your instruments with soap and water, rinse, and dry them.

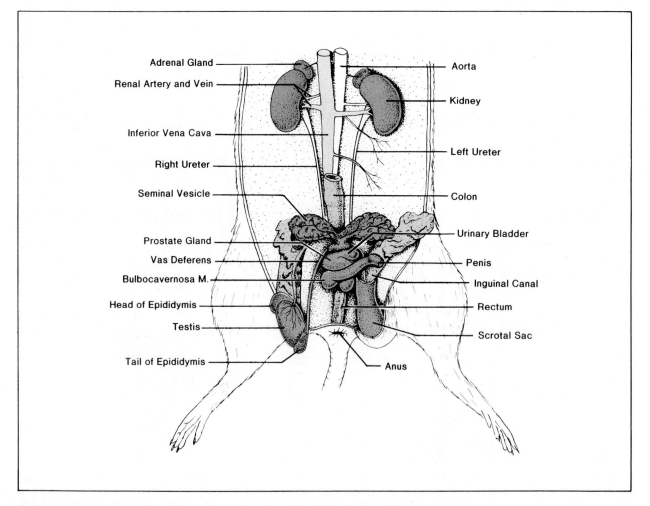

Figure 2.15 Abdominal cavity of male rat (intestines and liver removed).

The Microscope

Two essential prerequisites for successful cytological studies are (1) a well-designed, well-constructed microscope and (2) a knowledge of how to use this instrument. Assuming that the first criterion has been satisfied in the laboratory, it is the purpose of this exercise to provide the expertise necessary to utilize the instrument properly. Prior to attending the first laboratory period in which the microscope is to be used, read this exercise and answer the questions on the Laboratory Report.

All compound microscopes have certain things in common, yet they differ somewhat in mechanical operation. An attempt is made here to point out the similarities and differences of microscopes so that you will know how to use the instrument that is available to you.

Transport and Placement

In most laboratories, the microscopes are numbered and stored in a locked cabinet. You will be assigned a specific instrument for which you will be responsible. If it fails to function properly, notify the instructor immediately. If the instructor is unable to remedy the malfunction, another microscope will be available to you. Don't use a different instrument without authorization.

A microscope is a precision instrument that can be seriously damaged if dropped or accidentally bumped against hard objects. For this reason, it is recommended that the instrument be held as illustrated in figure 3.1 when carried from the cabinet to your desk. Note that the right hand has a firm grip on the arm and the left hand supports it from underneath. If the microscope is allowed to hang suspended from one hand as shown in figure 3.2, the chances of dropping it or colliding with furniture are considerably increased. *Under no circumstances should you attempt to carry two microscopes at one time.*

Before the microscope is placed on the desk, ample space must be provided for it. All books, purses, and other unneeded paraphernalia should be put away. Accidents with microscopes are more frequent when you work in crowded quarters.

Microscopes that have an inclined eyepiece and rotatable head can be used in two different ways. Figure 3.3 illustrates two students using the same microscope in two different ways. The stu-

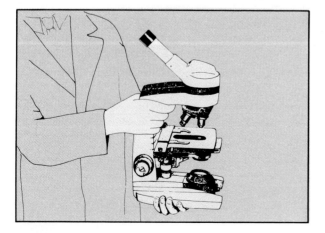

Figure 3.1 The microscope should be held firmly with both hands while carrying it.

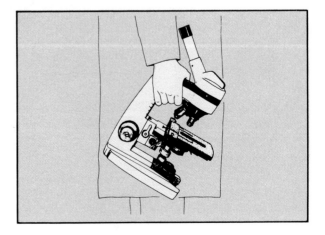

Figure 3.2 Carrying a microscope this way may result in costly accidents.

dent on the left is using the instrument in the conventional position. Note that the arm of the microscope is near the student. The student on the right, however, has the arm of the microscope away from her. This provides her with more direct access to the stage for positioning the slide. It also allows her arm to rest on the table as she focuses the instrument. This particular microscope was designed to be used in this manner. It is generally the preferred method. If the microscope does not have a rotatable head, however, it must be used as shown on the left.

Illumination

Most microscopes today have built-in lamps that provide the necessary illumination. Older microscopes and less expensive ones often have only a mirror under the stage to provide reflected light from a lamp or sunlight. If the mirror is flat on one side and concave on the other side, it is best to use the flat side if the microscope has a condenser; use the concave side if a condenser is lacking. The reason for this is that condensers accept parallel light rays and focus them on the slide. Only a flat surfaced mirror produces parallel light rays that can be used by a condenser.

Built-in lamps on microscopes may produce constant or variable amounts of light. Variable intensity lamps are regulated by changing the voltage. On some microscopes, this is accomplished with a separate transformer; on others, this is achieved with electronic circuitry that is incorporated into the base of the microscope. In the latter case, a small knob near the lamp controls the

voltage. Figure 3.4 illustrates the housing of such a lamp. The knob on the left is for voltage control.

The amount of voltage used should always be minimal. Remember in most instances the bulb of a microscope lamp that is used at different voltages will last much longer when used at minimum voltage. *A bulb used at 4 volts will last twenty times longer than an identical bulb that is used constantly at 5 or 5.5 volts!* When you need to brighten the microscopic field, it is best to open the diaphragm completely before increasing the voltage.

Components

Before we discuss the procedures for using a microscope, identify the various parts of the instrument as illustrated in figures 3.5 and 3.6. The **arm, base,** and **lamp** have already been mentioned to some extent. The lamp housing of this microscope has a lever for positioning a **neutral density filter** over the light bulb. In most instances, it is desirable to have the filter in position over the lamp to reduce excessive brightness. It should be moved out of position only when the microscope is used at maximum magnification. One mistake students often make is to allow the filter to be tilted at 45° over the bulb. This malposition causes the edge of the filter to scatter light rays and greatly impair the image seen through the microscope. Figure 3.4 illustrates a lamp housing that utilizes a knob instead of a lever to move the neutral density filter in and out of position.

The **stage** is the platform on which the microscope slide is placed. The **mechanical stage** is the device attached to the top surface of the stage that

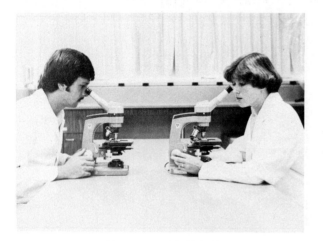

Figure 3.3 The microscope position on the right has the advantage of stage accessibility.

Figure 3.4 The left knob controls voltage; the other knob controls the neutral density filter.

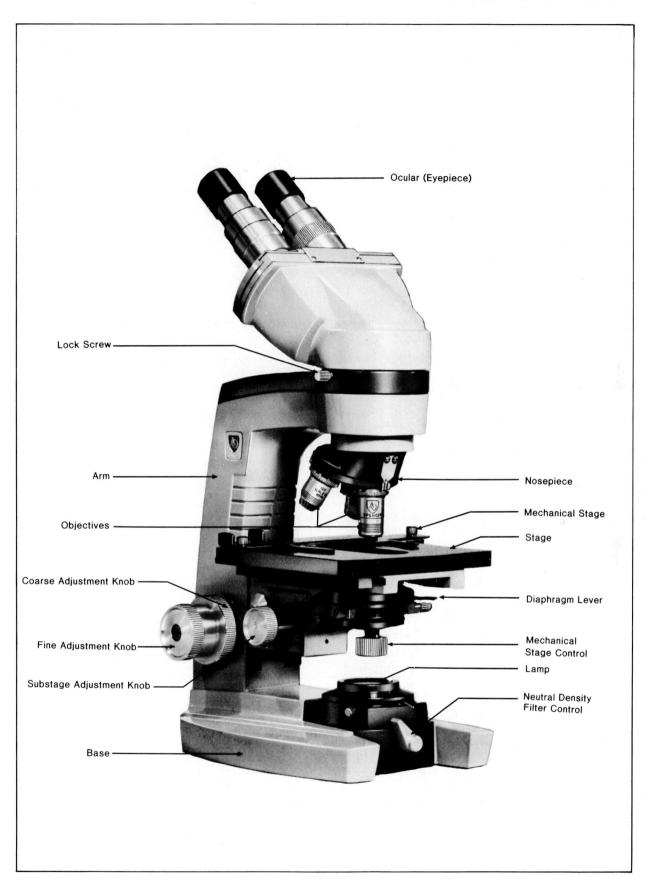

Figure 3.5 The compound microscope.

grips the slide and holds it in place. Figure 3.7 illustrates how a microscope slide is positioned within the confines of the mechanical stage. The mechanical stage can be moved in two directions by control knobs that may be located under the stage or on top of the stage; the substage location is more common.

Beneath the stage is a group of lenses that comprise the **condenser.** Since it is not visible in figure 3.5, refer to figure 3.6 where it can be seen clearly. The condenser causes light rays from the light source to converge on the surface of the microscope slide. It can be moved up and down with the **substage adjustment knob.** For most microscopic work, *it is best to keep the condenser at its highest level.* Only rarely is it desirable to lower it slightly. When the condenser is used at a lowered position, the resolving power of the microscope is greatly reduced.

Below the condenser is the **diaphragm,** which regulates the amount of light that passes up through the condenser. The diaphragm position is represented as a jagged line across the light path in figure 3.6. The **diaphragm lever** that controls it is labeled in figure 3.5.

The primary lens systems of the microscope are the objectives and ocular. The **objectives** are the lenses at the lower end of the microscope and are attached to a revolving plate called the **nosepiece.** The shortest objective is the **low power objective,** which is inscribed 10X to designate its power. The longest objective has 100X inscribed on its barrel. This lens is called the **oil immersion objective** because it must have a special oil interposed between the lens and slide when it is used. The intermediate length lens, or **high-dry objective,** has a power of 45X. The oil immersion and high-dry objectives may have slightly different powers, depending on who manufactures them. The **ocular** is usually 10X. Lower and higher powered oculars are available, but they are not widely used.

The concentric knobs on the side of the microscope are used for focusing the instrument. Note in figure 3.5 that the large knob is designated for **coarse adjustment** and the small knob is for **fine adjustment.**

Resolution and Magnification

The resolution limit, or **resolving power,** of a microscope lens is a function of the wavelength of light, the design of the condenser, and the use of

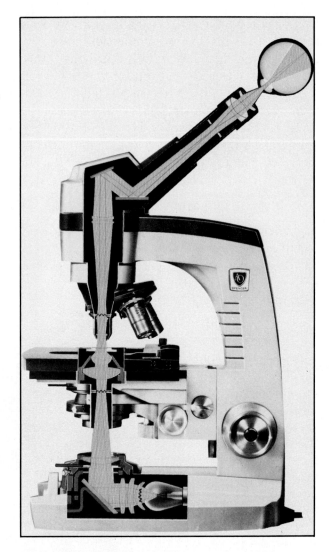

Figure 3.6   The light pathways of a microscope.

immersion oil with the 100X objective. The shortest wavelengths of visible light provide maximum resolution. It is for this reason that all microscopes use blue filters over the light source.

The best compound microscope lenses have a resolving power of approximately $0.2\mu m$. This means that two small objects that are $0.2\mu m$ apart will be seen as separate entities under the oil immersion lens. If they are closer than this, a single object is seen due to the fusion of images.

To achieve maximum resolution, it is essential that the condenser be kept in its highest position, the diaphragm be kept open, and immersion oil be used. It is also important that all lenses and other glass surfaces be clean.

Magnification of an object seen through a microscope is a function of the power of the ocular and objective. If the 10X low power objective is used with a 10X ocular, the magnification is 100X. When the oil immersion lens is used, the magnification is 10X × 100X, or 1000X. If a 15X ocular is used with the oil immersion lens, the magnification becomes 1500X. Although this might seem to be a 50% advantage over the use of a 10X ocular, it really is of no advantage. Keep in mind that the resolution limit of the microscope (0.2μm) negates the increase in magnification beyond 1000X. If the resolution would improve with increased magnification, then using the 15X ocular would be desirable.

Lens Care

To utilize the maximum amount of light entering the microscope, it is important that the mirror, condenser, and lenses be kept clean. The proper kinds of cleansing tissues and solutions must be used to avoid damage to the lenses.

Cleaning Tissues

Only lint-free, optically safe tissues should be used to clean lenses. Tissues free of abrasive grit fall in this category. Although booklets of lens tissue are most commonly used in laboratories, there are several brands of boxed tissues, such as Kimwipes, that are also safe to use. An advantage of boxed tissues is that they are better protected against dust accumulation than open books of tissue. It might be well to store lens tissue in a paper envelope to prevent dust contamination. A supposedly clean handkerchief or other piece of cloth should never be used.

Objectives

Objective lenses frequently become soiled with materials from slides or fingers and must be removed. A piece of lens tissue moistened with green soap and water may remove grease and other materials, but a final wiping with alcohol, acetone, or xylene is usually necessary. Since objective lenses are usually seated in cement, it is important that you not expose the lens to a solvent that will remove the cement. While one manufacturer (American Optical) might recommend the use of alcohol, another (Leitz or Zeiss) might disapprove of its use. Acetone is a very powerful solvent, but it evaporates rapidly and leaves no film. Xylene is not disallowed by any of the manufacturers and is an excellent cleansing agent. The exposed lens of an objective can be cleaned very conveniently with a cotton swab soaked in xylene. A blast of air from an air syringe will remove any remaining lint.

Eyepiece

To test the eyepiece for cleanliness, rotate it between the thumb and forefinger as you look through the microscope. A rotating pattern will be evidence of dirt. If cleaning the eye lens of the eyepiece fails to remove all debris, try cleaning the lower lens with lens tissue and blowing off lint with an air syringe. Be sure to cover the open end of the microscope with lens tissue when the eyepiece is removed (see figure 3.8).

If these efforts fail to get the eyepiece clean, it will be necessary for your instructor to disassemble the unit and clean the individual lenses. *Under no circumstances should a student attempt this operation independently.* Lens elements may be damaged or reassembled incorrectly by the novice.

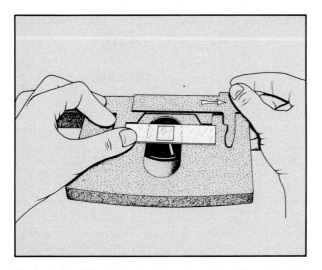

Figure 3.7 The slide must be properly positioned as the retainer level is moved to the right.

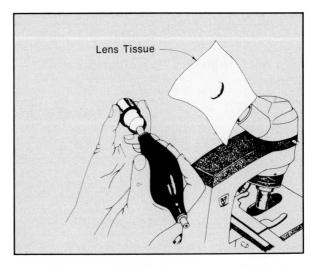

Figure 3.8 After cleaning the lenses, a blast of air from an air syringe removes residual lint.

Low Power Examination

The low power objective of the microscope is used for exploring the slide to find the desired material and to study objects that are quite large. Before the slide is examined, however, it must be properly mounted on the stage. It must be flat on the stage with the material to be studied on the upper surface.

The methods of focusing under low power may vary slightly from one make of microscope to another, depending on the presence or absence of an automatic stop. Some microscopes, such as those built by American Optical, have a built-in stop. For this type of microscope, the best procedure to use is to close the distance between the low power objective and the slide with the coarse adjustment knob until it stops; then, the fine adjustment knob is used to bring the image into focus.

On microscopes that lack stops, the objective is brought close to the slide until it stops; then, while peering through the microscope and turning the coarse adjustment knob slowly, the student increases the distance between the slide and the objective until the image is brought into focus.

Regardless of the type of microscope used, the following points should be kept in mind:

1. Except for a few microscopes, the condenser should be at its highest point. If a pattern shows in the field, lower the condenser until the pattern disappears.
2. The diaphragm should be closed somewhat to prevent an overly bright field.
3. The low power objective should be locked into position on the nosepiece.
4. When the image is in focus, the slide can be moved around on the stage by turning the appropriate mechanical stage knobs.

High-Dry Examination

When the microscopic field has been surveyed to locate a desired object, you can change the magnification by utilizing the high-dry objective. This lens will increase the magnification four or five times more than low power, making cellular detail much clearer. All that is necessary is to rotate the nosepiece so that the high-dry objective is locked in place over the slide. Since most good laboratory microscopes are of parfocal design, it can be assumed that the specimen under high-dry will be in focus (or nearly so) if it was in focus under low power. Only inexpensive microscopes or instruments that are out of adjustment will lack parfocalization. When changing from low power to high-dry, keep these points in mind:

1. For optimum clarity, the slide should have a cover glass on it.
2. Be sure the microscope is in focus under low power before changing to high-dry. Often students will increase the distance between the objective and slide just before changing to high-dry for fear that they might strike the slide with the objective. This is a mistake. It is important to remember that *when the objective is in focus under low power, the high-dry can be safely rotated into place.*
3. Open the diaphragm sufficiently to increase the illumination and bring it into sharp focus with the fine adjustment knob.
4. Keep the condenser at its highest point, unless instructed otherwise.

Oil Immersion Techniques

An oil immersion objective is a luxury that is not always available in the anatomy and physiology laboratory. However, it can be a very helpful aid in the study of different kinds of cells. This is particularly true in blood studies. The greatest difficulty students have with this lens is that its working distance (0.14 mm.) is so small that cover glasses are often broken when it is used the first

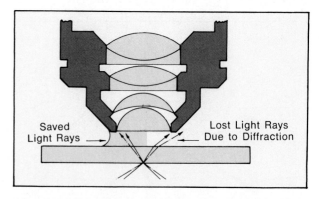

Figure 3.9 Immersion oil, having the same refractive index as glass, prevents light loss due to diffraction.

time. These objectives are the most expensive ones on microscopes, so it is essential that certain safeguards be observed. Two methods for using the oil immersion lens follow:

Easiest Method

The easiest (and safest) procedure is to progress from low power to oil immersion. This may be accomplished by bypassing the high-dry objective, if desired. Microscopes of parfocal design make it as easy to progress from low power to oil immersion as from high-dry to oil immersion. In either case, when the microscope has been in focus at one magnification, the oil immersion lens can be safely rotated into position. Before moving it into position, however, a drop of immersion oil should be placed on the slide. Slight adjustment of the fine adjustment knob is usually necessary to sharpen the focus. It will also be necessary to open up the diaphragm to its maximum aperture.

Direct Method

It is sometimes desirable to omit the lower magnifications and start directly with the oil immersion lens. Whether or not the instrument has an automatic stop will determine the procedure.

With Automatic Stop If the microscope has an automatic stop, place a drop of immersion oil on the slide and lower the objective into the oil by turning the coarse adjustment knob until it stops; then, bring the image into focus by turning the fine adjustment knob very slowly.

Without Automatic Stop The oil immersion objective of a microscope that lacks an automatic stop should be lowered into the oil on the slide until the objective just barely touches the slide. It is necessary to watch the objective from the side as it approaches the slide. Bringing the image into focus is achieved by turning the fine adjustment knob very slowly to increase the distance between the slide and lens.

Returning Microscope to Cabinet

When you take a microscope from the cabinet at the beginning of the period, you expect it to be clean and in proper working condition. The next person to use the instrument after you have used it will expect the same consideration. A few moments of care at the end of the period will insure these conditions. Check over this list of items at the end of the period before you return the microscope to the cabinet:

1. Remove the slide from the stage.
2. If oil has been used, wipe it off the lens and stage with lens tissue.
3. Rotate the lower power objective into position.
4. If the microscope has been inclined, return it to an erect position.
5. If the microscope has a movable body tube, lower it to its lowest position.
6. If the microscope has a built-in movable lamp, raise the lamp to its highest position and wrap the electric cord around the base.
7. Adjust the mechanical stage so that it does not project too far on either side.
8. Replace the dust cover on the instrument.
9. If the microscope has a separate transformer, return it to its designated place.

Laboratory Report

Before entering the laboratory for your first session with the microscope, answer the questions on the Laboratory Report. Preparation on your part prior to going to the laboratory will greatly facilitate your understanding. Your instructor may wish to collect this report at the beginning of the period on the first day that the microscope is used in class.

Cell Study

Proceed to the cellular studies of Exercise 4.

4 Cytology: Basic Cell Structure

As a prelude to the study of cellular physiology, it is essential to review cellular anatomy. An understanding of the nature of the plasma membrane and the roles of organelles such as mitochondria, ribosomes, and endoplasmic reticulum is of fundamental importance.

In the first portion of this exercise, a summary of the present knowledge of cellular structure and function is presented. Then laboratory studies of epithelial cells and living microorganisms help you observe the most visible elements of cell structure. Since our laboratory studies will be limited to observations through a compound microscope, much that is reviewed in the text will not be observable. To get the most out of this exercise *it is recommended that the questions on the Laboratory Report be answered prior to entering the laboratory.*

The Basic Design

Techniques in modern microscopy and cellular physiology have enabled us to study the structure and function of a living cell as never before possible. The old notion of the cell as a "sac of protoplasm" has been supplanted by a much more dynamic model. A cell is now viewed as a highly complex miniature computer with an integrated, compartmentalized ultrastructure capable of communicating with its environment, altering its shape, processing information, and synthesizing a great variety of substances. It is largely through our advancing knowledge of the cellular organelles that this new viewpoint has emerged.

Figure 4.1 is a view of a pancreatic cell as it might appear if magnified by an electron microscope. This cell has been chosen for our study because it lacks the degree of specialization that is seen in many other cells, and so it can be considered illustrative of generalized cell structure. It contains the majority of organelles present in body cells and is also representative of an actively secretory cell.

Although it is not possible to see the fine structure of many organelles with our ordinary light microscope, we are including descriptions of their appearance and current theories of function.

As the various cell structures are discussed in the following test, identify them in figure 4.1.

Plasma Membrane

The outer surface of every cell consists of an extremely thin, delicate *plasma membrane* or *cell membrane*. Chemical analyses of plasma membranes indicate that phospholipids, glycolipids, and proteins are the principal constituents. Lipids account for one-half the mass of plasma membranes, proteins the other half.

Examination with an electron microscope reveals that all cell membranes have a similar basic trilaminar structure, the so-called *unit membrane*. The current interpretation of the molecular organization of this membrane proposes that the **lipid molecules** form a double fluid layer in which **protein** and **glycoprotein molecules** are embedded. Figure 4.1 depicts an enlarged view of this fluid mosaic model (Singer-Nicholson, 1972). Note that some of the proteins protrude externally, others protrude internally, and some extend through both sides of the membrane. This asymmetrical architecture supposedly accounts for the external and internal differences in receptor sites that affect the permeability characteristics of the membrane.

Cell membranes are *selectively permeable* in allowing certain molecules to pass through easily and preventing other molecules from gaining entrance to the cell. Although cell membranes play both active and passive roles in molecular movements, experimental evidence seems to indicate that the movement of all molecules, including water, are assisted by the membrane. Exercise 8 pertains to the forces involved in permeability.

Nucleus

Near the basal portion of the cell is a large spherical body, the *nucleus*. It is shown in figure 4.1 with a portion of its outer surface cut away. The substance of this body *(nucleoplasm)* is sur-

rounded by a double-layered **nuclear envelope** which, unlike other membranes, is perforated by pores of significant size. These openings provide a viable passageway between the cytoplasm and nucleoplasm.

A cell that has been stained with certain dyes will exhibit darkly stained regions in the nucleus called *chromatin granules*. These granules are visible even with a compound microscope. They

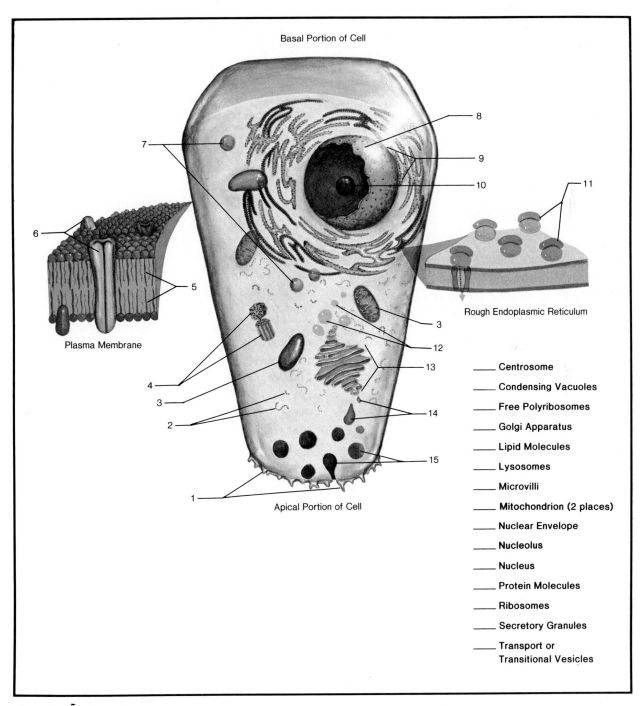

Basal Portion of Cell

Plasma Membrane

Rough Endoplasmic Reticulum

Apical Portion of Cell

____ Centrosome

____ Condensing Vacuoles

____ Free Polyribosomes

____ Golgi Apparatus

____ Lipid Molecules

____ Lysosomes

____ Microvilli

____ Mitochondrion (2 places)

____ Nuclear Envelope

____ Nucleolus

____ Nucleus

____ Protein Molecules

____ Ribosomes

____ Secretory Granules

____ Transport or Transitional Vesicles

Figure 4.1 Cellular anatomy (pancreatic cell).

represent highly condensed DNA molecules that comprise part of the chromosomes. Current theories suggest that increased condensed chromosomal material indicates a metabolically less active cell.

The most conspicuous substructure of the nucleus is the **nucleolus.** Although nucleoli of different cells vary considerably in number and structure, they all consist primarily of RNA, and function chiefly in the production of ribonucleoprotein for the ribosomes.

Cytoplasmic Matrix

Between the plasma membrane and nucleus is an area called the *cytoplasmic matrix,* or *cytoplasm.* This region is a heterogenous aggregation of many components involved in cellular metabolism. The cytoplasm is structurally and functionally compartmentalized by the numerous organelles, many of which are hollow and enclosed by unit membranes. A description of each of these cytoplasmic organelles follows:

Endoplasmic Reticulum The most extensive structure within the cytoplasm is a complex system of tubules, vesicles, and sacs called the *endoplasmic reticulum (ER).* It is a unit membrane organelle somewhat similar to the nuclear and plasma membranes and, in certain cases, continuous with these two membranes. It appears to function, in part, as a microcirculatory system for the cell, providing a passageway for intracellular transport of molecules.

Detailed studies of cells have shown that ER may have surfaces that are smooth or rough. ER that is designated as **rough endoplasmic reticulum (RER)** is a fluid-filled canalicular system studded with ribosomes. It is the type illustrated in the magnified view on the right side of figure 4.1. RER is believed to be an important site of protein synthesis and storage. It is more developed in cells that are primarily secretory in function. **Smooth endoplasmic reticulum (AGER)** also consists of multilayered unit membrane compartments, but lacks ribosomes. It is thought that AGER is active in various types of biosynthesis.

Ribosomes As stated above, ribosomes are small bodies attached to the surface of the RER. They are also scattered throughout the cytoplasm in chains and rosettes called free **polyribosomes.** Since they are only 170 Angstrom units (17 nanometers) in diameter they can be resolved only by electron microscopy.

The components for each ribosome originate in the nucleolus of the nucleus. They pass from the nucleolus through the pores of the nuclear membrane into the cytoplasm where they unite to form the completed ribosomal particle. Each ribosome consists of about 60% RNA and 40% protein. Their proposed structure is shown in figure 4.1.

Ribosomes are sites of protein synthesis and are particularly numerous in actively synthetic cells. There is evidence to indicate that free polyribosomes are involved in synthesis of proteins for endogenous use; attached (RER) ribosomes are implicated in the synthesis of protein to be transported extracellularly (i.e., secretion).

Mitochondria The double-walled bean-shaped organelles that have inward projecting partitions are *mitochondria.* These unit membrane structures are approximately 1 × 2–3 micrometers in size. The infoldings of the inner membrane are called *cristae* and serve to increase the inner surface area.

Mitochondria contain DNA and are able to replicate themselves. They also contain their own ribosomes. They are frequently referred to as the "power plants" of the cell since it is here that the energy-yielding reactions of oxidative respiration occur. The complex series of enzymatic reactions that yield stored energy in the form of ATP occurs here.

Golgi Apparatus This unit membrane organelle is quite similar in basic structure to AGER. In figure 4.1 it appears as a layered structure extending from the end of the RER. Although quite variable in structure in different cells, it is usually seen as a concave stack of vesicles, often many layers thick.

Its primary role relates to the packaging, movement, and completed synthesis of products to be released by secretory cells. Proteins synthesized by the RER are pinched off into ovoid sacs or **transitional vesicles** (label 12) which then coalesce with the cisternae of the Golgi apparatus proper. Here, further synthesis and completed processing of the secretory product take place. Completely processed materials accumulate at the apical end of the organelle and form **condensing vacuoles.** These membrane-bound bodies move to the apex of the cell where, as **secretory granules,** they are ultimately exocytosed. Figure 4.1 illustrates this process and how the granules fuse with the plasma membrane. The fate of the exocytosed product depends upon the cell type.

Lysosomes Oval sacs, such as the one in the upper left part of the cell in figure 4.1, contain digestive enzymes and are called *lysosomes.* These sacs consist of a single unit membrane and are believed to form by the pinching off of sacs from the Golgi apparatus. The contents of these organelles vary from cell to cell, but typically they include enzymes that hydrolyze proteins and nucleic acids.

The function of these organelles is not known for certain. It appears, however, that they act as disposal units of the cytoplasm, for they often contain fragments of mitochondria, ingested food particles, dead microorganisms, worn-out red blood cells, and any other debris that may have been taken into the cytoplasm. When the lysosome has performed its function, it is expelled from the cell through the plasma membrane.

In dying cells, the membrane of the lysosome disintegrates to release the enzymes into the cytoplasm. The hydrolytic action of the enzymes on the cell hastens the death of the cell. It is for this reason that these organelles have been called "suicide bags."

Centrosome At the far left of figure 4.1 is a nonmembranous organelle, the *centrosome.* It consists of two bundles of fine microtubules called *centrioles* that are oriented at right angles to each other. Because of their small size, they appear as very small dots when observed under a light microscope. Centrioles are most evident during cell division (mitosis) as they help form asters during chromatid separation.

Cytoskeletal Elements

It now appears that a living cell has an extensive permeating network of fine fibers that contribute to cellular properties such as contractility, support, shape, and translocation of materials. Microtubules and microfilaments fall into this category.

Microtubules Molecules of protein arranged in submicroscopic cylindrical hollow bundles, the *microtubules,* are found dispersed through the cytoplasm of most cells. They converge especially around the centrioles and are the same as the aster fibers that function during cell division. Other apparent functions of microtubules are to influence cell shape by imparting stiffness to certain areas and to act as pathways for the transport of particles into cell processes.

Microfilaments All cells show some degree of contractility. Cytoskeletal elements responsible for this phenomenon are the *microfilaments.* These filaments are composed of protein molecules arranged in solid parallel bundles. They are most highly developed in muscle cells that are adapted for contraction. In other types of cells, they are present as interwoven networks of the cytoplasm and are relatively inconspicuous even in electron micrographs. Besides aiding in contractility, they also appear to be involved in cell motility and support.

Cilia and Flagella

Hair-like appendages of cells that function in providing some type of movement are either *cilia* or *flagella.* While cilia are usually less than 20 micrometers long, flagella may be thousands of micrometers in length. Cells that line the respiratory tract have cilia. Human spermatozoa have flagella to propel them toward the ova for fertilization. A flagellum is an extension of one or two centrioles of the cell. Cilia originate from basal bodies that are derived from many extra centrioles.

Microvilli

Certain cells, such as the one in figure 4.1, have tiny protuberances known as *microvilli* on their apical or free surfaces. They appear as a fine *brush border* with the compound microscope. Each microvillus is an extension of cytoplasm enclosed by the plasma membrane. Microvilli are most numerous in intestinal and kidney cells and probably function to increase surface area for absorption of important substances such as water and nutrients.

Laboratory Assignment

After labeling figure 4.1 and answering the questions on the Laboratory Report pertaining to cell structure and function, do the following two cellular studies in the laboratory. The epithelial cell will be removed from the inner surface of the cheek and examined under high-dry magnification. The study of microorganisms (primarily protozoa) is performed here to observe the activity of living cells. Ciliary and flagellar action, amoeboid movement, secretion and excretion of cellular products, and cell division can be observed in viable cultures containing a mixture of protozoans.

Materials:

> microscope slides
> cover glasses
> toothpicks
> IKI solution
> mixed cultures of protozoa
> medicine dropper
> microscope

Epithelial Cell Prepare a stained wet mount slide of some cells from the inside surface of your cheek as follows:

1. Wash a microscope slide and cover glass with soap and water.

2. Gently scrape some cells loose from the inside surface of your cheek with a clean toothpick. It is not necessary to draw blood.
3. Mix the cells in a drop of IKI solution on the slide.
4. Cover with a cover glass.
5. After locating a cell under low power of your microscope, make a careful examination of the cell under high-dry magnification. Identify the *nucleus, cytoplasm,* and *cell membrane.*
6. Draw a few cells on the Laboratory Report, labeling the above structures.

Microorganisms Prepare a wet mount of some protozoans by placing a drop of the culture on a slide and covering it with a cover glass. Be sure to insert the medicine dropper all the way to the bottom of the jar for your study sample since most protozoans will be found there.

Examine the slide first under low power, then under high-dry. Study individual cells carefully, looking for nuclei, cilia, flagella, vacuoles of ingested food, excretory vacuoles, etc. If you wish to identify the organisms being observed, consult Appendix D for representative types. Since algae are often present in protozoan cultures, there are also some illustrations of algae in the appendix.

Laboratory Report

Answer the questions on Laboratory Report 4 that pertain to this exercise.

Histology: Epithelial, Connective and Supportive Tissues

Although all cells of the body share common structures such as nuclei, centrosomes, Golgi, etc., they differ considerably in size, shape, and structure according to their specialized functions. An aggregate of cells that are similar in structure and function is called a **tissue.** The science which relates to the study of tissues is called **histology.**

The kinds of tissues found in the body are grouped into four distinct categories: epithelial, connective, muscular, and nerve. Only the epithelial and connective tissues will be studied here. Nerve and muscle tissues will be studied in Exercises 13 and 14.

The histological descriptions provided here are necessarily brief. It is assumed that you will also utilize histology textbooks that are available in the library and laboratory. References at the end of this exercise and at the back of the book (Reading References) should be helpful. Reading over the exercise and answering the questions on the Laboratory Report prior to examining the prepared slides in the laboratory is recommended.

Epithelial Tissues

Tissues that cover the body, line cavities, and form glands fall into this category. All epithelial tissues are characterized by (1) a free surface, (2) the absence of intercellular substances, and (3) the presence of a basement membrane. The *basement membrane,* which makes up the basal surface, consists of fibers and ground substance. Its function is to anchor the tissue to underlying connective tissues. Figure 5.1 illustrates the four kinds of epithelial tissue.

Squamous Epithelium These cells are characterized by their flattened appearance. Figure 5.1 reveals that there are two types of squamous epithelium: simple and stratified. When the cells exist as a single layer the tissue is *simple squamous.* Layered flat cells are called *stratified squa-*

mous. The peritoneum, pleural membranes, and linings of blood vessels consist of simple squamous. The skin and lining of the mouth (mucosa) are good examples of stratified squamous. The stratified squamous in figure 5.1 is of the oral mucosa.

Plain Columnar Epithelium Elongated cells that line the digestive tract are of this type. Flask-like cells, called *goblet cells,* are interspersed among them. These goblet cells contain large amounts of mucus which they constantly secrete over the exposed surfaces of adjacent cells, affording protection to the tissue.

Ciliated Columnar Epithelium Columnar cells that have delicate hair-like projections, called *cilia,* on their exposed surfaces are of this type. If the tissue has some deep cells, as in figure 5.1, that do not extend to the surface, it is said to be *pseudostratified.* The trachea, bronchial tubes, and pharynx are lined with this tissue. Goblet cells are also present for the production of mucus. Movement of the cilia in the respiratory tract moves a sheet of mucus over the surface of the cells. Foreign objects, such as dust and bacteria, become entangled in the mucus, protecting the cells.

Cuboidal Epithelium These cubical cells are seen in various glands of the body. The function of this tissue is secretion. The thyroid, pancreas, and salivary glands are typical examples.

Connective Tissues

These tissues provide support to various organs, fill up spaces in the body, attach organs to each other and provide protection where needed.

Connective tissues are characterized by an abundance of non-living intercellular substance *(matrix).* The nature of this matrix determines the type of connective tissue. The cells of many connective tissues are called *fibroblasts* because of their ability to produce reinforcing fibers.

Three types of fibers may be present: *white* (collagenous), *yellow* (elastic), and *reticular* (argyrophilic) fibers. The white fibers are tough, nonelastic and composed of many parallel submicroscopic collagenous fibrils. The yellow fibers consist of elastin and are elastic and branching. Reticular fibers are somewhat similar to white fibers but differ in that they stain more readily with silver dyes *(argyrophilic)*.

Connective tissues are often grouped into two categories: (1) *connective tissue proper* and (2) *supportive tissues*. The distinction between the two is based on the nature of the matrix.

Connective Tissue Proper

Connective tissues that have a gelatinous to fibrous matrix fall into this category. The four principal types are areolar, adipose, fibrous, and reticular.

Areolar Connective Tissue This tissue is also known as *loose connective tissue*. It is characterized by the presence of both white and yellow fibers between the fibroblasts. It attaches the skin to the body and lends strength to the walls of the blood vessels and many other organs. Refer to illustration 1 in figure 5.2.

Adipose Tissue The cells of adipose tissue are specialized for the storage of fat. Each cell consists of a large vacuole filled with lipid. The cytoplasm is reduced to a thin outer layer that makes up about one-fiftieth of the total volume of the cell. Silver stains reveal that delicate reticular fibers lie between the cells. Adipose tissue lies under the skin, in the mesenteries, around the kidneys, and in many other places throughout the body. In addition to providing an excellent storage medium of energy, fat insulates the body and provides protection to certain organs. It is seen in illustration 2, figure 5.2.

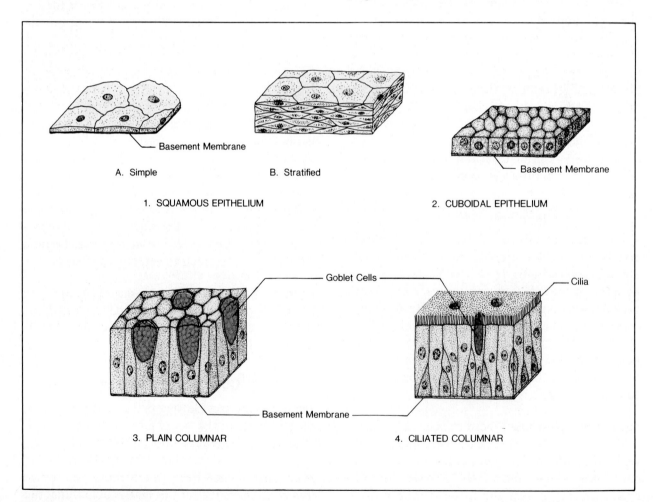

Figure 5.1 Epithelial tissues.

Fibrous Connective Tissue This tissue is silvery-white, pliant, and very strong. Its strength is derived from the masses of white fibers that are packed between rows of fibroblasts. Ligaments, tendons, and fibrous membranes around the heart and kidneys are of this material.

Reticular Tissue This type of tissue furnishes support in lymph nodes. Interspersed among the fibroblasts of this tissue are seen lymphoblasts, macrophages, and other blood cells. Reticular fibers are also present in the basement membrane of epithelial cells. The *reticulin* of reticular fibers is probably chemically identical to collagen of white fibers as evidenced by the fact that reticular fibers, in many instances, mature into collagenous fibers.

Supportive Connective Tissue

This type of connective tissue has a rigid or semi-solid matrix which imparts considerable strength to the tissue. Bone and cartilage fall in this category.

Cartilage This tissue is commonly referred to as *gristle*. It provides rigid support to such structures as the outer ear, trachea, and other structures throughout the body. There are three distinct kinds: (1) hyaline cartilage, (2) elastic cartilage, and (3) fibrocartilage.

Hyaline Cartilage Illustration 4 in figure 5.2 reveals a tissue that has a homogenous-appearing matrix. This intercellular substance actually con-

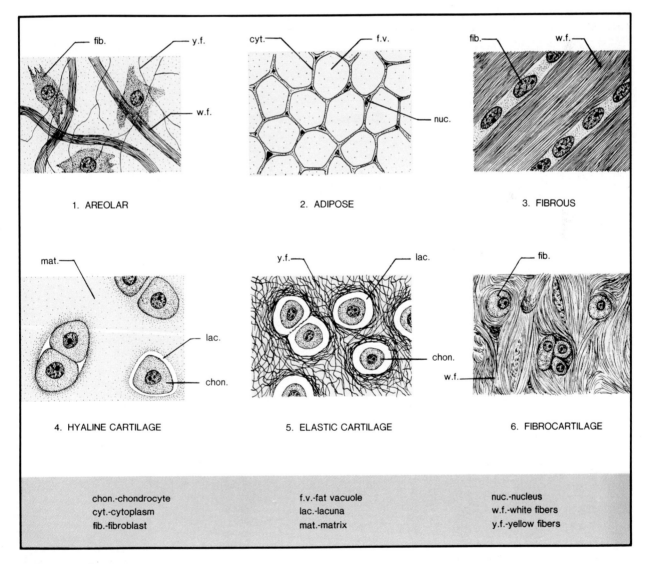

Figure 5.2 Connective tissues.

sists of very fine white fibers densely packed together, forming a moist, glassy matrix. Note that the cells are often paired and lie in spaces called **lacunae.** Hyaline cartilage covers the articulating (contact) surfaces of bones, reinforces the nose, and provides attachment of the ribs to the sternum.

Elastic Cartilage This tissue is somewhat similar in appearance to hyaline cartilage except that numerous yellow fibers are visible between the cells. This tissue is found in the external ear, Eustachean tube, and parts of the larynx.

Fibrocartilage The matrix in this type of cartilage is packed with white fibers. Fibrocartilage is found between the vertebrae as intervertebral disks and between the pubic bones of the pelvis. This tissue has a cushioning effect wherever it is located.

Bone The matrix of this type of connective tissue is extremely hard due to the presence of mineral salts. These salts, calcium carbonate and tricalcium phosphate, are secreted by young bone cells, the osteoblasts. Once ossification is complete, the osteoblasts mature into osteocytes.

In figure 5.3 a portion of the mandible has been enlarged in three steps to show the nature of bony tissue at different degrees of magnification. Two kinds of bone tissue, compact and cancellous, are

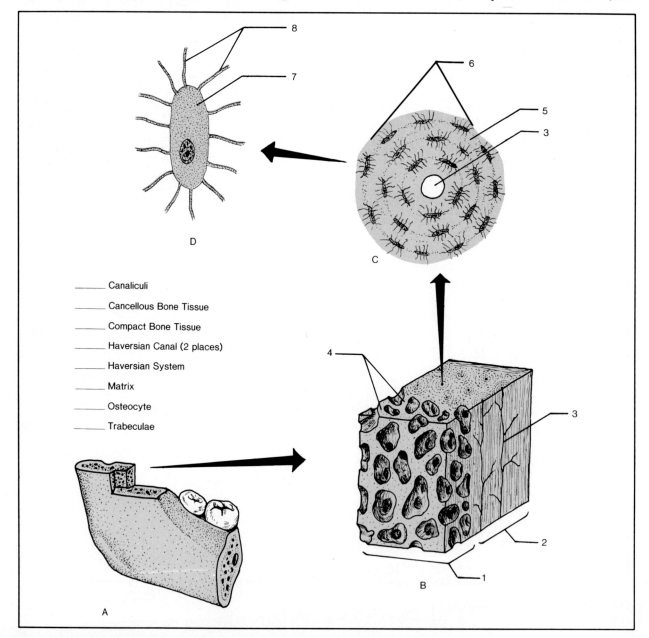

_____ Canaliculi

_____ Cancellous Bone Tissue

_____ Compact Bone Tissue

_____ Haversian Canal (2 places)

_____ Haversian System

_____ Matrix

_____ Osteocyte

_____ Trabeculae

Figure 5.3 Bone tissue.

seen in illustration B. The *compact bone tissue* is the dense outer portion. The inner part of the bone, which is porous or sponge-like, is the *cancellous bone tissue.* The bony processes between the open spaces in cancellous tissue are called **trabeculae.**

Permeating the compact bone tissue is an intricate system of passageways, the **Haversian canals,** which carry nourishment throughout the bone to the living osteocytes imbedded in the mineral salts. These canals are shown in the longitudinal cut face of the bone section in illustration B. Illustration C is an enlargement of an Haversian canal with a cluster of osteocytes arranged concentrically in the matrix around it. An Haversian canal with its surrounding cells and layers of bone is called an *Haversian system.* Illustration D shows an enlarged **osteocyte.** The tiny canals, **canaliculi,** that radiate from it connect with adjacent cells.

Laboratory Assignment

Labeling: Label figures 5.1, 5.2, and 5.3 and transfer the label numbers to the appropriate columns on the Laboratory Report.

Microscopic Studies Study available slides of the various kinds of tissues under high-dry and oil immersion objectives. Look for the various distinguishing characteristics. If drawings are required, label all differentiating structures (cilia, vacuoles, fibroblasts, etc.) that are recognizable.

Laboratory Report

Answer the questions on Laboratory Report 5,6 that apply to this exercise.

Supplemental Readings

Arey Leslie B. *Human Histology,* 4th ed. Philadelphia, Pa.: W. B. Saunders Co., 1974.
Bevelander, G. *Outline of Histology,* 7th ed., St. Louis, Mo.: C. V. Mosby Co., 1975.
Bloom, William and D. W. Fawcett. *A Textbook of Histology,* 10th ed. Philadelphia, Pa.: W. B. Saunders Co., 1975.
DiFiore, Mariono S. *An Atlas of Human Histology,* 4th ed. Philadelphia, Pa.: Lea and Febiger, 1974.

6 The Integument

The entire surface of the body is enveloped in a unique multifunctional covering, the **skin.** It is composed of epithelial, connective, nerve, and muscle tissues. Although its primary function is to protect deeper tissues against abrasion, desiccation, and microbial invasion, it also serves as an excretory organ and temperature regulating mechanism. The secretion of sweat by glands of the skin serves to dispose of metabolic wastes as well as cool the body.

In this exercise, prepared slides of the skin will be available for microscopic examination. Prior to examining the slides, however, label figure 6.1 from the descriptive text that follows.

The skin consists of two distinct layers: the epidermis and dermis. The outer renewable portion which consists of several layers is called the **epidermis.** Beneath it lies the **dermis,** or *corium,* which is considerably thicker.

The Epidermis

This outer layer of the skin has been enlarged in figure 6.1 to reveal its detailed structure. The entire enlarged section is the epidermis. It consists of four distinct layers: an outer **stratum corneum,** a thin translucent **stratum lucidum,** a darkly stained **stratum granulosum,** and a multilayered **stratum spinosum** (mucosum.)

All four layers of the epidermis originate from the deepest layer of cells of the stratum spinosum. This layer, which lies adjacent to the dermis, is called the **stratum germinativum,** or *stratum basale.* The columnar cells of this deep layer are constantly dividing to produce new cells that move outward to undergo metamorphosis at different levels. As they move outward the cells continue to divide, become polyhedral, and accumulate varying amounts of a brown pigment called *melanin.* This pigment is produced by stellate *melanocytes* which occur among the cells of the stratum germinativum. The marked differences of skin color in races is due to the amount of melanin present.

The stratum corneum of the epidermis consists of many layers of the scaly remains of dead epithelial cells. This protein residue of dead cells is primarily *keratin,* a water-repellent material. As the cells of the stratum spinosum are pushed outward they move away from the nourishment of the capillaries, die, and undergo *keratinization.* The *eleidin* granules of the stratum granulosum are believed to be an intermediate product of keratinization. The translucent stratum lucidum consists of closely packed cells with traces of flattened nuclei.

The Dermis

This layer is sometimes referred to as the "true skin." It varies in thickness from less than a millimeter to over six millimeters. It is highly vascular and provides most of the nourishment for the epidermis. It consists of two strata, the papillary and reticular layers.

The outer portion of the dermis, which lies next to the epidermis, is the **papillary layer.** This portion derives its name from the presence of numerous projections, the **papillae,** which extend into the upper layers of the epidermis. In most regions of the body these papillae form no pattern; however, on the palms and soles of the feet they form regularly arranged patterns of parallel ridges. These ridges improve the frictional characteristics of the soles and palms.

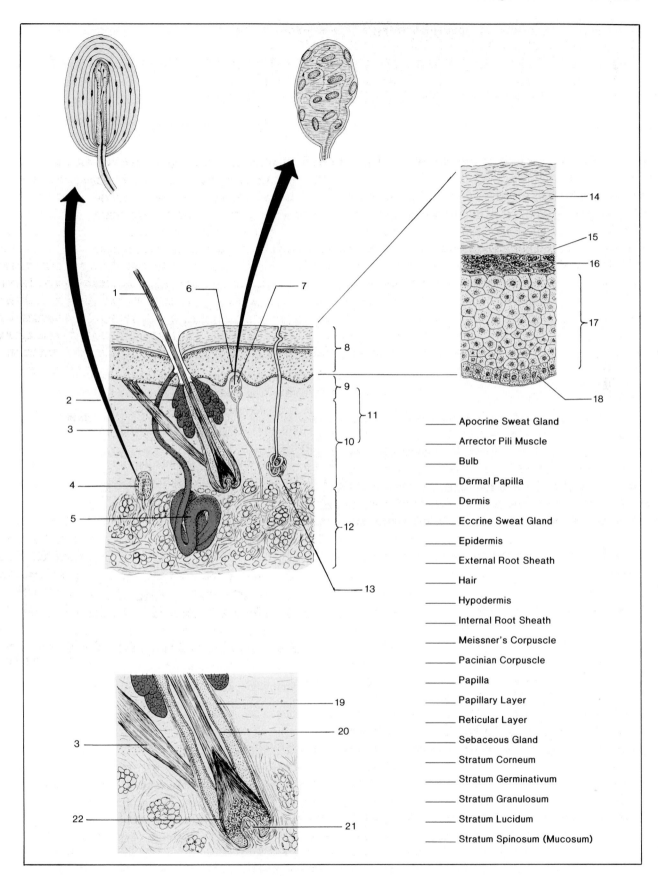

Figure 6.1 Skin structure.

_____ Apocrine Sweat Gland

_____ Arrector Pili Muscle

_____ Bulb

_____ Dermal Papilla

_____ Dermis

_____ Eccrine Sweat Gland

_____ Epidermis

_____ External Root Sheath

_____ Hair

_____ Hypodermis

_____ Internal Root Sheath

_____ Meissner's Corpuscle

_____ Pacinian Corpuscle

_____ Papilla

_____ Papillary Layer

_____ Reticular Layer

_____ Sebaceous Gland

_____ Stratum Corneum

_____ Stratum Germinativum

_____ Stratum Granulosum

_____ Stratum Lucidum

_____ Stratum Spinosum (Mucosum)

The deeper portion, or **reticular layer,** of the dermis contains more collagenous fibers than the papillary layer. These fibers form a dense felt-like network that provides most of the reenforcement in the skin. Suede leather, which is used in clothing manufacture, is derived from the reticular layers of hides of certain animals.

Subcutaneous Tissue

Beneath the dermis is the subcutaneous tissue, or **hypodermis.** It consists of loose connective tissue and is also called *superficial fascia.* Although it is variable in thickness throughout the body, it is generally thicker than the dermis. Its principal function is to bind the skin loosely to underlying structures. It contains many blood vessels and nerves that supply the skin.

Hair Structure

Except for a few areas of the body such as the palms of the hands and soles of the feet, **hair** *(pili)* is well distributed over the surfaces of the body. A single shaft of hair is seen in figure 6.1. It is composed of keratinized cells, compactly cemented together. Each hair is surrounded by a tube of epithelial cells, the **hair follicle.** The terminal end of the hair, or **root,** is enlarged to form an onion-shaped region called the **bulb.** Within the bulb is an involution of loose connective tissue called the **dermal papilla.** It is through this latter structure that nourishment enters the hair shaft. The root of the hair is encased in an **internal root sheath** and an **external root sheath.** These two layers are shown in the enlargement of the root in figure 6.1. Extending diagonally from the wall of the hair follicle to the epidermis is a bundle of smooth muscle tissue, the **arrector pili muscle.** Contraction of these fibers causes the hair to move to a more perpendicular position, causing elevations on the skin surface, commonly referred to as "goose pimples" or "gooseflesh."

Glands

Two kinds of glands are present in the skin: sebaceous and sweat. Both types are seen in figure 6.1.

The **sebaceous glands** are located within the epithelial tissue surrounding each hair follicle. Each gland produces an oily secretion, *sebum,* which fills the hair follicle and spreads over the surface of the skin. Its function is to keep the hair pliable and assist in waterproofing the skin. It also contains antimicrobial agents that play an important part in resistance to infection. When the arrector muscle pulls on the hair follicle, pressure is exerted on the sebaceous gland to force sebum up the follicle.

Sweat glands are of two types: eccrine and apocrine. Both kinds are seen in figure 6.1. The small sweat glands that empty directly out through the surface of the skin are **eccrine sweat glands.** These glands are simple tubular structures that have their coiled basal portions located deep in the dermis. Except for the lips, glans penis, and clitoris, they are widely distributed over the body.

Although all eccrine glands produce essentially the same secretion, they differ in nervous control. Glands on the palms of the hands and the soles of the feet respond primarily to emotional stimuli; glands on the forehead, neck, back, and front of the body are affected by thermal stimuli. It is these latter eccrine glands that produce the sweat that cools the body during exercise and hot weather.

Apocrine sweat glands are larger glands that have their secretory coiled portions in the hypodermis. Instead of emptying out onto the surface of the epidermis, all apocrine glands empty directly into a hair follicle canal. These glands are found in such places as the axillae (armpits), scrotum of the male, the perigenital area of the female, the perianal region, the external ear canal, and the nasal passages. While eccrine sweat is watery, the secretion of apocrine glands is a thick white, gray, or yellowish secretion. Malodorous substances in apocrine sweat are the principal contributors to "body odors." The production of apocrine sweat is influenced by psychic factors rather than ambient temperature.

Receptors

Receptors seen in figure 6.1 are Meissner's and Pacinian corpuscles. **Meissner's corpuscles** are located in the papillary layer of the dermis, projecting up into papillae of the epidermis. They are receptors of touch. **Pacinian corpuscles** are spherical receptors with onion-like laminations and lie deep in the reticular layer of the dermis. Pacinian corpuscles are sensitive to variations in sustained pressure.

Tissues

A review of the structure of the skin reveals that many different kinds of tissues are represented here. The epidermis, with its flattened cells in the stratum lucidum, is a good example of stratified squamous epithelium. Note that the basement membrane of this tissue is on the cells of the stratum germinatum.

A section through the secretory portions of the various glands of the skin would reveal the presence of cuboidal epithelium. The hypodermis region consists of adipose and areolar connective tissue. Smooth muscle tissue (arrector pili), blood in capillaries, and nerve tissue can also be seen.

Laboratory Assignment

After labeling figure 6.1 and answering the questions on Laboratory Report 5,6, proceed as follows:

Materials:

microscope slides of the skin showing hair structure, glands, and receptors.

Examine a prepared slide of a section through the skin under low and high power objectives. Identify the various layers and structures by referring to figure 6.1. Make a drawing of a section as seen under high power and label as many structures as are recognizable.

7 Osmosis and Cell Membrane Integrity

Every cell of the body is bathed in a watery fluid that contains a mixture of molecules that are essential to its survival. This fluid may be the plasma of blood or the tissue fluid in the interstitial spaces. In either case, these molecules, whether water, nutrients, gasses, or ions, pass in and out of the cell through the plasma membrane. Some molecules, usually of small size, are able to diffuse passively through the cell membrane. Organic molecules such as glucose and amino acids and certain ions move through either with or against a concentration gradient by active transport which requires energy expenditure by the cell. The movement of all these molecules takes place only as long as the plasma membrane is viable and undisturbed.

The integrity of cell membranes throughout the body is due largely to the fact that body fluids are isotonic. In this exercise, a study will be made of diffusion and osmosis and the effects of hypertonic and hypotonic solutions on the plasma membrane of red blood cells.

Demonstration Setups:
   Brownian Movement (India Ink Preparation)
   Osmosis (Thistle Tube Setup).

Molecular Movement

All molecules, whether in a gas or liquid state, are in constant motion. As molecules bump into each other, directions are changed, causing random dispersal. Although direct observation of molecular movement is impossible (invisibility of molecules), one can observe this activity indirectly by both microscopic and macroscopic techniques. Two such observations will be made here.

Brownian Movement

A Scottish botanist, Robert Brown, in 1827 observed that extremely small particles in suspension in the protoplasm of plant cells were in constant vibration. He erroneously assumed, at first, that this movement was a characteristic peculiar to living cells. Subsequent studies revealed, however, that this vibratory movement was caused by invisible water molecules bombarding the small visible particles. This type of movement became known as *Brownian movement.*

Demonstration:

The simplest way to observe this phenomenon is to examine a wet mount slide of India ink under the oil immersion objective of a microscope. India ink is a colloidal suspension of carbon particles in water, alcohol, and acetone.

Examine the demonstration setup that has been provided in the laboratory. This type of movement is of particular interest to microbiologists, since it must be differentiated from true motility in bacteria. Nonmotile bacteria are small enough to be displaced in this manner by water molecules.

Diffusion

If a crystal of some soluble compound is placed on the bottom of a container of water, molecules will disperse from the crystal to all parts of the container. The molecules are said to have spread throughout the water by *diffusion.* Movement of the molecules away from the original site of high concentration to low concentration is due to the fact that the reduced number of obstructing molecules in the lower concentration area favors movement of molecules into that area. The rate of diffusion is variable and depends on ambient temperature and molecular size. Given sufficient time, diffusion eventually achieves even dispersal of all molecules throughout the container.

To observe this type of molecular activity, one can use crystals of colored compounds such as potassium permanganate (purple) and methylene blue. Figure 7.1 illustrates how to set up such a demonstration. Methylene blue has a molecular weight of 320. Potassium permanganate has a molecular weight of 158. A Petri plate of 1.5% agar-agar will be used. This agar medium is 98.5% water and allows freedom of movement for molecules. Prepare such a plate as follows and record your observations on the Laboratory Report.

Materials:

crystals of potassium permanganate and
  methylene blue
Petri plate with about 12 ml. of 1.5% agar-
  agar

1. Select one crystal of each of the two different chemicals and place them on the surface of the agar medium about 5 cm. apart. Try to select crystals of similar size.
2. Every fifteen minutes, for one hour, examine the plate to measure the extent of diffusion.
3. Record the measurements on the Laboratory Report.

Osmotic Effects

Water molecules, because of their small size, move freely through cell membranes into and out of the cell. Water acts as a vehicle which enables ions and other molecules to move through the mem-

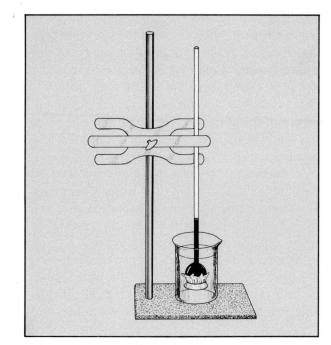

Figure 7.2 Osmosis setup.

brane. This movement of water molecules through a semipermeable membrane is called *osmosis*.

Although water molecules are always moving both ways through the membrane, the predominant direction of flow is determined by the concentrations of solutes in each side of the membrane. If a funnel containing a sugar solution is immersed in a beaker of water and separated from the water with a semipermeable membrane, as in figure 7.2, more water molecules will enter the funnel through the membrane than will leave

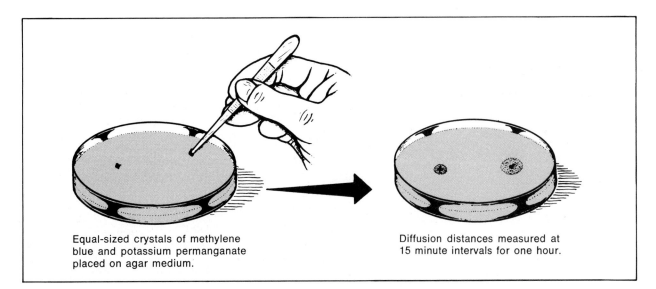

Equal-sized crystals of methylene
blue and potassium permanganate
placed on agar medium.

Diffusion distances measured at
15 minute intervals for one hour.

Figure 7.1 Comparing rates of diffusion.

the sugar solution. As was observed in the diffusion experiment, molecules tend to move from high concentration to low concentration areas. In this case, the water molecules are of greater concentration in the beaker than in the sugar solution. The inward flow of water molecules creates an upward movement of the sugar solution. The force that would be required to restrain this upward flow is called the *effective osmotic pressure.*

Solutions that contain the same concentration of solutes as cells are said to be **isotonic solutions.** Blood cells immersed in such a solution gain and lose water molecules at the same rate, establishing an *osmotic equilibrium.* See figure 7.3.

If a solution contains a higher concentration of solutes than is present in the cells, water leaves the cells faster than it enters and the cells shrink. This shrinkage is called *crenation.* A solution that contains such a high solute concentration is called a **hypertonic solution.** These solutions are said to have a higher osmotic potential than isotonic solutions.

A solution that has a lower solute concentration than is present in cells is said to be a **hypotonic solution.** When red blood cells are immersed in this type of solution, water flows rapidly into them, causing the cells to burst, or *lyse,* as the plasma membrane disintegrates. Lysis of red blood cells is called *hemolysis.*

To observe the effects of the various types of solutions on red blood cells, we will follow the procedures outlined in figure 7.4. Blood cells will be added to various concentrations of solutions. The effects of the solutions on the cells will be determined macroscopically and microscopically. Proceed as follows:

Materials:

    5 serological test tubes (13mm.
        dia. × 100mm.)
    test tube rack (Wassermann type)
    small beaker of distilled water (50 ml. size)
    depression slides (2 per student)
    cover glasses
    vaseline
    toothpicks
    serological pipettes, 5 ml. size (1 per
        student)
    cannister for used pipettes
    mechanical pipetting device (optional)
    .15M NaCl
    .30M NaCl
    .28M glucose
    .30M glycerine
    .30M urea
    syringes and needles
    fresh blood

1. Label five clean serological tubes 1 to 5 and arrange them sequentially in a test tube rack.
2. With a 5 ml. pipette, deliver 2 ml. of each solution to the appropriate tube, rinsing out the pipette with distilled water between each de-

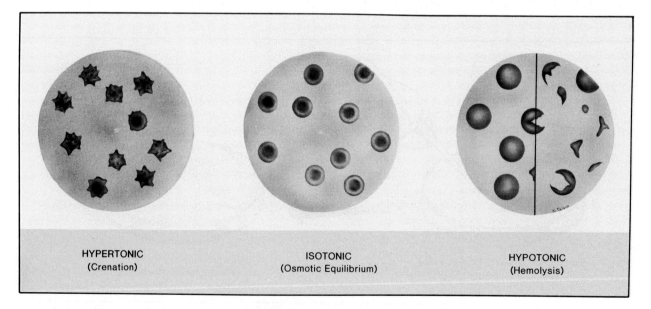

HYPERTONIC
(Crenation)

ISOTONIC
(Osmotic Equilibrium)

HYPOTONIC
(Hemolysis)

Figure 7.3   Effects of solutions on red blood cells.

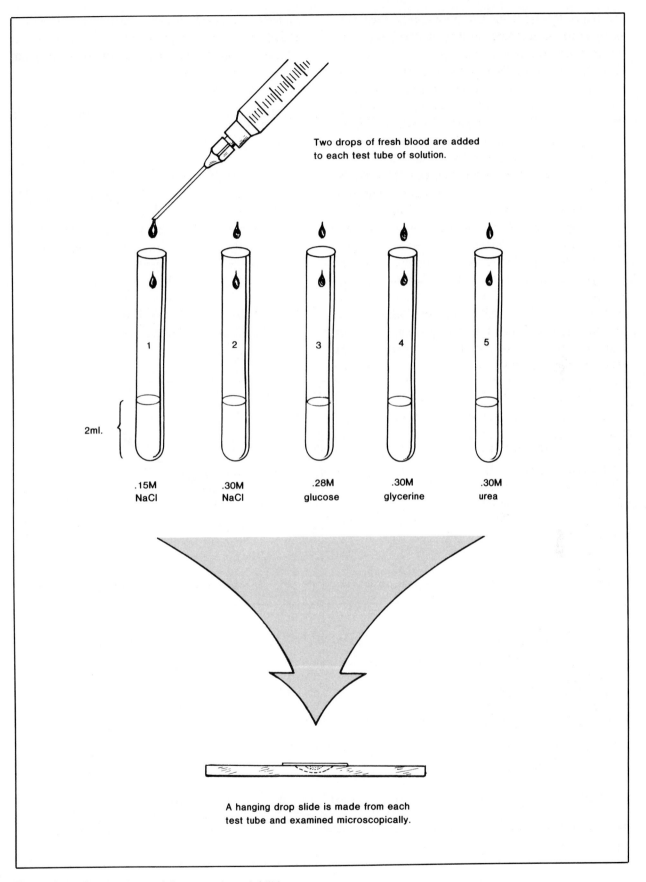

Two drops of fresh blood are added
to each test tube of solution.

2ml.

1	2	3	4	5
.15M NaCl	.30M NaCl	.28M glucose	.30M glycerine	.30M urea

A hanging drop slide is made from each
test tube and examined microscopically.

Figure 7.4   Routine for studying osmotic variabilities.

livery. Make certain that all residual water is forced out of the tip of the pipette before filling with the next solution.

Method of Delivery. If mechanical pipetting devices, such as the one shown in figure 7.6, are available, use the thumb to control delivery. If no mechanical device is available, use your index finger to control the delivery, holding the pipette as shown in figure 7.7.

When using sterile pipettes that are to be placed in your mouth, avoid contaminating the end of the pipette with your fingers prior to inserting it into your mouth. Also take care not to contaminate other pipettes when removing your pipette from the cannister. *Other students don't want your germs on their pipettes!*

Once a pipette has been used and is to be discarded, place it in the proper cannister designated by the instructor. Used pipettes are usually placed in a cannister that is partially filled with a detergent solution prior to washing. **Never return a used pipette to the original sterile cannister.**

When drawing fluid up into the pipette with your mouth draw it up slowly. Excessive speed might surprise you with an unpleasant mouthful. Draw up slightly more than you need and place the tip of your tongue over the end of the pipette so that the fluid does not drop back down. The tongue is then quickly replaced with the index finger over the end of the pipette to hold the fluid level. The fluid is allowed to ease down to the desired volume by gently releasing the pressure of the index finger on the end of the pipette. When discharging the fluid into a container it is often necessary to blow the last drop out through the end of the pipette. This should be done gently.

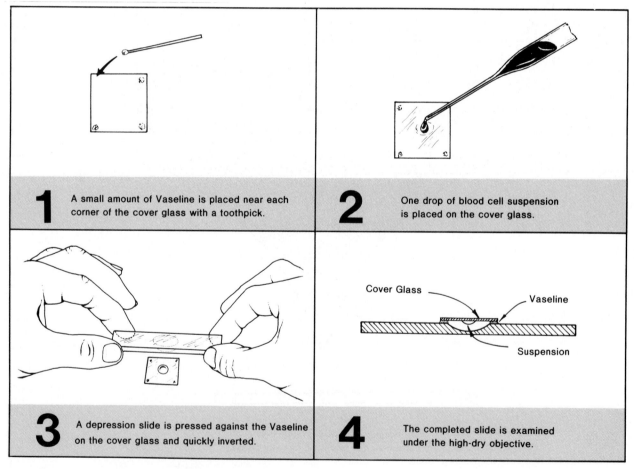

1 A small amount of Vaseline is placed near each corner of the cover glass with a toothpick.

2 One drop of blood cell suspension is placed on the cover glass.

3 A depression slide is pressed against the Vaseline on the cover glass and quickly inverted.

4 The completed slide is examined under the high-dry objective.

Cover Glass

Vaseline

Suspension

Figure 7.5 The hanging-drop slide procedure.

3. Dispense 2 drops of blood to each tube, using a syringe. Shake each tube from side to side to mix, then let stand for 5 minutes.
4. Hold the rack of tubes up to the light and compare them. If the solution is transparent, *hemolysis* has occurred. If you are unable to see through the tube, *no hemolysis* has occurred and the cells should be intact. If crenation has occurred, the appearance will be somewhat between the clarity of hemolysis and the opacity of osmotic equilibrium. Record your results on the Laboratory Report.
5. Make a hanging drop slide from each tube and examine under the high-dry objective of your microscope. Follow the procedure in figure 7.5.

Laboratory Report

Complete the Laboratory Report for this exercise.

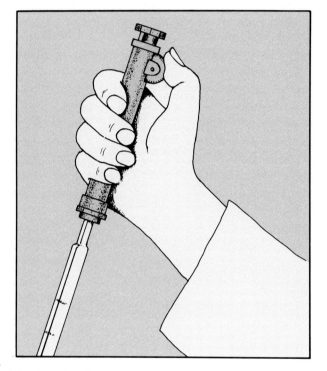

Figure 7.6 Fluid volume delivery is best controlled with the thumb when using a mechanical pipetting device.

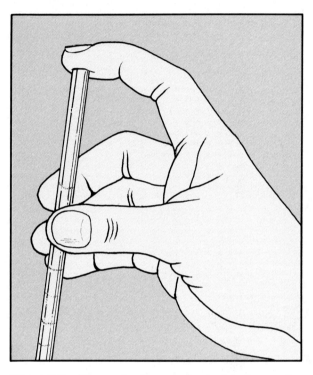

Figure 7.7 When not using a mechanical device use the index finger instead of thumb for delivery control.

Part 2 The Skeletal System

A thorough understanding of the structure and function of the skeletal system is invaluable to students of many phases of medical science. For example, if a knowledge of the muscular system is important, then one must know the names of the processes, ridges, and grooves of the bones to which muscles attach, for it is the exact points of attachment that determine how specific muscles function. In addition, studies of the pathways followed by nerves and of blood vessels throughout the body reveal that one must know the names of the openings (foramina) in the bones that provide passageways for these structures. This is particularly true for the study of the nerves and blood vessels of the skull.

The X-ray technologist, in particular, relies heavily on his or her knowledge of osteology. A thorough knowledge of the shape and normal position of bones is essential for proper positioning of the patient for X-rays. In dentistry the structure of the mandible and surrounding facial bones is of particular importance in taking X-rays of the teeth. In most branches of medicine all parts of the skeleton should be well understood.

 **The Skeletal Plan**

In this exercise we will study the structure of the skeleton as a whole and the anatomy of a typical long bone. Detailed examination of individual parts will follow in subsequent exercises.

Materials:

> fresh beef bones, sawed longitudinally
> articulated human skeleton

The adult skeleton is made up of 206 named bones and many smaller unnamed ones. They are classified as being long, short, flat, irregular, or sesamoid. The *long* bones include the bones of the arm, leg, metacarpals, metatarsals, and phalanges. The *short* bones are seen in the wrist and ankle. In addition to being shorter, the short bones differ from the long ones in another respect: they are filled with cancellous bone instead of having a medullary cavity. The *flat* bones are the protective bones of the skull. Those bones that are neither long, short, or flat are classified as being *irregular*. The vertebrae and bones of the middle ear fall into this category. Round bones embedded in tendons are called *sesamoid* bones (shape resembling sesame seeds). The kneecap is the most prominent one of this type.

Bone Structure

Bones contain cavities, holes, processes, depressions, and other variations which serve different purposes. Terms that pertain to the several kinds of **depressions** or **cavities** are:

> *Foramen.* An opening in a bone which provides a passageway for nerves and blood vessels.
> *Fossa.* A shallow depression in a bone. In some instances the fossa is a socket into which another bone fits.
> *Sulcus.* A groove or furrow.
> *Meatus.* A canal or long tube-like passageway.
> *Fissure.* A narrow slit.
> *Sinus (antrum).* A cavity in a bone.

Any prominence on a bone may be referred to as a **process.** They exist in various shapes and sizes. Different types of processes are:

> *Condyle.* A rounded knuckle-like eminence on a bone which articulates with another bone.
> *Tuberosity.* A large roughened process on a bone which serves as a point of anchorage for a muscle.
> *Tubercle.* A small rounded process.
> *Trochanter.* A very large process on a bone.
> *Head.* A portion supported by a constricted part, or *neck.*
> *Crest.* A narrow ridge of bone.
> *Spine.* A sharp slender process.

Figure 8.1 shows a long bone, the *femur,* which has been sectioned to reveal its internal structure. Linearly, it consists of an elongated shaft, the **diaphysis,** and two enlarged ends, the **epiphyses.** Where the epiphyses meet the diaphysis are growth zones called **metaphyses.** During the growing years a plate of hyaline cartilage, the **epiphyseal disk,** exists in this area. As new cartilage forms on the epiphyseal side it is destroyed and then replaced by bone on the diaphyseal side. The metaphysis, thus, consists of the epiphyseal disk, calcified cartilage, and bone during the growing years. At maturity the area becomes completely ossified and linear growth ceases.

Note that the central portion of the diaphysis is a hollow chamber, the **medullary cavity.** The compact bone tissue of the shaft provides ample strength, obviating the need for central bone tissue. This cavity is lined with a membrane called the **endosteum** that is continuous with the Haversian canals. The entire medullary cavity and much of the cancellous bone of the extremities contain a fatty material, the **yellow marrow.** The cancellous bone of the epiphyses of the humerus and femur contain **red marrow** in the adult. The epiphyses of other long bones in the adult contain yellow marrow. Most red marrow in adults is found in the ribs, sternum, and vertebrae.

A tough covering, the **periosteum,** envelops the surfaces of the entire bone except for the areas of articulation. This covering consists of fibrous con-

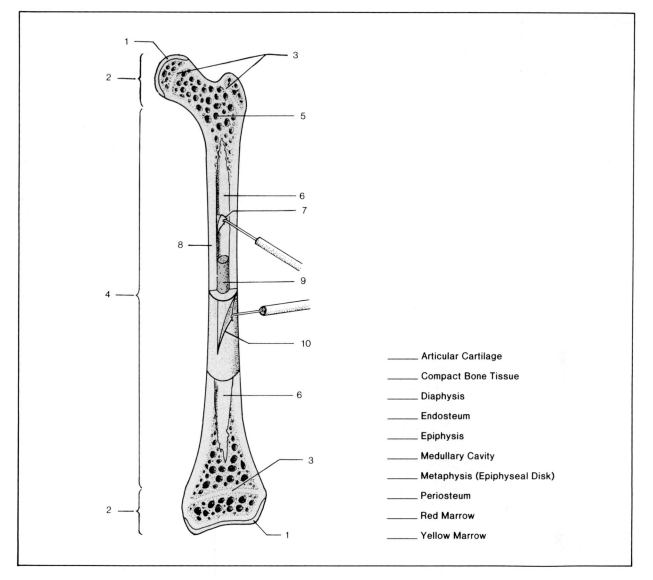

Figure 8.1 Long bone structure.

_____ Articular Cartilage

_____ Compact Bone Tissue

_____ Diaphysis

_____ Endosteum

_____ Epiphysis

_____ Medullary Cavity

_____ Metaphysis (Epiphyseal Disk)

_____ Periosteum

_____ Red Marrow

_____ Yellow Marrow

nective tissue which is quite vascular. The surfaces of each epiphysis which contact adjacent bones are covered with smooth **articular cartilage.**

Assignment:

Figure 8.1: Identify the labels in this illustration.

Beef Bone Study: Examine a freshly cut section of bone. Identify all structures shown in Figure 8.1. Probe into the periosteum near a torn ligament or tendon; note the continuity of fibers between the periosteum and these structures. Probe into the marrow and note its texture.

Bone Names

The bones of the skeleton fall into two main groups: those that make up the axial skeleton and those forming the appendicular skeleton.

Axial Skeleton The parts of the axial skeleton are the **skull, hyoid bone, vertebral column** (spine), and **rib cage.** The hyoid bone is a horseshoe-shaped bone that is situated under the lower jaw. The rib cage consists of twelve pairs of **ribs** and a **sternum** (breastbone).

Appendicular Skeleton This portion of the skeleton includes the upper and lower extremities. The upper extremities consist of the shoulder girdles and arms. Each **shoulder girdle** consists of a **scapula** (shoulder blade) and **clavicle** (collarbone). Each arm consists of an upper portion, the **humerus,** two forearm bones, the **radius** and **ulna,** and the **hand.** The radius is lateral to the ulna.

The lower extremities consist of the pelvic girdle and legs. The **pelvic girdle** is formed by two bones, the **os coxae,** which are attached to the base of the vertebral column (sacrum) and to each other on their anterior surfaces. The joint where the os coxae are united on the median line is the **symphysis pubis.** Each leg consists of a femur, tibia, fibula, patella, and foot. The **femur** is the long bone of the upper part (thigh) of the leg. The **tibia** (shinbone) is the largest bone of the lower portion of the leg. The **fibula** (calfbone) parallels the tibia, lateral to it. The **patella** is the kneecap.

Assignment:

Label the parts of the skeleton in figure 8.2.

Bone Fractures

When bones of the body are subjected to excessive stress, various kinds of fractures occur. The type of fracture that results will depend on the nature and direction of forces that are applied. Some of the more common types are illustrated in figure 8.3.

If the fracture is contained in the soft tissues and does not communicate in any way with the skin or mucous membranes it is considered to be a **closed,** or **simple,** fracture. A majority of the fractures in figure 8.3 would probably fall in this category. If the fracture does communicate with the external surfaces, however, it is called an **open,**

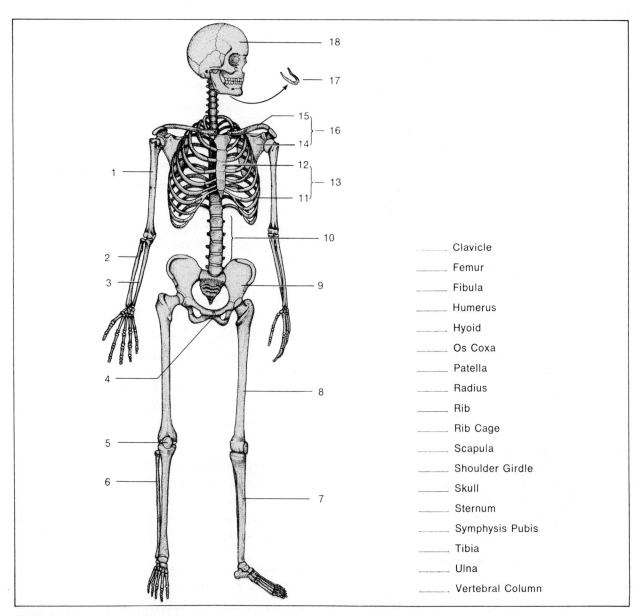

_____ Clavicle
_____ Femur
_____ Fibula
_____ Humerus
_____ Hyoid
_____ Os Coxa
_____ Patella
_____ Radius
_____ Rib
_____ Rib Cage
_____ Scapula
_____ Shoulder Girdle
_____ Skull
_____ Sternum
_____ Symphysis Pubis
_____ Tibia
_____ Ulna
_____ Vertebral Column

Figure 8.2 The human skeleton.

or **compound,** fracture. Fractures of this type may become considerably more difficult to treat because bone marrow infections (osteomyelitis) may result.

If the forces applied to a bone are of insufficient magnitude to accomplish a complete fracture, thus resulting in a splitting or splintering of the bone, the break is called an **incomplete** fracture. Illustrations A, B, and C in figure 8.3 are of this type. If the fractured portion of the bone is only on the convex surface, as in illustration B, it is called a **greenstick** fracture. These fractures are most common among the young. An incomplete break which is characterized by a linear splitting of the bone is called a **fissured** fracture.

Complete fractures may be transverse, segmental, oblique, spiral, or comminuted. In a **transverse** fracture the break is at right angles to the long axis. If a piece of bone is broken out of the shaft it is referred to as a **segmental** fracture. An **oblique** fracture is one in which the fracture is at an angle to the long axis. **Spiral** fractures are caused by torsional forces that twist the bone to produce a spiral-like break. If the bone is broken

to produce more than two fragments at the break it is classified as being a **comminuted** fracture.

Fractures characterized by pieces of bone that are forced out of alignment are called **displaced** fractures. A fractured pelvis of this type, as seen in figure 8.3, is often accompanied by considerable damage to soft tissues in the area.

If a portion of a bone is forced into another part of the bone so that the trabeculae of the cancellous tissue are compacted, a **compacted** fracture has occurred. This type is frequently seen in the hip joint where the neck of the femur is broken and driven into the upper end of the diaphysis. A **compression** fracture is a type sometimes seen in the vertebral column. In this type of fracture vertical forces crush the vertebrae so that the thickness of the bone is compressed.

Assignment:

Identify the types of fractures illustrated in figure 8.3.

Complete the Laboratory Report for this exercise.

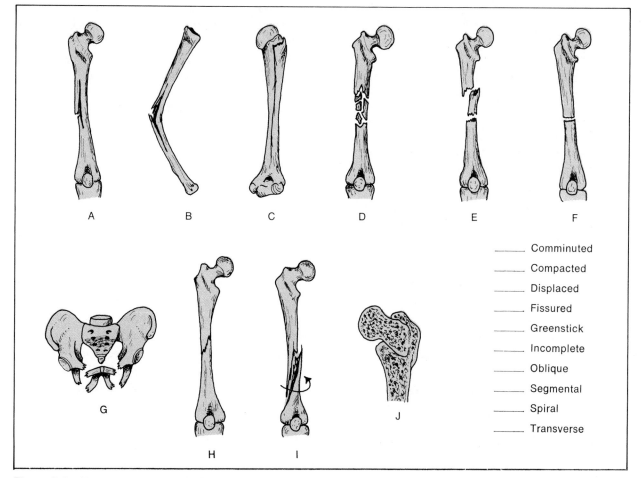

_____ Comminuted
_____ Compacted
_____ Displaced
_____ Fissured
_____ Greenstick
_____ Incomplete
_____ Oblique
_____ Segmental
_____ Spiral
_____ Transverse

Figure 8.3 Types of bone fractures.

9 The Skull

For this study of the skull, specimens will be available in the laboratory. As you read through the discussion of the various bones identify them first in the illustrations and then on the specimens. Compare the specimens with the illustrations to note the degree of variance.

Care of Skulls

When handling laboratory skulls be very careful to avoid damaging them. **Never use a pencil as a pointer.** Pencil marks must not be made on the bones. A metal probe or a pipe cleaner should be used instead. If a metal probe is used, **touch the bones very gently to avoid bone perforation** where bone is thin.

Materials:

>   whole and disarticulated skulls
>   fetal skulls
>   metal probe or pipe cleaner

The Cranium

The portion of the skull that encases the brain is called the *cranium.* It consists of the following bones: a single frontal, two parietals, one occipital, two temporals, and an ethmoid. All these bones are joined together at their margins by irregular interlocking joints called *sutures.* The lateral and inferior aspects of the cranium are illustrated in figures 9.1 and 9.2. A sagittal section of the cranium is seen in figure 9.8.

Frontal The anterior superior portion of the skull consists of the frontal bone. It forms the eyebrow ridges and the ridge above the nose. The most inferior edge of this bone extends well into the orbit of the eye to form the **orbital plates** of the frontal bone. On the superior ridges of the eye orbits are a pair of foramina, the **supraorbital foramina.** See figure 9.4.

Parietals Directly posterior to the frontal bone on the sides of the skull are the parietal bones. The lateral view of the skull actually shows only the left parietal bone. The right parietal is on the other side of the skull. The right and left parietals meet on the midline of the skull to form the **sagittal suture.** Between the frontal and each parietal bone is another suture, the **coronal suture.** Two semicircular bony ridges that extend from the forehead (frontal bone) and over the parietal bone are the **superior temporal line** and **inferior temporal line.** These ridges form the points of attachment for the longest muscle fibers of the *temporalis* muscle. Reference to figure 22.1 shows the position of this muscle (label 1). It is the upper extremity of this muscle that falls on the superior temporal line.

Temporals On each side of the skull, inferior to the parietal bones, are the temporals. These bones are colored yellow in figure 9.1. Each temporal is joined to its adjacent parietal by the **squamosal suture.** A depression, the **mandibular** *(glenoid)* **fossa,** on this bone provides a recess into which the lower jaw articulates. Pull the jaw away from the skull to note the shape of this fossa. The rounded eminence of the mandible that fits into this fossa is the **mandibular condyle.** Just posterior to the mandibular fossa is the ear canal, or **external acoustic meatus** (*acoustic:* hearing; *meatus:* canal or passage).

The temporal bone has three significant processes: the zygomatic, styloid and mastoid processes. The **zygomatic process** is a long slender process that extends forward, anterior to the external acoustic meatus, to form a bridge to the cheekbone of the face. The **styloid process** is a slender spine-like process that extends downward from the bottom of the temporal bone to form a point of attachment for some muscles of the tongue and pharyngeal region. This process is often broken off in laboratory specimens. The **mastoid process** is a rounded eminence on the inferior surface of the temporal just posterior to the styloid process. It provides anchorage for the *sternocleidomastoideus* muscle of the neck. Middle ear infections which spread into the cancellous bone of this process are referred to as *mastoiditis.*

Sphenoid The pink colored bone seen in the lateral view of the skull, figure 9.1, is the sphenoid bone. Note in the bottom view, figure 9.2, that this bone extends from one side of the skull to the other. Examine your laboratory skull carefully to see if you can follow its margins from one side to the other. If an "exploded" or disarticulated skull is available examine this bone to note its extreme complexity. The part that is seen on the side of the skull is called the **greater wing of the sphenoid.** Note in figure 9.4 that this bone makes up the posterior wall of the eye orbit. Here it is called the **orbital surface of the sphenoid.**

Ethmoid On the medial surface of each orbit of the eye is seen the ethmoid bone (label 6, figure 9.1). This bone forms a part of the roof of the nasal cavity and closes the anterior portion of the cranium.

Examine the upper portion of the nasal cavity of your laboratory skull. Note that the inferior portion of the ethmoid has a downward extending **perpendicular plate** on the median line. Refer to figure 9.8 (label 3). This portion articulates anteriorly with the nasal and frontal bones. Posteriorly, it articulates with the sphenoid and vomer. On each side of the perpendicular plate are irregular curved plates, the **superior** and **middle nasal**

conchae. They provide bony reinforcement for the upper nasal conchae of the nasal cavity.

Occipital The posterior inferior portion of the skull consists, primarily, of the occipital bone. It is joined to the parietal bones by the **lambdoidal suture.** Examine the inferior surface of your laboratory skull and compare it with figure 9.2. Note the large **foramen magnum** which surrounds the brain stem in real life. On each side of this opening is seen a pair of **occipital condyles** (label 16). These two condyles rest on fossae of the *atlas,* the first vertebra of the spinal column.

Assignment:

Label all bones of the cranium in figure 9.1. Facial bones will be labeled later.

Label all cranial bones in figure 9.2. Bones of hard palate and all foramina will be labeled later.

Floor of Cranium

The inside surface of the skull reveals other structural details of the ethmoid, sphenoid and temporal bones of significance. Remove the top of your laboratory skull and compare the floor of the cranium with figure 9.3 to identify the following structures.

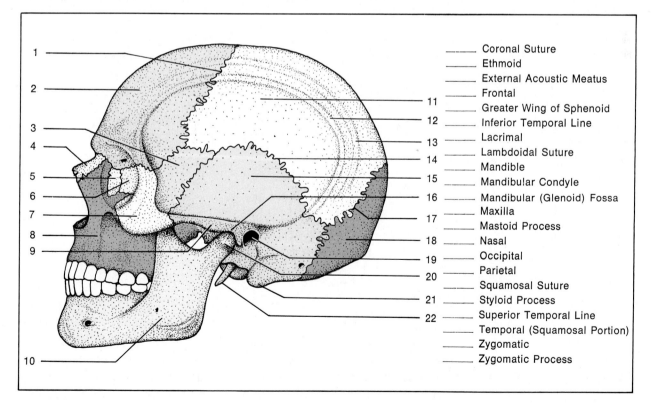

_____ Coronal Suture	
_____ Ethmoid	
_____ External Acoustic Meatus	
_____ Frontal	
_____ Greater Wing of Sphenoid	
_____ Inferior Temporal Line	
_____ Lacrimal	
_____ Lambdoidal Suture	
_____ Mandible	
_____ Mandibular Condyle	
_____ Mandibular (Glenoid) Fossa	
_____ Maxilla	
_____ Mastoid Process	
_____ Nasal	
_____ Occipital	
_____ Parietal	
_____ Squamosal Suture	
_____ Styloid Process	
_____ Superior Temporal Line	
_____ Temporal (Squamosal Portion)	
_____ Zygomatic	
_____ Zygomatic Process	

Figure 9.1 Lateral view of skull.

Cranial Fossae As you look down on the entire floor of the cranium note that it is divided into three large depressions called *cranial fossae*. The one formed by the orbital plates of the frontal is called the **anterior cranial fossa.** The large depression made up mostly of the occipital bone is the **posterior cranial fossa.** It is the deepest fossa. In between these two fossae is the **middle cranial fossa** which is at an intermediate level. The latter involves the sphenoid and temporal bones.

Ethmoid The ethmoid bone in the anterior cranial fossa is seen as a pale yellow structure between the orbital plates of the frontal bone. Note that it consists of a perforated horizontal portion, the **cribriform plate,** and an upward projecting process, the **crista galli** (cock's comb). The holes in the cribriform plate allow branches of the olfactory nerve to pass from the brain into the nasal cavity. The crista galli serves as an attachment for the *falx cerebri* (label 6, figure 27.1).

Sphenoid Observe that on the median line of the sphenoid there is a deep depression called the **hy-**pophyseal fossa. This depression contains the pituitary gland *(hypophysis)* in real life. Posterior to this fossa is an elevated ridge called the **dorsum sella.** The two spine-like processes anterior and lateral to the hypophyseal fossa that project backward are the **anterior clinoid processes.** The outer spiny processes of the dorsum sella are the **posterior clinoid processes.** The hypophyseal fossa, dorsum sella and clinoid processes, collectively, make up the **sella turcica,** or Turkish saddle.

Temporals The significant parts of the temporal bone to identify in figure 9.3 are the petrous and squamous portions, the carotid canal and the internal acoustic meatus. The **squamous portion** of the temporal is that thin portion that forms a part of the side of the skull. The **petrous portion** (label 8) is probably the hardest portion of the skull. It contains the hearing mechanism of the ear.

On the medial sloping surface of the petrous portion is seen an opening to the **internal acoustic meatus.** This canal contains the facial and stato-acoustic cranial nerves. The latter pass from the inner ear region to the brain.

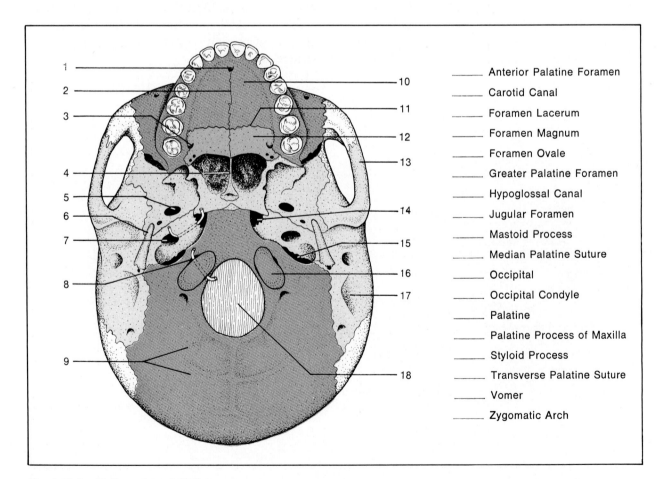

_____ Anterior Palatine Foramen
_____ Carotid Canal
_____ Foramen Lacerum
_____ Foramen Magnum
_____ Foramen Ovale
_____ Greater Palatine Foramen
_____ Hypoglossal Canal
_____ Jugular Foramen
_____ Mastoid Process
_____ Median Palatine Suture
_____ Occipital
_____ Occipital Condyle
_____ Palatine
_____ Palatine Process of Maxilla
_____ Styloid Process
_____ Transverse Palatine Suture
_____ Vomer
_____ Zygomatic Arch

Figure 9.2 Bottom view of skull.

The **carotid canal** is a passageway running through the petrous portion of the temporal bone which allows the internal carotid artery of the neck to pass through the skull into the brain. The brainside opening to this canal can be seen on the anterior margin of the petrous portion of the temporal. Insert a piece of wire (straightened-out paper clip) into this foramen and note where it comes out on the ventral side of the skull.

Other Foramina In addition to the above foramina the following major foramina of the floor of the cranium should be identified: foramen lacerum, foramen ovale, optic foramen, jugular foramen and hypoglosal canal.

The **foramen lacerum** is a large jagged-edged foramen located on each side of the hypophyseal fossa of the sphenoid bone. Lateral to this foramen in the sphenoid bone is an oval opening, the **foramen ovale**. This foramen provides a passageway for the mandibular nerve. Just anterior to the anterior clinoid processes are a pair of **optic foramina** (label 11) which provide passageways for the optic nerves to the eyes. Between the medial margin of the petrous portion of the temporal and the occipital bone is seen an irregular **jugular foramen**. Blood from the brain drains into the internal jugular vein through this opening. The inner openings to the **hypoglossal canals** are seen on each side of the foramen magnum. Insert a probe (paper clip) into one of these canals and note that it passes through the base of the occipital condyle. This canal provides an exit for the 11th and 12th cranial nerves (spinal accessory and hypoglossal). Be able to identify any of these foramina that pass through to the inferior surface of the skull.

Assignment:

Label figure 9.3.
Label the foramina in figure 9.2.

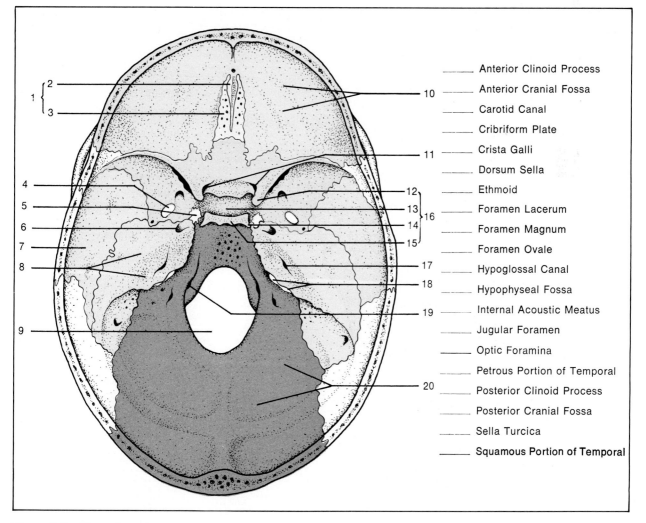

Anterior Clinoid Process

Anterior Cranial Fossa

Carotid Canal

Cribriform Plate

Crista Galli

Dorsum Sella

Ethmoid

Foramen Lacerum

Foramen Magnum

Foramen Ovale

Hypoglossal Canal

Hypophyseal Fossa

Internal Acoustic Meatus

Jugular Foramen

Optic Foramina

Petrous Portion of Temporal

Posterior Clinoid Process

Posterior Cranial Fossa

Sella Turcica

Squamous Portion of Temporal

Figure 9.3   Floor of cranium.

The Face

The bones that make up the anterior portion of the skull constitute the facial bones. Except for two bones, the vomer and mandible, all of them paired. Figure 9.4 reveals the majority of the facial bones.

Maxillae The upper jaw consists of two maxillary bones (maxillae) that are joined by a suture on the median line. Remove the mandible from your laboratory skull and examine the hard palate. Compare it with figure 9.2. Note that the anterior portion of the hard palate consists of two **palatine processes of the maxillae.** A **median palatine suture** joins the two bones on the median line.

The maxillae of an adult support sixteen permanent teeth. Each tooth is contained in a socket, or *alveolus.* That portion of the maxillae that contains the teeth is called the **alveolar process.**

Three significant foramina are seen on the maxillae: two infraorbital and one anterior palatine. The **infraorbital foramina** are situated on the front of the face under each eye orbit. Nerves and blood vessels emerge from each of these foramina to supply the nose. The **anterior palatine foramen** is seen in the anterior region of the hard palate just posterior to the central incisors.

Palatines In addition to the palatine processes of the maxilla, the hard palate also consists of two palatine bones. These bones form the posterior third of the palate. Locate them on your laboratory specimen. Note that each palatine bone has a large **greater palatine foramen** and two smaller **lesser palatine foramina.**

Assignment:

Label the parts of the hard palate in figure 9.2.

Zygomatics On each side of the face are two zygomatic *(malar)* bones. They form the prominence of each cheek and the inferior, lateral surface of each eye orbit. Each zygomatic has a small foramen, the **zygomaticofacial foramen.**

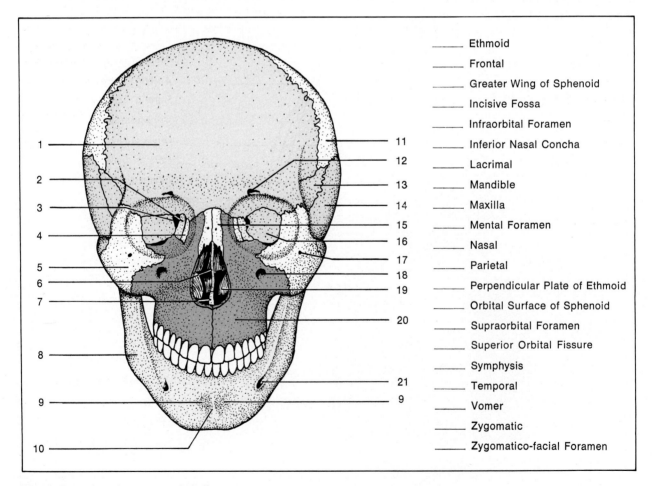

_____ Ethmoid

_____ Frontal

_____ Greater Wing of Sphenoid

_____ Incisive Fossa

_____ Infraorbital Foramen

_____ Inferior Nasal Concha

_____ Lacrimal

_____ Mandible

_____ Maxilla

_____ Mental Foramen

_____ Nasal

_____ Parietal

_____ Perpendicular Plate of Ethmoid

_____ Orbital Surface of Sphenoid

_____ Supraorbital Foramen

_____ Superior Orbital Fissure

_____ Symphysis

_____ Temporal

_____ Vomer

_____ Zygomatic

_____ Zygomatico-facial Foramen

Figure 9.4 Anterior aspect of skull.

Lacrimals Between the ethmoid and upper portion of the maxillary bones are a pair of lacrimal (*lacrima:* tear) bones—one in each eye orbit. Each of these small bones has a groove which allows the tear ducts from the lacrimal glands of the eye to pass down into the nasal cavity.

Nasals The bridge of the nose is formed by a pair of thin, rectangular nasal bones.

Vomer This thin bone is located in the nasal cavity on the median line. Its posterior upper edge articulates with the back portion of the perpendicular plate of the ethmoid and the rostrum of the sphenoid. The lower border of the vomer is joined to the maxillae and palatines. The *septal cartilage* of the nose extends between the anterior margin of the vomer and the perpendicular plate of the ethmoid. Locate this bone on figures 9.2, 9.4 and 9.8 as well as on your laboratory specimen.

Inferior Nasal Conchae The inferior nasal conchae are curved bones attached to the walls of the nasal fossa. They are situated beneath the superior and middle nasal conchae which are part of the ethmoid bone.

Mandible The only bone of the face that is not fused as an integral part of the skull is the lower jaw, or *mandible.* Figure 9.5 reveals the anatomical details of this bone.

It consists of a horizontal portion, the **body,** and two vertical portions, the **rami.** Embryologically, the mandible forms from two centers of ossification, one on each side of the face. As the bone develops toward the median line, the two halves finally meet and fuse to form a solid ridge. This point of fusion on the midline is called the **symphysis** (label 10, figure 9.4). On each side of the symphysis are two depressions, the **incisive fossae.**

The superior portion of each ramus has a condyle, coronoid process and notch. The **mandibular condyle** occupies the posterior superior terminus of the ramus. The process on the superior anterior portion of the ramus is the **coronoid process.** This tuberosity provides attachment for the *temporalis* muscle. Between the mandibular condyle and the coronoid process is the **mandibular notch.** At the posterior inferior corners of the mandible, where the body and rami meet, are two protuberances, the **angles.** The angles provide attachment for the *masseter* and *internal pterygoid* muscles.

A ridge of bone, the **oblique line,** extends at an angle from the ramus down the lateral surface of the body to a point near the mental foramen. This bony elevation is strong and prominent in its upper part, but gradually flattens out and disappears, as a rule, just below the first molar. On the internal

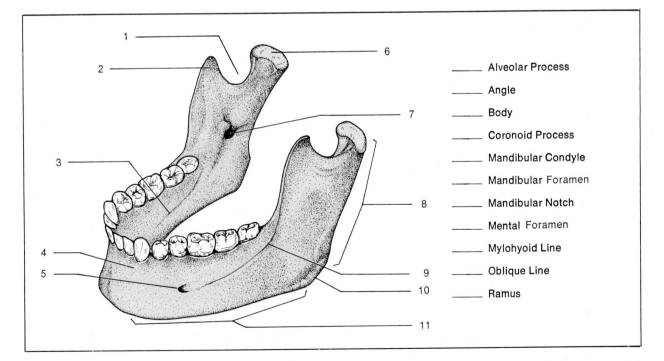

_____ Alveolar Process

_____ Angle

_____ Body

_____ Coronoid Process

_____ Mandibular Condyle

_____ Mandibular Foramen

_____ Mandibular Notch

_____ Mental Foramen

_____ Mylohyoid Line

_____ Oblique Line

_____ Ramus

Figure 9.5 The mandible.

(medial) surface of the mandible is another diagonal line, the **mylohyoid line.** It extends from the ramus down to the body. To this crest is attached a muscle, the *mylohyoid,* which forms the floor of the oral cavity. The bony portion of the body that exists above this line makes up a portion of the sides of the oral cavity proper.

Each tooth lies in a socket of bone called an *alveolus.* As in the case of the maxilla, the portion of this bone that contains the teeth is called the **alveolar process.** The alveolar process consists of two compact tissue bony plates, the *external* and *internal alveolar plates.* These two plates of bone are joined by partitions, or *septae,* which lie between the teeth and make up the transverse walls of the alveoli.

On the medial surfaces of the rami are two foramina, the **mandibular foramina.** On the external surface of the body are two prominent openings, the **mental foramina** (*mental:* chin).

Assignment:

Label figures 9.4 and 9.5.
Label the bones of the face in figure 9.1.

The Paranasal Sinuses

Some of the bones of the skull contain cavities, the *paranasal sinuses,* which reduce the weight of the skull without appreciably weakening it. All of the sinuses have passageways leading into the nasal cavity and are lined with a mucous membrane similar to the type that lines the nasal cavities. The paranasal sinuses are named after the bones in which they are situated. Figures 9.6 and 9.8 show the location of these cavities. Above the eyes in the forehead are the **frontal sinuses.** The largest sinuses are the **maxillary sinuses,** which are situated in the maxillary bones. These sinuses are also called the *antrums of Highmore.* The **sphenoidal sinus** is the most posterior sinus seen in figure 9.6. It is also shown in figure 9.8 (label 8). Between the frontal and sphenoidal sinuses are a group of small spaces called the **ethmoid air cells.**

Assignment:

Label figures 9.6 and 9.8.
Answer the questions on the Laboratory Report that pertain to the bones of the face.

The Fetal Skull

The human skull at birth is incompletely ossified. Figure 9.7 reveals its structure. These unossified membranous areas, called *fontanels,* facilitate compression of the skull at childbirth. During labor, the bones of the skull are able to lap over each other as the infant passes down the birth canal without causing injury to the brain.

There are six fontanels. The **anterior fontanel** is somewhat diamond-shaped and lies on the median line at the junction of the frontal and parietal bones. The **posterior fontanel** is somewhat smaller

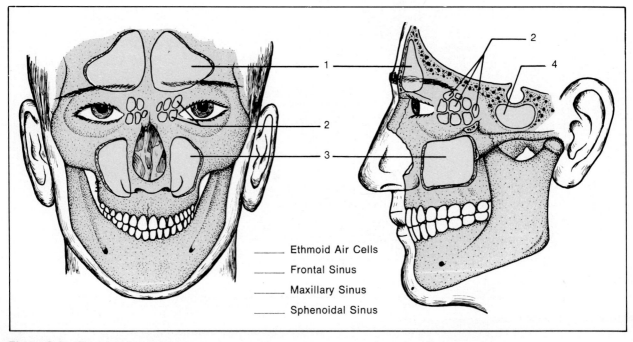

_____ Ethmoid Air Cells

_____ Frontal Sinus

_____ Maxillary Sinus

_____ Sphenoidal Sinus

Figure 9.6 The paranasal sinuses.

and lies on the median line at the junction of the parietal and occipital bones. The **sagittal suture** extends between these two fontanels. On each side of the skull, where the frontal, parietal, sphenoid, and temporal bones come together, is the **anterolateral fontanel.** The **coronal suture** extends from the anterior fontanel to the anterolateral fontanels on each side of the skull. The **posterolateral fontanels** lie at the junction of the parietal, temporal and occipital bones on each side of the skull.

The **squamosal suture** lies between the parietal and temporal bones. Ossification of these fontanels is usually completed in the two-year-old child.

Assignment:

Label figure 9.7.

Examine a fetal skull, identifying all of the structures in figure 9.7.

Complete the Laboratory Report for this exercise.

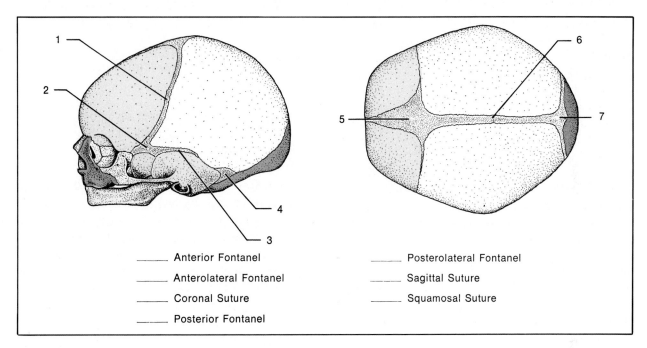

_____ Anterior Fontanel

_____ Anterolateral Fontanel

_____ Coronal Suture

_____ Posterior Fontanel

_____ Posterolateral Fontanel

_____ Sagittal Suture

_____ Squamosal Suture

Figure 9.7 The fetal skull.

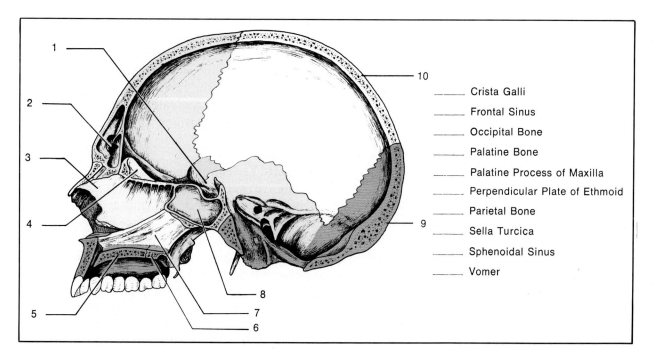

_____ Crista Galli

_____ Frontal Sinus

_____ Occipital Bone

_____ Palatine Bone

_____ Palatine Process of Maxilla

_____ Perpendicular Plate of Ethmoid

_____ Parietal Bone

_____ Sella Turcica

_____ Sphenoidal Sinus

_____ Vomer

Figure 9.8 Sagittal section of skull.

10 The Vertebral Column and Thorax

The vertebral column, ribs, and sternum form the skeletal structure of the trunk of the body.

Materials:

> skeleton, articulated
> skeleton, disarticulated
> vertebral column, mounted

The Vertebral Column

The vertebral column consists of thirty-three bones, twenty-four of which are individual movable vertebrae. Figure 10.1 illustrates its structure. Note that the individual vertebrae are numbered from the top.

The Vertebrae

Although the vertebrae in different regions of the vertebral column vary considerably in size and configuration, they do have certain features in common. Each one has a structural mass, the **body,** which is the principal load-bearing contact area between adjacent vertebrae. The space between the surfaces of adjacent vertebral bodies is filled with a fibrocartilaginous **intervertebral disk.** The collective action of these twenty-four disks imparts a vital cushion effect to the spinal column.

In the center of each vertebra is an opening, the **vertebral** or **spinal foramen,** which contains the spinal cord. Projecting out from the posterior surface of each vertebra is a **spinous process.** On each side is a **transverse process.** These processes of the spinal column are joined to each other by ligaments to form a unified flexible structure. Various muscles of the body are anchored to them.

Extending backward from the body of each vertebra are two processes, the **pedicles,** which form a portion of the bony arch around the vertebral foramen. These structures are labeled in illustration C. The opening formed between the pedicles of adjacent vertebrae allows spinal nerves to emerge from the spinal cord. These openings

are the **intervertebral foramina** (label 26) of the vertebral column.

The posterolateral portions of each vertebra consists of two broad plates, the **laminae** (label 7). The two pedicles and two laminae constitute the **neural arch.**

Cervical Vertebrae The upper seven bones are the cervical vertebrae of the neck. The first of these seven is the **atlas.** Illustration A, figure 10.1, reveals the superior surface of this bone. Note that the spinal foramen is much larger here than on the lumbar vertebrae. Its larger size is necessary to accommodate a short portion of the brain stem which extends down into this space. Note that on each side of the spinal foramen is a depression, the **superior articular surface (facet),** which articulates with the skull. Which processes of the skull fit into these depressions?

Within each transverse process is a small **transverse foramen.** These foramina are seen only in the cervical vertebrae. Collectively, they form a passageway on each side of the spinal column for the vertebral artery and vertebral vein.

The second cervical vertebra is called the **axis.** It differs from all other vertebrae in having a vertical protrusion, the **odontoid process,** which provides a pivot for the rotation of the atlas. When the head is turned from side to side, movement occurs between the axis and atlas around this process.

Thoracic Vertebrae Below the seven cervical vertebrae are twelve thoracic vertebrae. Observe that these bones are larger and thicker than the ones in the neck. The superior surface of a typical thoracic vertebra is seen in illustration C. A distinguishing feature of these vertebrae is that all twelve of them have facets on their transverse processes for articulation with ribs.

Lumbar Vertebrae Inferior to the thoracic vertebrae lie five lumbar vertebrae. The bodies of

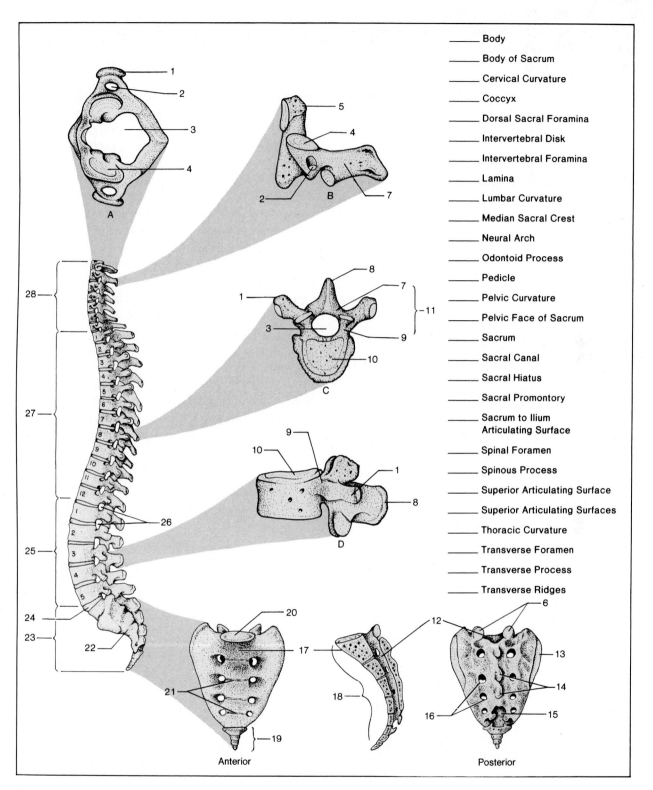

Body
Body of Sacrum
Cervical Curvature
Coccyx
Dorsal Sacral Foramina
Intervertebral Disk
Intervertebral Foramina
Lamina
Lumbar Curvature
Median Sacral Crest
Neural Arch
Odontoid Process
Pedicle
Pelvic Curvature
Pelvic Face of Sacrum
Sacrum
Sacral Canal
Sacral Hiatus
Sacral Promontory
Sacrum to Ilium
Articulating Surface
Spinal Foramen
Spinous Process
Superior Articulating Surface
Superior Articulating Surfaces
Thoracic Curvature
Transverse Foramen
Transverse Process
Transverse Ridges

Figure 10.1 The vertebral column.

these bones are much thicker than those of the other vertebrae due to the greater stress that occurs in this region of the vertebral column. Illustration D is of a typical lumbar vertebra.

The Sacrum

Inferior to the fifth lumbar vertebra lies the sacrum. It consists of five fused vertebrae. Note that there are two oval **superior articulating surfaces** (facets) which provide contact with articulating facets on the fifth lumbar vertebra. On its lateral surfaces are a pair of **sacrum to ilium articulating surfaces.** The intervertebral disk between the fifth lumbar vertebra and the sacrum contacts the flat surface of the **body** of the sacrum.

Note how the anterior aspect, or **pelvic face,** of the sacrum curves backward and that the body of the first sacral vertebra forms a protrusion called the **sacral promontory.** Observe, also, that four **transverse ridges** can be seen on the pelvic face which reveal where the five vertebrae are fused together.

Identify the **median sacral crest** (label 14) and the **dorsal sacral foramina** on the posterior surface. The neural arches of the fused sacral vertebrae form the **sacral canal,** which exits at the lower end as the **sacral hiatus.**

The Coccyx

The "tailbone" of the vertebral column is the coccyx. It consists of four or five rudimentary vertebrae. It is triangular in shape and is attached to the sacrum by ligaments.

Spinal Curvatures

Four curvatures of the vertebral column, together with the intervertebral disks, impart considerable springiness along its vertical axis. Three of them are identified by the type of vertebrae in each region: the **cervical, thoracic,** and **lumbar curves.** The fourth curvature, which is formed by the sacrum and coccyx, is the **pelvic curve.**

Assignment:

Label figure 10.1.

The Thorax

The sternum, ribs, costal cartilages, and thoracic vertebrae form a cone-shaped enclosure, the *thorax.* It is illustrated in figure 10.2. This portion of the skeleton supports the shoulder girdles and forms a protective shield for the heart and lungs.

Ribs There are twelve pairs of ribs. The first seven pairs attach directly to the sternum by **costal cartilages** and are called **vertebrosternal** or **true ribs.** The remaining five pairs are called **false ribs.** The upper three pairs of false ribs, the **vertebrochondral ribs,** have cartilaginous attachments on their anterior ends, but do not attach directly to the sternum. The lowest false ribs, the **vertebral** or **floating ribs,** are unattached anteriorly.

Sternum The sternum, or breastbone, consists of three separate bones: the upper **manubrium,** the middle **body,** or **gladiolus,** and the lower **xiphoid (ensiform) process.** On both sides of the sternum are notches where the sternal ends of the costal cartilages are attached.

Assignment:

Label figure 10.2.

Complete the Laboratory Report for this exercise.

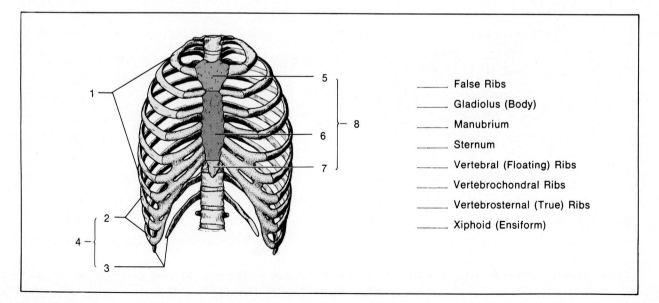

_____ False Ribs

_____ Gladiolus (Body)

_____ Manubrium

_____ Sternum

_____ Vertebral (Floating) Ribs

_____ Vertebrochondral Ribs

_____ Vertebrosternal (True) Ribs

_____ Xiphoid (Ensiform)

Figure 10.2 The thorax.

The Appendicular Skeleton

In this exercise a study will be made of the individual parts of the upper and lower extremities. For this exercise the following materials should be available:

skeleton, articulated
skeleton, disarticulated
male pelvis and female pelvis

The Upper Extremities

The upper extremities consist of the shoulder girdle and arm. As you locate the following structures in figures 11.1 and 11.2 compare the illustrations with the bones on a complete skeleton.

Shoulder Girdle The articulation of the arm and shoulder is seen in figure 11.2. The **clavicle** of the shoulder girdle is a slender S-shaped bone which articulates with the manubrium, medially, and with the acromion on its lateral end.

The **scapula** of the shoulder girdle is a triangular bone which has a socket, the **glenoid cavity,** into which the head of the humerus fits. The scapula is not attached directly to the axial skeleton;

rather, it is loosely held in place by muscles providing more mobility to the shoulder. Figure 11.1 illustrates the anatomical details of this bone. Examination of its lateral aspect reveals its two most prominent processes, the coracoid and acromion. The **coracoid process** lies superior and anterior to the glenoid cavity. The **acromion process** is posterior and superior to the cavity. This process forms the point of attachment for the clavicle.

The margins of the scapula are best seen in the posterior aspect, figure 11.1. The **medial,** or **vertebral margin,** is the semicircular margin on the left side of this view. This border extends from the **superior angle** at its upper extremity to the **inferior angle** at the bottom. The **lateral,** or **axillary margin,** is the opposite border which extends from the glenoid cavity down to the inferior angle. The **superior margin** of the scapula extends from the superior angle to a depression, the **scapular notch.** A ridge of bone, the **spine,** extends from the medial border to the acromion.

Assignment:

Label figure 11.1.

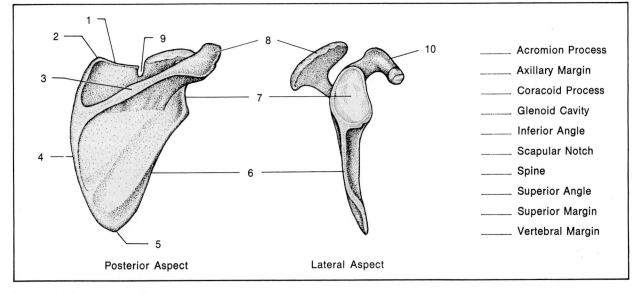

_____ Acromion Process
_____ Axillary Margin
_____ Coracoid Process
_____ Glenoid Cavity
_____ Inferior Angle
_____ Scapular Notch
_____ Spine
_____ Superior Angle
_____ Superior Margin
_____ Vertebral Margin

Posterior Aspect   Lateral Aspect

Figure 11.1 The scapula.

Upper Arm The skeletal structure of the upper arm consists of a single bone, the **humerus.** It consists of a shaft with two enlarged extremities. The smooth rounded upper end which fits into the glenoid cavity of the scapula is the **head.** Inferior to the head are two eminences, the greater and lesser tubercles. The **greater tubercle** is the larger process which is lateral to the **lesser tubercle.** Below these two tubercles is the **surgical neck,** so-named because of the frequency of bone fractures in this area.

The surface of the shaft of the humerus has a roughened raised area near its mid-region which is the **deltoid tuberosity.** In this same general area is also an opening, the **nutrient foramen.**

The distal terminus of the humerus has two condyles, the capitulum and trochlea, which contact the bones of the forearm. The **capitulum** is the lateral condyle that articulates with the radius. The **trochlea** is the medial condyle that articulates with the ulna. Superior and lateral to the capitulum is an eminence, the **lateral epicondyle.** On the opposite side (the medial surface) is a larger tuberosity, the **medial epicondyle.** Above the trochlea on the anterior surface is a depression, the **coronoid fossa.** The posterior surface of this end of the humerus has a depression, the **olecranon fossa.**

Forearm The radius and ulna constitute the skeletal structure of the forearm. Figure 11.2 shows the relationship of these two bones to each other and the hand.

The **radius** is the lateral bone of the forearm. The proximal end of this bone has a disk-shaped **head** which articulates with the capitulum of the humerus. The disk-like nature of the head makes it possible for the radius to rotate at the upper end when the palm of the hand is changed from one position to another (pronation-supination, see figure 21.1). A few centimeters below the head on the medial surface is an eminence, the **radial tuberosity.** This process is the point of attachment for the *biceps brachii,* a muscle of the arm. The region between the head and the radial tuberosity is the **neck** of the radius. The distal lateral prominence of the radius which articulates with the wrist is the **styloid process.**

The **ulna,** or elbow bone, is the largest bone in the forearm. The proximal posterior prominence of this bone is the **olecranon process.** Within this process is a depression, the **semilunar notch,** which

articulates with the trochlea of the humerus. The eminence just below the semilunar notch on the anterior surface is the **coronoid process** (label 24). Where the head of the radius contacts the ulna is a depression, the **radial notch.** The lower end of the ulna is small and terminates in two eminences: a large portion, the **head,** and a small **styloid process.** The head articulates with a fibrocartilage disk which separates it from the wrist. The styloid process is a point of attachment for a ligament of the wrist joint.

Hand Each hand consists of a carpus, metacarpus, and phalanges. The **carpus,** or wrist, consists of eight small bones arranged in two rows of four bones each. The **metacarpus,** or palm, consists of five metacarpal bones. They are numbered from one to five, the thumb side being one. The **phalanges** are the skeletal elements of the fingers. They are distal to the metacarpal bones. There are three phalanges in each finger and two in the thumb.

Assignment:

Label figure 11.2.

The Lower Extremities

Pelvic Girdle The two hip bones (os coxae), which articulate in front, form an arch called the *pelvic girdle.* This arch is completed behind by the sacrum and coccyx to form a rigid ring of bone called the **pelvis.** Figure 11.4 illustrates one half of the pelvis and the right leg.

Figure 11.3 illustrates the anatomical details of the right hip bone. The large circular depression into which the head of the femur fits is the **acetabulum.** Lines of ossification of three parts of the hip bone meet in this fossa. Although these ossification lines are readily discernable in the os coxa of a young child they are generally obliterated in the adult.

The three parts of the os coxa are the ilium, ischium, and pubis. The **ilium** is the broad flaring upper portion of the os coxa. It forms the prominence of the hip. The **ischium** is the lower posterior portion. The **pubis** is the part that is most anterior and forms half of the pubic arch. The point of union of the pubic bones is the **symphysis pubis.** The line of juncture between the ilium and sacrum is the **sacroiliac joint.**

The os coxa has many protuberances, fossae and landmarks of importance. Reference to figure

11.3 will reveal these structures. The upper margin of the ilium is called the **iliac crest.** This crest terminates on its anterior surface in an eminence, the **anterior superior spine** (label 8), and on its posterior surface as the **posterior superior spine** (label 2). Just below this latter process is another eminence, the **posterior inferior spine.** Similarly,

on the anterior margin of the ilium is the **anterior inferior spine** which is below the anterior superior spine.

Two eminences of the ischium are of significance: the ischial spine and the tuberosity of the ischium. The **ischial spine** is the uppermost smaller process. The lower, more prominent process is the

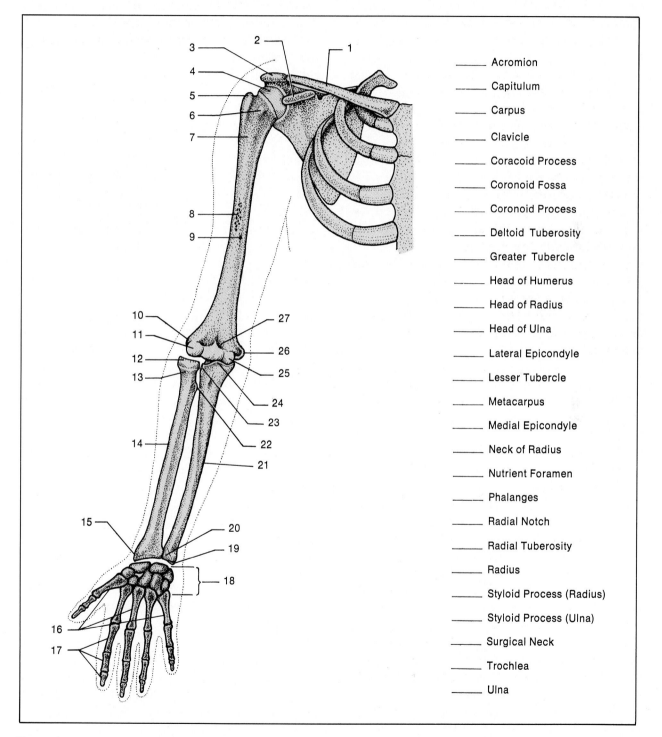

_____ Acromion
_____ Capitulum
_____ Carpus
_____ Clavicle
_____ Coracoid Process
_____ Coronoid Fossa
_____ Coronoid Process
_____ Deltoid Tuberosity
_____ Greater Tubercle
_____ Head of Humerus
_____ Head of Radius
_____ Head of Ulna
_____ Lateral Epicondyle
_____ Lesser Tubercle
_____ Metacarpus
_____ Medial Epicondyle
_____ Neck of Radius
_____ Nutrient Foramen
_____ Phalanges
_____ Radial Notch
_____ Radial Tuberosity
_____ Radius
_____ Styloid Process (Radius)
_____ Styloid Process (Ulna)
_____ Surgical Neck
_____ Trochlea
_____ Ulna

Figure 11.2 The arm and shoulder girdle.

tuberosity of the ischium. Just below the ischial spine is a depression, the **lesser sciatic notch.** The larger notch between the ischial spine and posterior inferior spine is the **greater sciatic notch.** The large opening surrounded by the pubis and ischium is the **obturator foramen.**

Assignment:

Label figure 11.3.

Compare a male pelvis with a female pelvis and answer questions on the Laboratory Report that pertain to their anatomical differences.

Upper Leg Skeletal support of the thigh is achieved with one bone, the **femur.** Its upper end consists of a hemispherical **head,** a **neck,** and two eminences, the greater and lesser trochanters. The **greater trochanter** is the large process on the lateral surface. The **lesser trochanter** is located further down on the medial surface. A ridge, the **intertrochanteric line,** lies obliquely between the two trochanters on the anterior surface. The ridge between these two trochanters on the posterior surface of the femur is the **intertrochanteric crest.**

The lower extremity of the femur is larger than the upper end and is divided into two condyles: the **lateral** and **medial condyles.**

Lower Leg The tibia and fibula constitute the skeletal structure of the lower leg. The **tibia,** or shinbone, is the stronger bone of this part of the leg. Its upper portion is expanded to form two condyles and one tuberosity. The condyles (labels 7 and 19, figure 11.4) are named according to their location: **medial** and **lateral condyles.** The **tibial tuberosity** is located just below these condyles on the anterior surface of the tibia. The distal extremity of the tibia is smaller than the upper portion. A strong process, the **medial malleolus,** of the distal end forms the inner prominence of the ankle. The tibia is somewhat triangular in cross section, with a sharp ridge on its anterior surface, the **anterior crest.**

The **fibula** is lateral to the tibia and parallel to it. The upper extremity, or **head,** articulates with the tibia, but it does not form a part of the knee joint. Below the head is the **neck** of the fibula. The lower extremity of the fibula terminates in a pointed process, the **lateral malleolus,** which lies just under the skin forming the outer ankle bone. Like the tibia, the fibula has an **anterior crest** extending down its anterior surface.

Foot Each foot consists of a tarsus, metatarsus and phalanges. The **tarsus,** or ankle, consists of seven tarsal bones. The *calcaneous,* or heelbone, is the largest tarsal bone. The tibia of the leg articulates with the *talus,* a tarsal bone on top of the foot.

The **metatarsus,** or instep, consists of five elongated metatarsal bones. They are numbered

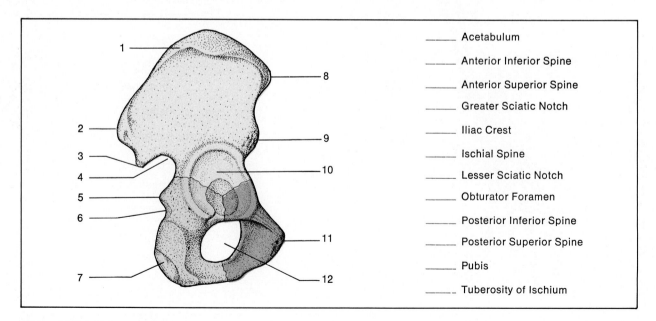

_____ Acetabulum

_____ Anterior Inferior Spine

_____ Anterior Superior Spine

_____ Greater Sciatic Notch

_____ Iliac Crest

_____ Ischial Spine

_____ Lesser Sciatic Notch

_____ Obturator Foramen

_____ Posterior Inferior Spine

_____ Posterior Superior Spine

_____ Pubis

_____ Tuberosity of Ischium

Figure 11.3 The os coxa.

one through five, number one being on the medial side of the foot.

The **phalanges** are the bones of the toes. There are two phalanges in the great toe and three in each of the other ones.

Assignment:

Label figure 11.4.

Answer the questions on the first part of combined Laboratory Report 11,12.

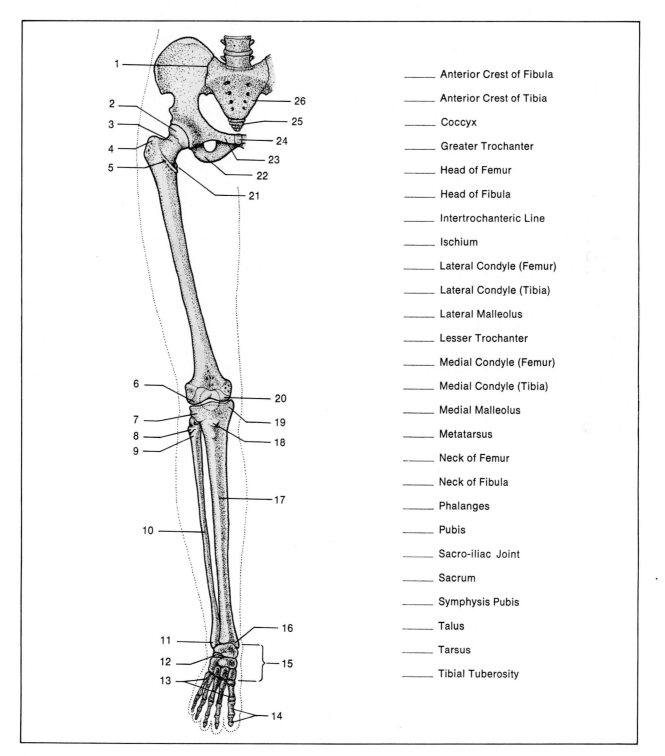

Figure 11.4   The leg and pelvic girdle.

_____ Anterior Crest of Fibula

_____ Anterior Crest of Tibia

_____ Coccyx

_____ Greater Trochanter

_____ Head of Femur

_____ Head of Fibula

_____ Intertrochanteric Line

_____ Ischium

_____ Lateral Condyle (Femur)

_____ Lateral Condyle (Tibia)

_____ Lateral Malleolus

_____ Lesser Trochanter

_____ Medial Condyle (Femur)

_____ Medial Condyle (Tibia)

_____ Medial Malleolus

_____ Metatarsus

_____ Neck of Femur

_____ Neck of Fibula

_____ Phalanges

_____ Pubis

_____ Sacro-iliac Joint

_____ Sacrum

_____ Symphysis Pubis

_____ Talus

_____ Tarsus

_____ Tibial Tuberosity

12 Articulations

The various bones of the skeleton are attached to each other with different degrees of rigidity. Whereas some articulations are freely movable, others are only slightly movable or completely rigid. In this exercise we will study, in general, the basic characteristics of the three types of joints, and the details of the predominant freely movable joints.

Materials:

Fresh knee joint of cow or lamb, sawed through longitudinally.

Three Types

The three basic types of joints in the body are illustrated in figure 12.1. They are classified according to the degree of movement that occurs in the joint.

Immovable Joints

Joints in which the adjacent bones are rigidly held together are known as *synarthrotic* or immovable joints. The rigidity of these joints is due to the bonding effect of fibrous connective tissue or cartilage. The two most common types of immovable joints are sutures and synchondroses.

Sutures are the irregular joints seen between the flat bones of the cranium. Illustration A, figure 12.1, illustrates a sectional view of such a joint. Instead of the bone edges being completely fused to each other, **fibrous connective tissue** lies between them. This tissue is continuous with the **periosteum** (label 2) on the external surface of the bone and the **dura mater** on the inner surface.

Synchondroses are rigid joints that have cartilage between the bone segments. Typical synchondroses are the metaphyses of the long bones in children. These growth zones which lie between the epiphyses and diaphyses of long bones (label 9, figure 12.1) contain cartilage during growth but become completely ossified with maturity.

Slightly Movable Joints

Joints in which there is a small amount of movement between adjacent bones are called *amphiarthrotic,* or slightly movable joints. Two types exist:

Symphyses are joints in which a pad of fibrocartilage provides a cushion between two bones. Illustration B, figure 12.1 of an intervertebral joint is typical. The ends of the bones which contact the fibrocartilage are covered with hyaline **articular cartilage.** The joint is held together with a fibroelastic **capsule.** The symphysis pubis is another example of a joint of this type.

Syndesmoses are joints in which adjacent bones are held together by an interosseous ligament. A good example is the point of articulation between the tibia and fibula at their distal ends. See figure 12.4.

Freely Movable Joints

Joints that are freely movable are also known as *diarthrotic joints.* Most of the articulations of the body are of this type. They are also referred to as *synovial joints.* Illustration C, figure 12.1, is of a typical freely movable joint. As in slightly movable joints, the bone ends of freely movable joints are covered with smooth **articular cartilage.** Surrounding the joint is a fibrous articular capsule which consists of an outer layer of **ligaments** and an inner lining of **synovial membrane.** This latter membrane produces a viscous fluid, *synovium,* which lubricates the joint. These joints also contain sacs, or **bursae,** of synovial tissue. Fibrocartilaginous pads may also be present.

There are six different types of freely movable joints: gliding, hinge, condyloid, saddle, pivot, and ball and socket.

Gliding joints are seen between the carpals of the wrist, the tarsals of the ankle, and the articular processes of the vertebrae. The articular surfaces are nearly flat, or slightly convex, and allow gliding movement only.

Hinge joints allow bending movement in one direction only, much like the hinge on a door. The elbow, ankle, and knee joints are of this type.

Condyloid joints permit angular movement in two directions. The wrist is such a joint. These joints have an oval-shaped head, or condyle, in an elliptical cavity.

Saddle joints resemble condyloid joints in having angular movement in two planes but their anatomical structures differ. The articulation of the thumb metacarpal bone with the trapezium of the carpus is a good example. The articular surface of each of the bones is convex in one direction and concave in another.

Pivot joints allow rotary movement in one axis. The articulation of the axis to the atlas is a good example. In this case rotation of the atlas results in rotation of the head.

Ball-and-socket joints have angular movement in all directions combined with pivotal rotation. The shoulder and hip joints are of this type.

Assignment:

Label figure 12.1.

The Shoulder, Hip, and Knee

A comparative study of the three largest diarthrotic joints reveals that although they all share the common anatomical features illustrated in illustration C, figure 12.1, they differ considerably in structure. Their differences are due essentially to the types of loads and stresses encountered. Detailed anatomy of one hinge joint, the knee, and two ball-and-socket joints, the shoulder and hip, follows.

The Shoulder Joint

The shoulder joint is the most freely movable articulation of the body. Its structure is revealed in figure 12.2. Note that the articulating surfaces of both the humerus and glenoid cavity are covered

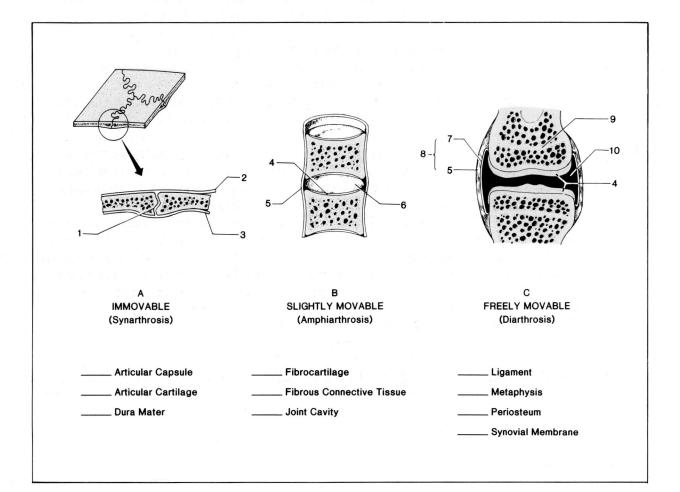

A
IMMOVABLE
(Synarthrosis)

B
SLIGHTLY MOVABLE
(Amphiarthrosis)

C
FREELY MOVABLE
(Diarthrosis)

_____ Articular Capsule

_____ Articular Cartilage

_____ Dura Mater

_____ Fibrocartilage

_____ Fibrous Connective Tissue

_____ Joint Cavity

_____ Ligament

_____ Metaphysis

_____ Periosteum

_____ Synovial Membrane

Figure 12.1 Types of articulations.

with articular cartilage. Several muscles are attached to the upper part of the humerus to effect multiple movements, but only one, the **supraspinatus,** is shown in this section. Its tendon is attached to the greater tubercle of the humerus. Another muscle, the **deltoid** (label 1), is a major muscle of the shoulder. The upper end of this muscle is attached to the **acromion** of the scapula. Its lower end attaches further down on the humerus at a spot not shown in figure 12.2. Between the deltoid and the humerus is the **subdeltoid bursa.** It minimizes friction between the deltoid and the humerus. Below the epiphysis of the humerus is seen another synovial sac, the **subacromial bursa.**

Assignment:

Label figure 12.2.

The Hip Joint

The hip joint, another ball-and-socket joint, is shown in figure 12.3. The rounded head of the femur is confined in the acetabulum of the os coxa by the acetabular labrum and the transverse acetabular ligament. The **acetabular labrum** (label 8) is a fibrocartilaginous rim attached to the margin of the acetabulum. The **transverse acetabular ligament** (label 11) is an extension of the acetabular labrum. Note in the sectional view that a structure, the **ligamentum teres femoris,** is attached to the middle of the curved condylar surface of the femur. The other end of this ligament is attached to the surface of the acetabulum. This structure adds nothing to the strength of the joint; instead, it contributes to nourishment of the head of the femur and supplies synovial fluid to the joint.

The entire joint, including the above three structures, is enclosed in an **articular capsule.** This capsule consists of longitudinal and circular fibers surrounded by three external accessory ligaments. The three accessory ligaments are the iliofemoral, pubocapsular, and ischiocapsular. The **iliofemoral ligament** (label 7) is a broad band on the anterior surface of the joint. This ligament is attached to the **anterior inferior iliac spine** at its upper margin and the **intertrochanteric line** on its lower margin.

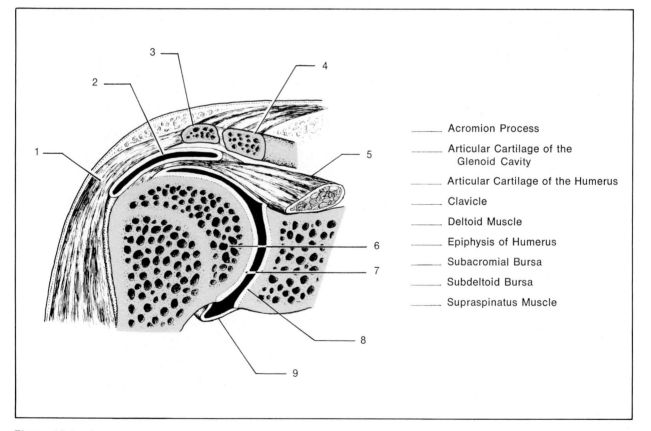

_____ Acromion Process

_____ Articular Cartilage of the
        Glenoid Cavity

_____ Articular Cartilage of the Humerus

_____ Clavicle

_____ Deltoid Muscle

_____ Epiphysis of Humerus

_____ Subacromial Bursa

_____ Subdeltoid Bursa

_____ Supraspinatus Muscle

Figure 12.2 The shoulder joint.

Adjacent and medial to the iliofemoral ligament lies the **pubocapsular ligament.** The posterior surface of the capsule is reinforced by the **ischiocapsular ligament.** Lining the inside of the capsule is the **synovial membrane** which provides the lubricant for the joint.

Movements of flexion, extension, abduction, adduction, rotation, and circumduction of the thigh are readily achieved through this joint. This is made possible by the unique angle of the neck of the femur and the relationship of the condyle to the acetabulum. Excessive backward movement of the body at the joint is controlled to a great extent by the ileofemoral ligament, taking some of the strain from certain muscles.

Assignment:

Label figure 12.3.

Complete the combined Laboratory Report 11,12.

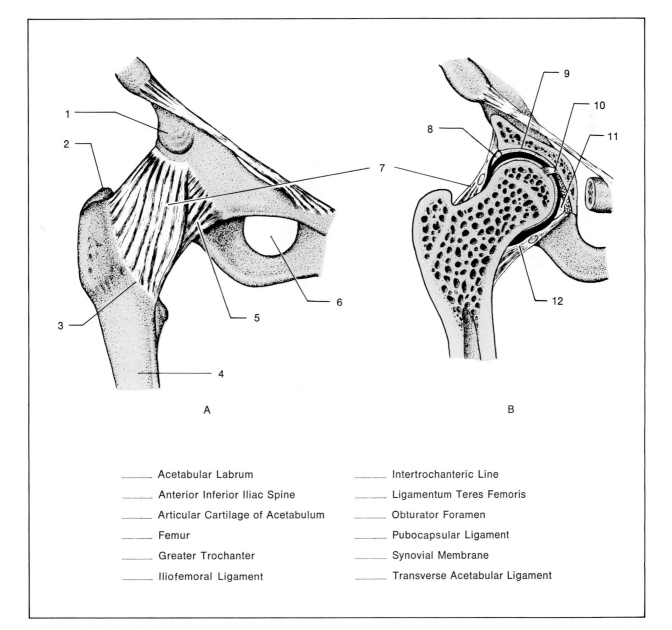

A   B

_____ Acetabular Labrum

_____ Anterior Inferior Iliac Spine

_____ Articular Cartilage of Acetabulum

_____ Femur

_____ Greater Trochanter

_____ Iliofemoral Ligament

_____ Intertrochanteric Line

_____ Ligamentum Teres Femoris

_____ Obturator Foramen

_____ Pubocapsular Ligament

_____ Synovial Membrane

_____ Transverse Acetabular Ligament

Figure 12.3   The hip joint.

The Knee Joint

Two views and a sagittal section of the knee are shown in figure 12.4. Although the action of this joint has been described above as being essentially hinge-like, it is, by no means, a simple hinge. The curved surfaces of the condyles of the femur allow rolling and gliding movements within the joint. There is also some rotary movement due to the nature of hip and foot alignment.

The knee joint is probably the highest stressed joint in the body. To absorb some of this stress are two *semilunar cartilages,* or *menisci,* in each joint. As revealed in the sagittal section, these fibrocartilaginous pads are thick at the periphery and thin in the center of the joint, providing a deep recess for the condyles. The **lateral meniscus** lies between the lateral condyle of the femur and the tibia; the **medial meniscus** is between the medial condyle and the tibia. These menisci are best seen in the anterior and posterior views of figure 12.4. Anteriorly and peripherally, they are connected by a **transverse ligament.**

The entire joint is held together by several layers of ligaments. Innermost are the two cruciate and two collateral ligaments. The **posterior** and **anterior cruciate ligaments** form an X on the median line of the posterior surface, with the posterior cruciate ligament being outermost. The anterior cruciate ligament is the outermost one seen on the anterior surface when the leg is flexed. The **fibular collateral ligament** is on the lateral surface, extending from the lateral epicondyle of the femur to the head of the fibula. The **tibial collateral ligament** extends from the medial epicondyle of the femur to the upper medial surface of the tibia. In addition to these four ligaments are the *oblique*

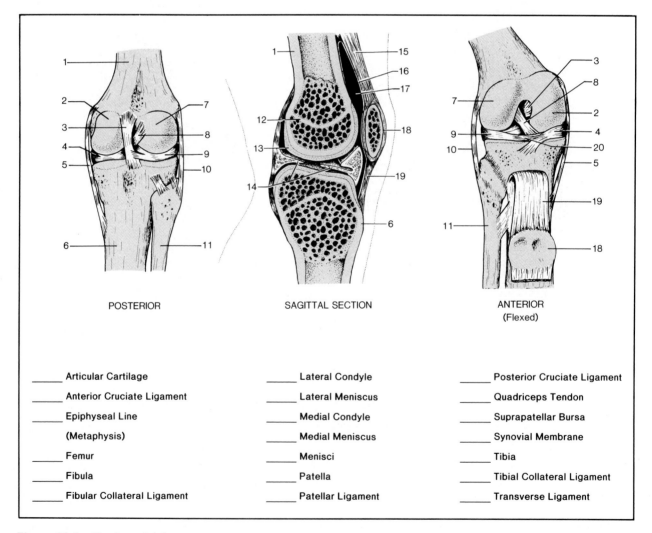

POSTERIOR

SAGITTAL SECTION

ANTERIOR
(Flexed)

_____	Articular Cartilage	_____	Lateral Condyle	_____	Posterior Cruciate Ligament
_____	Anterior Cruciate Ligament	_____	Lateral Meniscus	_____	Quadriceps Tendon
_____	Epiphyseal Line	_____	Medial Condyle	_____	Suprapatellar Bursa
	(Metaphysis)	_____	Medial Meniscus	_____	Synovial Membrane
_____	Femur	_____	Menisci	_____	Tibia
_____	Fibula	_____	Patella	_____	Tibial Collateral Ligament
_____	Fibular Collateral Ligament	_____	Patellar Ligament	_____	Transverse Ligament

Figure 12.4 The knee joint.

and *arcuate popliteal ligaments* on the posterior surface. These are not shown in figure 12.4.

Encompassing the entire joint is the *fibrous capsule*. It is a complicated structure of special ligaments united with strong expansions of the muscle tendons that pass over the joint.

Observe in the sagittal section that the kneecap is held in place by an upper **quadriceps tendon** and a lower **patellar ligament.** The inner surfaces of these latter structures are lined with **synovial membrane.** The space between the synovial membrane and femur is the **suprapatellar bursa.**

It is significant that although this joint is capable of sustaining considerable stress, it lacks bony reinforcement to prevent dislocations in almost any direction. It relies almost entirely on soft tissues to hold the bones in place. It is because of this fact that knee injuries are so commonplace today among the general populace, as well as professional athletes. It is a vulnerable joint, particularly to lateral and rotational forces. An understanding of its anatomy should alert one to its limitations.

Assignment:

Label figure 12.4.

Laboratory Assignment

Animal Joint Study. Examine the knee joint of a cow or lamb that has been sawed through longitudinally. Identify as many of the structures as possible that are shown in figure 12.4.

Laboratory Report. Complete the combined Laboratory Report 11,12.

Part 3 The Nerve-Muscle Relationship

The study of skeletal muscle physiology in Part 4 can be meaningful only if one understands the nature of neurons, muscle cells, and the neuromuscular junction. In this unit, the study of nerve and muscle tissue is broadly approached to include all three types of muscle tissue and various kinds of neurons to provide the proper basics for experiments in Parts 6, 7, 9, and 10, in addition to Part 4.

Although nerve and muscle cells are morphologically different, they do have one property in common that sets them apart from all other cells of the body: they are capable of transmitting electrochemical impulses along their membranes. This property is called *excitability.*

Excitability is a function of cell membrane electrical potential. Although virtually all cells of the body possess membrane potentials, only nerve and muscle cells demonstrate the ability to generate *action potentials* (impulses). Muscular function and neural coordination depend entirely on this property of excitability.

For our study of the different kinds of muscle and nerve cells in Exercises 13 and 14 we will examine prepared slides of various types to note the characteristic differences. The neuromuscular junction will be studied in Exercise 15.

The nervous system is composed of two kinds of cells: neurons and neuroglia. **Neurons** are the functional units of the nervous system in that they transmit nerve impulses. **Neuroglial cells,** or **glia,** are the connective, supportive, and nutritive cells of the nervous system and lack the ability to carry nerve impulses. Although both types of cells will be discussed in this exercise, only the neurons will be studied in the laboratory since our prime concern here is with the nerve-muscle relationship.

The first portion of this exercise is descriptive and should be studied prior to examining prepared slides in the laboratory. Figure 13.1 should be labeled and the questions on the Laboratory Report should be answered prior to laboratory studies. Since much of our knowledge of these cells is the result of electron microscopy, the text here will encompass cellular anatomy that cannot always be perceived with the compound microscope.

Basic Structure of Neurons

Neurons perform different roles in different parts of the nervous system. They exist in a variety of sizes and configurations. In spite of this diversity all neurons have much in common.

The large peripheral neurons illustrated in figure 13.1 are closely associated with skeletal muscle physiology. The cross-section through the spinal cord illustrates the relationship of these neurons to the spinal cord and to each other. Three kinds of neurons are shown: the sensory neuron, the motor neuron, and the interneuron. These three neurons, together with the receptor and effector, comprise a *reflex arc,* which is described more in detail in Exercise 25.

The Cell Body

The cell body or **perikaryon,** of a neuron usually contains a rather large nucleus, Nissl bodies, and neurofibrils. **Nissl bodies** are dense aggregations of rough endoplasmic reticulum scattered throughout the cytoplasm. They stain readily with basic aniline dyes such as toluidine blue or cresyl violet. They represent sites of protein synthesis. Some Nissl bodies are shown in the perikaryon of the motor neuron in figure 13.1. **Neurofibrils** are slender protein filaments that extend through the cytoplasm parallel to the long axis. They resemble roadways that circumvent Nissl bodies. It is significant that these minute fibers converge into the axon to form the core of the nerve fiber.

A most remarkable feature of neurons are their cytoplasmic processes: the axons and dendrites. Traditional terminology once categorized these processes purely from a morphological sense; however, numerous exceptions require more accurate and universal definitions that emphasize their functional attributes. A **dendrite,** or **dendritic zone,** functions in receiving signals from receptors or other neurons and plays an important part in integration of information. An **axon** is a single elongated extension of the cytoplasm that has the specialized function of transmitting impulses away from the dendritic zone. The cell body can be located at any position along the conduction pathway. In the motor neuron of figure 13.1 the dendrites are the short branching processes of the perikaryon; the axon is the long single process. Although a neuron will usually have several dendrites, there will be only one axon. Some neurons, such as the *amacrine cells* of the retina, have no axon at all.

Fiber Construction

The neural fibers of all peripheral neurons are enclosed by a covering of *Schwann cells.* This covering extends from the perikaryon to the peripheral termination of the fiber. In the larger peripheral neurons such as those seen in figure 13.1, the Schwann cells are wrapped around the axon in a unique manner to form a **myelin sheath.** External to the myelin sheath is a thin **sheath of Schwann,** or **neurilemma,** which is essentially the outer plasmalemma of the Schwann cells. The axons that have this thick myelin sheath with a sheath of Schwann are designated as being *myelinated.* The smaller peripheral neurons that have fibers enclosed in Schwann cells but lack the myelin sheath are said to have *unmyelinated* fibers. One of the most important functions of Schwann cells is to facilitate the regeneration of damaged fibers. When a fiber is damaged, regeneration occurs along the pathway formed by the Schwann cells.

The myelin sheath and sheath of Schwann are formed during embryological development by the folding of the Schwann cells around the axons. Figure 13.1 reveals an enlarged cross section of this lipid-protein sheath and neurilemma. The myelin sheath and sheath of Schwann are interrupted at regular intervals by **nodes of Ranvier,** which are points of discontinuity between successive Schwann cells; each internode consists of a single Schwann cell. Each node of Ranvier represents a place where the exposed axon lacks the myelin and neurilemmal sheath. These nodes enable the nerve impulses to travel with great rapidity by jumping from one node to another along the fiber. This form of nerve impulse transmission is known as *saltatory conduction.*

Types of Neurons

The variability in size, shape, and arborization of neurons is considerable. Neurons may be *multipolar, bipolar, unipolar,* or *pseudounipolar.* Some have perikarya as small as 4 micrometers; others may be as large as 150 micrometers. Some have perikarya that are encapsulated by satellite cells; others are not encapsulated. Many have myelinated fibers, others do not. A brief discussion of the principal types of neurons follows.

Motor Neurons As indicated in figure 13.1, these multipolar neurons have their perikarya located in the gray matter of the central nervous system

(CNS). They are also referred to as *efferent neurons* since they carry nerve impulses away from the CNS. The axons of these neurons are myelinated and terminate in *skeletal muscle fibers.* They function in the maintenance of voluntary control over skeletal muscles.

The perikarya of these neurons are somewhat angular or star-shaped rather than ovoid. The cytoplasm in a properly stained perikaryon reveals a nucleus, many Nissl bodies, and neurofibrils. Many short branching dendrites are present that make synaptic connections with axons of interneurons or sensory neurons. The myelinated axons often have **collaterals** that emanate at right angles from a node of Ranvier. Collaterals are branches that provide innervation to additional muscle fibers, other neurons, and even the same neuron, in some instances.

Sensory Neurons These neurons are also called *ganglion cells* because their perikarya are located in **spinal ganglia** (label 18, figure 13.1) outside the CNS. Since they carry nerve impulses from the receptor toward the CNS, they are also referred to as *afferent neurons.*

As illustrated in figure 13.1, the perikaryon of a sensory neuron is ovoid in shape and is covered with a capsule of **satellite cells.** These cells are continuous with Schwann cells and have the same relationship to these neurons that certain neuroglial cells have to the CNS. Satellite cells are not neuroglia, however, since they have a different embryological origin.

Although sensory neurons appear to be unipolar, histologists classify them as pseudounipolar on the basis of their embryological development. Note that the entire fiber is myelinated from the **receptor** in the skin to the synaptic connection in the spinal cord. The entire myelinated fiber is designated as an **axon** even though part of it functions like a dendrite in carrying nerve impulses toward the cell body. A **synapse** (label 1, figure 13.1) is the point where the terminus of an axon of one neuron adjoins the dendritic zone of another neuron.

Interneurons *(Association and Internuncial).* These multipolar unmyelinated neurons (figure 13.1) exist in large numbers in the gray matter of the CNS. They are essentially the *integrator cells* between sensory and motor neurons. Most reflexes involve one or more interneurons between the sen-

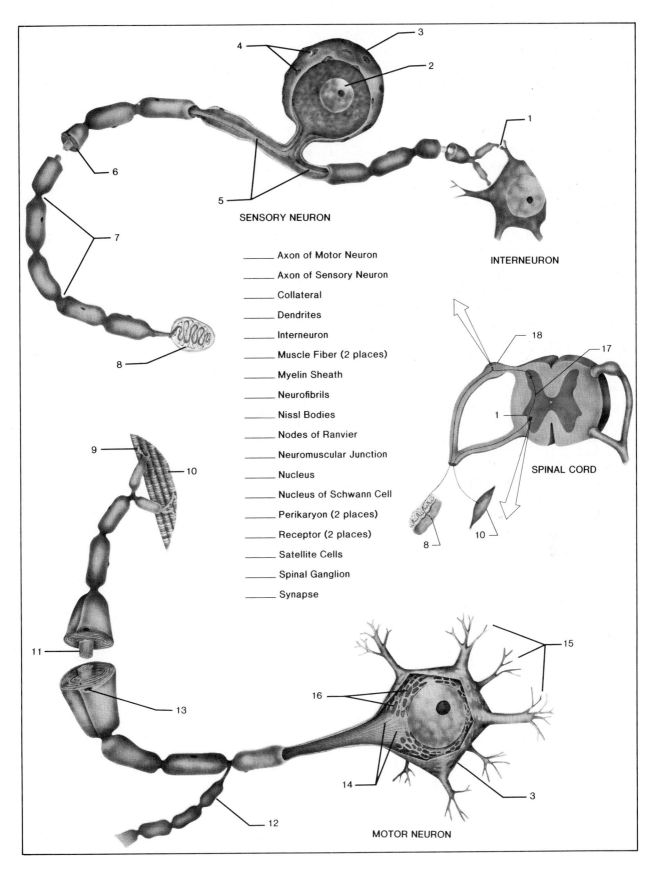

SENSORY NEURON

INTERNEURON

SPINAL CORD

MOTOR NEURON

_____ Axon of Motor Neuron

_____ Axon of Sensory Neuron

_____ Collateral

_____ Dendrites

_____ Interneuron

_____ Muscle Fiber (2 places)

_____ Myelin Sheath

_____ Neurofibrils

_____ Nissl Bodies

_____ Nodes of Ranvier

_____ Neuromuscular Junction

_____ Nucleus

_____ Nucleus of Schwann Cell

_____ Perikaryon (2 places)

_____ Receptor (2 places)

_____ Satellite Cells

_____ Spinal Ganglion

_____ Synapse

Figure 13.1   Sensory and motor neurons.

sory and motor neurons. However, some automatic responses, such as the knee jerk reflex, are monosynaptic in that they lack interneurons.

Assignment:

Label figure 13.1.

Pyramidal Cells These multipolar neurons are the principal cells of the cortex of the cerebrum. The upper neuron in figure 13.2 is of this type. Two strata of these cells are seen in the cerebral cortex. The perikarya of these neurons are somewhat triangular. A unique characteristic of these cells is that they have two sets of dendrites: (1) a long dendritic trunklike structure with many branches that ascends vertically in the cortex, and (2) several basal branching dendrites that extend outward from the surfaces of the perikaryon. The short axon shown in figure 13.2 is actually much longer than illustrated.

Stellate Cells *(Golgi Type II)* These neurons are associated with the pyramidal cells of the cerebral cortex. They are much smaller than pyramidal cells and exist in several different forms. Some are multipolar as the one shown in figure 13.2; others are bipolar. These cells provide linkage between pyramidal cells of different layers within the cerebral cortex.

Purkinje Cells The large multipolar neuron in the lower left-hand corner of figure 13.2 is a Purkinje cell. These cells are the predominant neurons located within the convoluted surface layer of the cerebellum of the brain. The large branching dendritic processes of these cells fill up most of the space in the outer molecular layer of the cerebellum. The axons of these neurons extend inward into the white matter of the brain.

Basket Cells These multipolar neurons (figure 13.2) are associated with the Purkinje cells in the cortex of the cerebellum. They have short, thick, branching dendrites and a long axon. The axons of these neurons usually have five or six collaterals that make synaptic connections with the dendrites of the Purkinje cells. These cells are considered to be association neurons.

Neuroglia

Numerically, the neuroglial cells of the nervous system exceed neurons by a ratio of approximately 5:1. The term *neuroglia* (nerve glue) is ap-

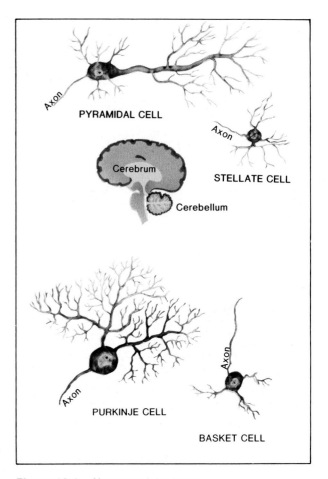

Figure 13.2 Neurons of the brain.

plied to the ependyma, which lines the ventricles of the brain and spinal cord, and to the neuroglial cells, which intermesh with neurons of the nervous system and retina. The satellite and Schwann cells, which function much like neuroglial cells but have a different origin, are often referred to as *peripheral neuroglia* to set them apart.

Although neuroglial cells do not carry nerve impulses, they do perform many important functions within the CNS. First, during embryological development, they provide a framework in the CNS for young developing neurocytes. As the nervous system develops, the neuroglial cells also seem to act as barriers to the formation of synaptic connections so that they control the development of functional reflex patterns in the CNS. They appear to play a role as mediators for the normal metabolism of neurons, and they are involved in the degeneration and regeneration of nerve fibers. They are also the chief source of tumors of the CNS.

The neuroglia are divided into three distinct morphological categories: astrocytes, oligoden-

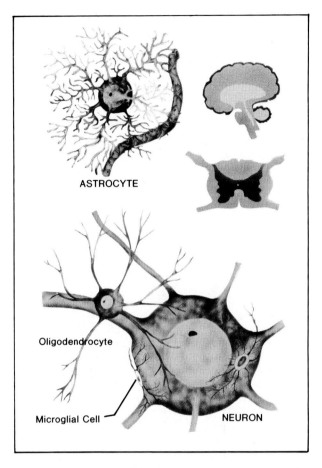

ASTROCYTE

Oligodendrocyte

Microglial Cell

NEURON

Figure 13.3 Neuroglial cells.

drocytes, and microglia. Representatives of each group are illustrated in figure 13.3.

Astrocytes These neuroglial cells have large cell bodies with numerous processes radiating out in all directions. There are two types: protoplasmic and fibrous astrocytes. The one shown in figure 13.3 is of the protoplasmic variety. These cells may be attached to blood vessels (figure 13.3), pia mater, or neurons, much like satellite cells.

Oligodendrocytes These cells have much in common with Schwann cells, since they develop in rows

along nerve fibers and produce myelin within the CNS. These cells are large like the astrocytes and together with astrocytes are often referred to as *macroglia*. In tissue culture, these cells exhibit rhythmic pulsatile movements, which are not well understood.

Microglia These cells are much smaller than macroglial cells. Two of them are seen attached to the perikaryon of a neuron in figure 13.3. The cytoplasmic processes on these cells are more twisted than the processes on astrocytes and oligodendrocytes. Microglia are scattered throughout the brain and spinal cord. They appear to function as CNS phagocytes.

Laboratory Assignment

After answering the questions on combined Laboratory Report 13,14 that pertain to cells of the nervous system, examine prepared slides of the nervous system using oil immersion optics whenever necessary. The following procedure will involve most of the cells discussed in this exercise so far.

Materials:

Prepared slides of cross-section of rat spinal cord, ganglion cells, cerebral cortex section, and cerebellar cortex section.

1. Examine a cross-section of the rat spinal cord with low power magnification. Scan the gray matter, looking for the large perikarya of motor neurons. When a cell is located, examine with the oil immersion objective. Look for **Nissl bodies, nucleus, neurofibrils,** and **processes.** Draw a few cells labeling all structures.
2. Follow the same procedure in examining ganglion cells, cerebral cortex, and cerebellar cortex slides. Look for all the different kinds of cells described in this exercise.

14 Muscle Tissue

The unique characteristic common to all muscle cells is **contractility.** This property is made possible by the presence of contractile protein fibers within the cells. The shortening of muscle cells, due to this property, is responsible for the various movements of appendages and organs of the body. Because of their elongated structure, muscle cells are often referred to as **muscle fibers.**

On the basis of cytoplasmic differences, there are two categories of muscle fibers: striated and smooth. **Striated muscle** is characterized by regularly spaced transverse bands called **striae.** **Smooth muscle** lacks these transverse bands and has other distinctions to set them apart. Striated muscle cells are further subdivided into skeletal and cardiac muscle. **Skeletal muscle** cells are multinucleated, or **syncytial.** Attached primarily to the skeleton, they are responsible for voluntary body movements. **Cardiac muscle** cells, which make up the muscle of the heart wall, are not syncytial and have less prominent striations than skeletal muscle.

In completing this exercise, follow the same procedures that were used in the last exercise. First, read over the text material and answer the questions on the Laboratory Report. Preparation in this manner will make the tissue studies in the laboratory more meaningful.

Smooth Muscle

Smooth muscle fibers are long, spindle-shaped cells as illustrated in figure 14.1. Each cell has a single elongated nucleus that usually has two or more nucleoli. When studied with electron microscopy, the cytoplasm reveals the presence of fine linear filaments of the same protein composition that are seen in striated muscle. The fact that both smooth and striated fibers contain actin, myosin, troponin, and tropomyosin indicates that the chem-

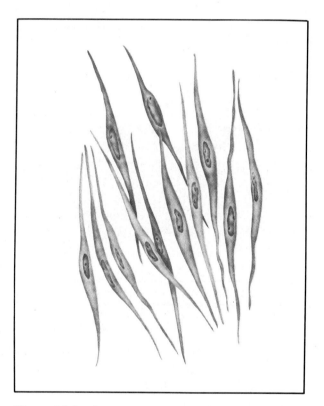

Figure 14.1 Smooth muscle fibers, teased.

istry of contraction is probably similar or identical in all types of muscle fibers.

Smooth muscle cells are found in the walls of blood vessels, walls of organs of the digestive tract, the urinary bladder, and other internal organs. They are innervated by the autonomic nervous system and, thus, are under *involuntary nervous control.*

Skeletal Muscle

Muscle fibers of this type are long, cylindrical, and multinucleated. Large numbers of these fibers are grouped together into bundles called **fasciculi.** The individual cells within each fascicle are separated from each other by a thin layer of connective tissue called **endomyosin.** This separateness of muscle fibers enables each cell to respond independently to nerve stimuli.

Examination of the cytoplasm, or **sarcoplasm,** with a compound microscope reveals a series of distinct dark and light striae. The detailed structure of these bands will be explored in a study of the chemistry of muscle contraction in Exercise 17. Surrounding each cell is a cell membrane called the **sarcolemma.** Note in figure 14.2 that the nuclei of each cell lie in close association with the sarcolemma.

Although striated muscle is referred to as skeletal and voluntary, there are certain muscles of this type that are not attached to bone or subject to voluntary control. Examples: (1) striated muscles of the tongue and external anal sphincter have no skeletal attachment; and (2) striated muscles of the pharynx and upper esophagus have no skeletal attachment and are not voluntarily controlled. In spite of these differences, these two aberrational examples are morphologically indistinguishable from other skeletal muscle.

Cardiac Muscle

Muscle cells of this type are found in the wall of the heart. A unique characteristic of cardiac muscle is its ability to contract rhythmically and con-tinuously, as a result of intrinsic cellular activity. Morphologically, cardiac muscle is readily distinguished from skeletal and smooth fibers; yet it shares some characteristics of each type.

Cardiac muscle differs from skeletal muscle in the following ways:

1. Instead of being anatomically syncytial, cardiac muscle is made up of separate cellular units that are separated from each other by **intercalated disks.** These disks stain somewhat darker than transverse striae.
2. Some fibers of cardiac cells are *bifurcated* to form a branched three-dimensional network instead of forming only long straight cylinders.
3. The elongated nuclei of each cellular unit lie deep within the cells instead of near the surface as in skeletal muscle.
4. Cardiac tissue is not normally subject to voluntary control.

Laboratory Assignment

Examine prepared slides of the various kinds of muscle tissue and make drawings of each type. Label the various structures discussed above. Complete Laboratory Report 13,14.

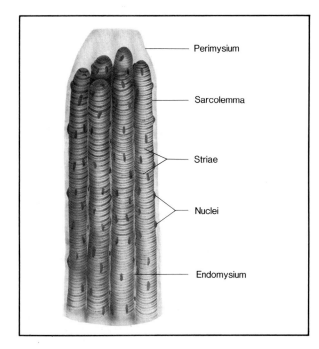

Figure 14.2 A fascicle of skeletal muscle fibers.

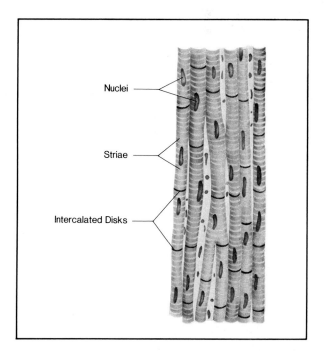

Figure 14.3 Cardiac muscle fibers.

15 The Neuromuscular Junction

It was observed in Exercise 13 that motor neurons emerge from the spinal cord through the anterior roots of spinal nerves to innervate skeletal muscle fibers throughout the body. Each large myelinated nerve fiber branches many times through collaterals to stimulate from 3 to 2000 skeletal muscle fibers. At the terminus of each branch, a junction called the *neuromuscular junction* is made with a muscle fiber. This juncture is usually made at approximately the midpoint of the fiber so that the action potential generated in the fiber will reach the opposite ends of the fiber at about the same time. Normally there is only one junction per muscle fiber.

Figure 15.1 reveals the structure of a neuromuscular junction as determined by electron microscopy. Chemical studies have uncovered a series of events that occur here. After labelling the structures in figure 15.1, answer the questions on the Laboratory Report.

Note in figure 15.1 that the end of the neuron consists of a cluster of knob-like structures that lie within invaginations of the **sarcolemma** (label 7) of the muscle fiber. The invaginations are called **synaptic gutters;** the knob-like projections are **axon terminal branches.**

Between the axon terminal branches and lining of the synaptic gutter is a gap, the **synaptic cleft,** which is approximately 20 μm wide and is filled with a gelatinous ground substance. The **presynaptic membrane** (label 5) and the **postsynaptic membrane** (label 4) form the surfaces on each side of this cleft. Note, also, that the surface area of the postsynaptic membrane is considerably in-creased by the presence of further infoldings to form **subneural clefts** (junctional folds).

The transmission of an impulse from the axon terminal across the synaptic cleft involves chemical rather than electrical transmission. The basic sequence is believed to include depolarization of the neurilemma, exocytosis of *acetyl choline (ACh)* transmitter, and the development of an action potential on the sarcolemma.

When the axonal action potential (AP) reaches the presynaptic membrane it initiates exocytosis of **synaptic vesicles** (label 3) that contain ACh. This coalescence of membranes results in release of ACh into the cleft. ACh quickly diffuses to its specific postsynaptic receptors, creating sarcolemmal depolarization and originating an action potential. This phenomenon is followed by (1) the inward propagation of the AP to the **T-tubule system** (label 2) and the **sarcoplasmic reticulum** (label 1); (2) the release of calcium ions by the sarcoplasmic reticulum, thus permitting contraction of the myofibrils; and (3) the inactivation of ACh by choline esterase within 1/500 of a second.

Energy for contraction is supplied by ATP from mitochondria in the area. When depolarization and contraction have occurred, repolarization of the membrane takes place. The muscle fiber is now ready for another nerve impulse to initiate contraction.

Laboratory Report

Complete the Laboratory Report for this exercise.

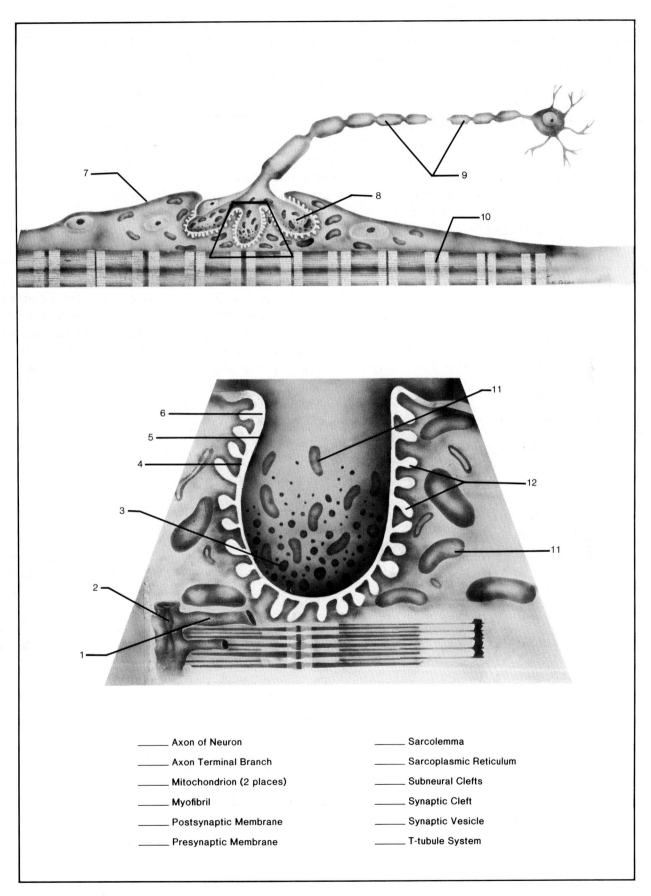

Figure 15.1 The neuromuscular junction.

_____ Axon of Neuron

_____ Axon Terminal Branch

_____ Mitochondrion (2 places)

_____ Myofibril

_____ Postsynaptic Membrane

_____ Presynaptic Membrane

_____ Sarcolemma

_____ Sarcoplasmic Reticulum

_____ Subneural Clefts

_____ Synaptic Cleft

_____ Synaptic Vesicle

_____ T-tubule System

Part 4 Skeletal Muscle Physiology

This unit contains three exercises pertaining to muscle physiology and one exercise on instrumentation. Exercise 16, the first one in this unit, presents an in-depth overview of the types of electronic instrumentation that are used throughout this book. Experiments in Exercises 18 and 19 employ most of the different types of equipment described in Exercise 16. Attempting to perform these experiments without a clear understanding of how these instruments can be effectively used would be absurd. It is anticipated that you will answer all the questions on Laboratory Report 16 prior to attempting Exercises 18 and 19.

Exercise 17 is a study of the microstructure of muscle fibers as related to the property of fiber contractility. The microscope will be used here to note the changes that occur when contraction is chemically induced in a muscle fiber.

In Exercise 18, frog muscle tissue will be used to study three phenomena of skeletal muscle contraction: muscle twitch, motor unit summation, and wave summation. Since this experiment is somewhat lengthy, students will have to work in teams of five or six to complete the experiment in the allotted time. Exercise 19, on the other hand, is relatively simple and will be performed by smaller teams of only three students per team.

Electronic Instrumentation

The study of physiological phenomena, such as muscle contraction, heart action, or brain activity, often requires the use of sophisticated electronic instrumentation. An understanding of the nature of this type of equipment is essential if one is to perform such experiments. Two exercises in this unit and many other experiments throughout this book utilize electronic instrumentation. It is the purpose of this exercise to provide you with some basic essentials that will enable you to proceed with a sense of confidence in these experiments.

Figure 16.1 reveals the arrangement of equipment that might be used in a typical instrumentation setup. First, one must use a device, or combination of devices, to pick up the biological signal that is being studied. If the signal is electrical this device will be an **electrode;** nonelectrical signals, such as temperature or pressure changes, however, require the utilization of an **input transducer.** Since biological phenomena often produce weak electrical signals that require augmentation, they are next fed into an **amplifier.** Finally, the amplified signals are converted to some form of visual or audio display with an **output transducer,** which may be a chart recorder, loud speaker, or some other such type of equipment.

Electrodes, transducers, amplifiers, and electronic stimulators are described here in general terms. Our main concerns here will be to learn about each component's functions, its limitations, and its vulnerability to damage. Since much of this equipment is costly, proper use and handling is of paramount importance.

Electrodes

Electrodes are devices that may induce an electrical current into a biological specimen as well as receive an electrical signal from a specimen. If an electrical current is carried to the specimen, it is anticipated that some measurable reaction will be produced in the tissue. On the other hand, if an electrode picks up an electrical signal from the specimen it must be fed directly to the amplifier and output transducer.

An electrode may be a bare wire, a metal disk to which a wire is attached, a wick saturated with conductive solution, or almost any conducting device. Figures 16.2 and 16.3 illustrate various kinds of electrodes. The type of electrode that is to be used depends on such factors as electrical conductivity, toxicity to tissue, size, chemical prop-

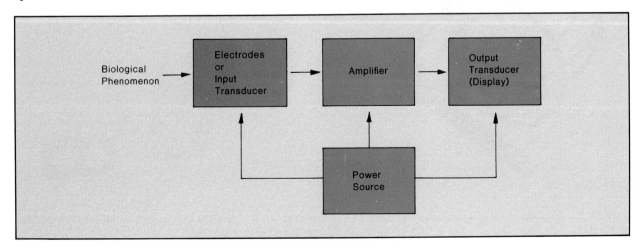

Figure 16.1 Equipment setup for monitoring biological phenomena.

erties, and cost. Metals such as platinum, palladium, gold, and nickel-silver alloys are most frequently used because of their relative inertness and good conductivity.

The greatest problem in using electrodes that pick up electrical signals through the skin is making positive conductive skin contact. The resistance of the cornified outer layer of skin cells to low voltage electrical transmission is considerable. To overcome this, it is often necessary to rub off some of the outer dead skin cells with an abrasive material and place a high conductance medium between the skin and electrode. Such a solution may be 0.05N saline, or a commercially prepared paste or jelly.

Input Transducers

Since electrodes are only able to pick up electrical currents from biological phenomena, they cannot be used as sensing devices for thermal changes, acoustical expression, movement, or any other form of energy. To accommodate these other energy forms, one must use a transducer.

A *transducer,* by definition, is any device that converts a nonelectrical signal to an electrical signal, or vice versa. An **input transducer** is a device that converts a nonelectrical form of energy to an electrical signal for amplification. Although there are many types of transducers that can be used in different applications, only force and pulse transducers will be discussed in detail here since they are the principal ones used in this manual.

Force Transducers

For our study of muscle contraction in Exercise 18 we will use a force transducer of the type shown in figure 16.4. Note that the transducer has a number of flexible steel leaves to which a string is attached to measure the force exerted by a contracting muscle. If the forces applied are weak, only a single leaf is used; strong forces are accommodated by using two or more leaves. If you study figure 18.1 you will see how this type of a transducer is set up. By varying the number of leaves, one can measure forces that range from 10 mg. to 10 kg., a ratio of 1:1,000,000.

This type of transducer utilizes a resistor which is affected by stress that is applied to the steel leaves. As deformation occurs in the spring leaves due to loading, proportional bending also occurs in the resistor. The induced stretching or compression of the resistor changes its resistivity. Resistivity changes, in turn, result in changing electrical signals that pass to the amplifier, accomplishing the conversion of mechanical activity to electrical activity.

One complication with this type of transducer is that the resistor is one of the variable resistors of a *Wheatstone bridge* (see figure 16.5). This fact makes it desirable to electronically balance the Wheatstone bridge of the transducer by making certain adjustments on the output transducer before any measurements are made. Specific instructions for balancing are outlined in Appendix C.

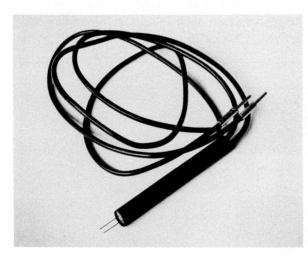

Figure 16.2 An electrode that works well in frog muscle experiments.

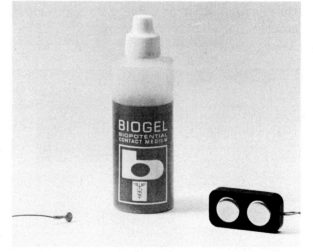

Figure 16.3 An EEG electrode (on the left), electrode jelly, and skin electrodes.

Pulse Transducers

Several different kinds of transducers have been developed for monitoring the human pulse. If you will refer to figure 42.1 in Part 9, you will see the setup for using this type of transducer. Figure 16.6 illustrates how this type of transducer works. Note that a small light source produces a beam of light that is transmitted through the tissues of the finger. A few millimeters away from the light source is a photosensitive resistor (photoresistor). When a surge of blood passes through the pad of the finger, the amplitude of light is altered, resulting in a change of reflected light that strikes the photoresistor. Light striking the photoresistor generates a signal which is fed to the amplifier and output transducer.

Amplifiers

Unless the weak electrical signals of biological systems are increased many times it is difficult to monitor them. Electronic amplifiers accomplish this task.

Amplifiers are usually designed to function within a limited signal to noise range. The signal received by the amplifier (input signal) must have sufficient intensity to overcome the electrical *noise* of the amplifier's electronic circuitry, yet not be so great that it overloads and distorts the amplifier output. It is very desirable that the amplifier allow signals to be increased in amplitude only, and not be distorted.

Types of Amplifiers

In some experimental setups the amplifier may be an independent unit; in other setups it may be incorporated into the output transducer, with its controls being on the control panel of the transducer. For equipment that is designed to do many things, such as the Gilson Unigraph (figure 16.8), a multipurpose amplifier is incorporated into the unit that is able to handle signals of considerable variability in strength.

Linearity

The sensitivity range of an amplifier will determine what size of input signal it can accept. If the input signal does not exceed the inherent noise of its circuitry, however, it is not within its minimum range. On the other hand, the input signal must not be so great that it distorts the amplifier output. The range of frequency in which an amplifier handles input without distortion is called its *linearity*. This information is important to the operator since it determines the fidelity of readout in the final analysis.

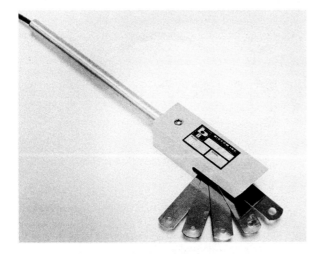

Figure 16.4 A strain gage force transducer, such as this one, utilizes a stress sensitive resistor to convert stress into measurable electrical signals. When stress is applied to the leaves the resistor is also bent, which affects its electrical properties. The resistor functions through a Wheatstone bridge (see figure 16.5).

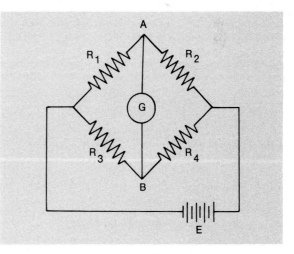

Figure 16.5 In this schematic of a Wheatstone bridge, the stress sensitive resistor is represented by R_1. Bending of this element causes unbalancing of the bridge due to increased resistance to electrical flow through it. When the bridge is not in balance an electrical signal is produced that is proportional to the stress produced.

Power Supply

Amplifiers need direct current (DC) for certain functions. Since their external source of electricity is usually conventional 110–120V alternating current (AC), they must have a power supply unit which converts the AC to DC. In addition to making this conversion, the power supply unit either increases or decreases the voltage required by the amplifier. Since most amplifiers are partially or entirely transistorized today, the voltage requirements are quite low; this results in physically smaller power supply units that are often an integral part of the amplifier unit.

Electrical Interference

Electrical currents through power cables, electrical appliances, and lighting systems generate magnetic and electrostatic fields that can induce interference in amplifier units. To prevent this type of electrical interference in electrical systems, interconnecting cables are usually *shielded*. This shielding usually consists of a braided metallic covering between the outer and inner insulation coatings, which is grounded to carry away spurious signals. Additional methods for minimizing electrical interference will be suggested in various experiments where the problem becomes more acute.

Output Transducers

As we have seen, the output transducer is the inverse counterpart of the input transducer. While the input transducer transforms some nonelectrical form of energy into an electrical signal, the output transducer converts an amplified electrical signal into a perceptible display. The type of output transducer that one might use in an experiment will depend, partially, on how the display record is to be used.

Although there are many different kinds of output transducers available (panel meters, oscilloscopes, audio systems, chart recorders, etc.) our focus here will be on the types of equipment that are used in experiments in this book.

The Audio Monitor

One very useful diagnostic procedure for the cardiologist is to listen to the heart beat of the patient. Usually, this is performed with a stethoscope; however, if the sound is to be heard by two or more individuals at the same time, a loud speaker, or *audio monitor* (figure 16.7), may be used. The input transducer for one of these units is a small microphone which is placed over specific auscultatory areas of the heart. A microphone is seen suspended on the left side of the audio monitor in figure 16.7.

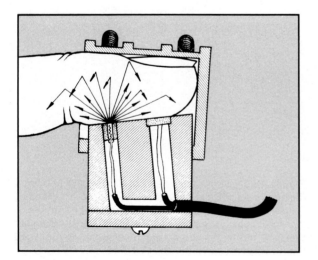

Figure 16.6 A pulse transducer uses light rays and a photoresistor to produce a signal.

Figure 16.7 The audio monitor with microphone may be used for ausculating heart and respiratory sounds.

Chart Recorders

Permanent traces on paper are achieved with chart recorders. These units record information on a moving chart which is graduated with lines that are one millimeter apart. The display is produced by a stylus, or pen, which may or may not utilize ink.

On most chart recorders the paper can be moved along at different speeds. For some experiments a slow speed is desirable; for others a faster speed may be required. The better recorders usually have an amplifier and recorder incorporated into the chasis.

Although there are many different recorders on the market we will focus our attention on only two here: the Unigraph and Duograph. Both are products of Gilson Medical Electronics, Inc. The reason for featuring these instruments instead of others is that they are the types used in this book. Your laboratory may be equipped with some other brand, but you will find that there is considerable similarity between these and other units. The following discussion can readily be adapted to other brands.

The Unigraph

There is probably no other recorder as small as the Unigraph that is able to monitor as many bi-ological phenomena as the Unigraph. It is only twelve inches long, six inches wide, nine inches high and can easily be transported from one place to another. With this single unit, one can monitor blood pressure, pulse rate, brain waves (EEG), heart action (ECG), muscle action (EMG), eye movements, respiratory rate, lung volume, organ secretion, and many other physiological activities. A study of the various controls will reveal how so many things can be accomplished with this unit. The control panel is illustrated in figure 16.9. The function of each control follows.

Mode Selection Control Any multipurpose chart recorder will have a control which can be set for specific functions. On the Unigraph the mode control is the yellow knob that can be set at six different options: EEG, ECG, CC-Cal, DC-Cal, DC, and Trans. Identify this control and its different modes in figure 16.9. Whether balancing a transducer or monitoring an EEG, the correct setting of this control is of paramount importance.

Stylus Heat Control There are two styluses on the Unigraph: the **recording stylus** and the **event marker.** These styluses are heated elements that produce a record on the chart by converting a heat sensitive chemical on the paper into a visible blue line. The width of the line is determined by the

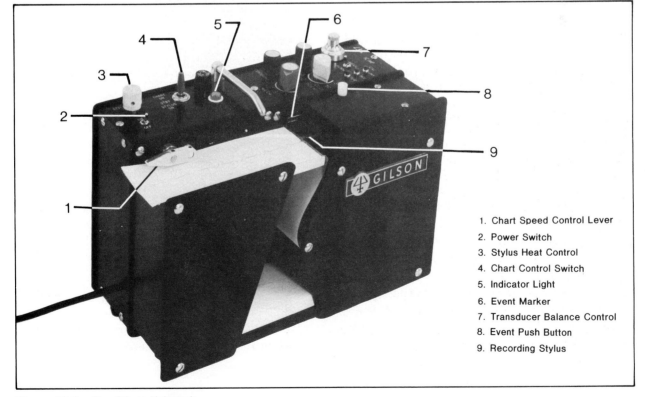

1. Chart Speed Control Lever
2. Power Switch
3. Stylus Heat Control
4. Chart Control Switch
5. Indicator Light
6. Event Marker
7. Transducer Balance Control
8. Event Push Button
9. Recording Stylus

Figure 16.8 The Gilson Unigraph.

amount of electrical current through the stylus. This regulation is made with the **stylus heat control,** label 3 in figure 16.8. To produce a visible line on the moving chart, the stylus heat control is usually placed first with its indicator line set at the two o'clock position.

While the recording stylus is used to record the amplified input signal, the event marker is used to record when an event begins or when some new pertinent influence affects the recorded event. The event may be recorded manually by depressing the white **event push button** at the beginning of the event, or it may be recorded automatically if an **event synchronization cable** is used.

Centering Control (Zero Offset Control) Note in figure 16.9 that on the left side, near the middle of the panel is a knob labeled CENTERING. This control is used for positioning the recording stylus in the most desirable position. If the tracing produced by the stylus is too near one edge of the chart, it can be moved closer to the center by turning this control. Sometimes it is desirable, however, to have the baseline of the tracing near one of the borders of the chart and not in the center. The position of the baseline will be determined by the direction and magnitude of excursion of the stylus in any particular event.

Chart Speed The chart speed can be adjusted so that it moves along at a **slow speed** of 2.5 mm. per second or a **fast speed** of 25 mm. per second. The chart speed control lever is label 1 in figure 16.8. When the pointed end of the lever is on the left, as shown, it is set for the fast speed. To slow down the chart, the lever is rotated clockwise 180° to the position shown in figure 16.9. For the chart to move at all, however, the chart control switch has to be in the Chart On position.

Toggle Switches The Unigraph has four toggle switches, two of which are extensively used in its operation. The main **power switch** (label 2, figure 16.8) is the small one at the left end of the unit. As shown in figure 16.9, this switch has ON and OFF settings, which enables one to turn the entire instrument on or off.

The large gray plastic switch to the right of the power switch is the **chart control switch.** It has three settings: STBY, Chart On, and Stylus On. See figure 16.9. In the STBY (standby) position

the chart does not move and the stylus is deactivated. When placed at Stylus On, the stylus heats up, but the chart does not move. At the Chart On position the chart moves and the stylus is activated.

There are two small toggle switches located near the TRANS-BAL control. Three labels, HI FILTER, NORM, and MEAN, identify their modes. Identify them in figure 16.9. For most experiments these switches should be set toward the NORM setting as shown in this illustration. If a

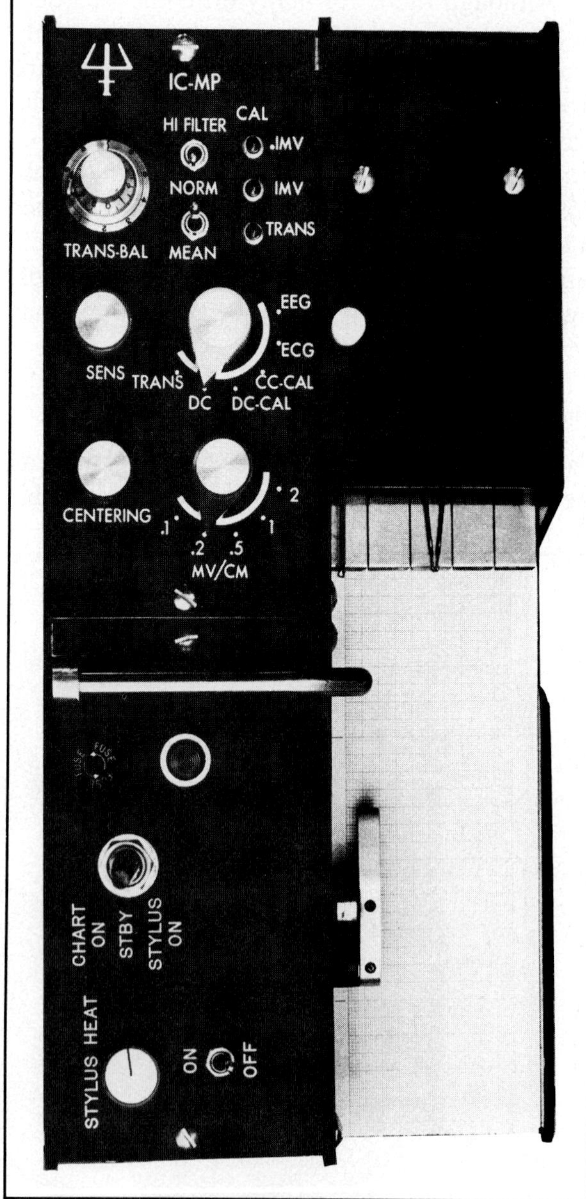

Figure 16.9 The Unigraph control panel.

different setting is required the instructions in the experiment will point this out.

Calibration Push Buttons Near the small toggle switches mentioned earlier are three small push buttons which have .1MV, 1MV, and TRANS labels near them. The millivolt buttons are used to calibrate certain types of measurements. The TRANS button is used in conjunction with transducer calibration. Detailed procedures for using these buttons are outlined in Appendix C.

Sensitivity Controls Sensitivity of the Unigraph is controlled by the red **gain control** and the blue **sensitivity knob.** The gain control is calibrated in MV/CM (millivolts per centimeter) and it has five ranges: 2, 1, .5, .2, and .1. The least sensitive position is 2; the greatest sensitivity is at .1. When the sensitivity knob is turned counterclockwise all the way, it is in the least sensitive position.

Operational Procedures

When setting up the Unigraph in an experiment one should follow these procedures:

1. Insert the electrode cable or transducer terminal into the receptacle located at one end of the Unigraph. For some transducers it is necessary to provide a special adapter between the Unigraph and the transducer.
2. If an electronic stimulator is going to be used, it is desirable to have an event synchronization cable between the stimulator and the Unigraph.
3. With the chart control switch at STBY and the power OFF, insert the power cord of the Unigraph into a grounded 110V electrical outlet. Now, turn on the power switch.
4. Rotate the stylus heat control knob to the two o'clock position.
5. After allowing a few minutes for the stylus to heat up, put the chart control switch (c.c.) in the Chart On mode to see if the stylus is producing a distinct visible line on the chart. For this test it is desirable to have the chart speed control lever set at slow speed. If the line is too light, increase its darkness by adjusting the stylus heat control knob. Return the c.c. switch to STBY.
6. If a force transducer is going to be used, balance the transducer, following the procedure that is outlined in Appendix C.
7. Calibrate the unit, if recommended. When

Figure 16.10 The Duograph.

monitoring an EEG, ECG, or EMG this procedure is recommended. Calibrating instructions are given in Appendix C.

8. Set the mode control on the proper setting.
9. Set the sensitivity and gain controls. These controls are often set at their lowest setting at the beginning of an experiment.

Once the above settings have been completed you are ready to proceed with the experiment. After the experiment is completed *all controls should be returned to their least sensitive settings and the switches should be in Off or STBY position.*

The Duograph

When it is necessary to monitor two physiological phenomena instead of one, the Unigraph cannot be used; it is a single channel recorder. To record two parameters simultaneously, such as respiratory and pulse rates, one must use a unit that has at least two channels. There are several polygraphs (Gilson, Narco, Grass, etc.) that can be used for this type of experiment. Most polygraphs have five channels, but they can be purchased with less. The Duograph has the advantage over the larger polygraphs in being less expensive and more portable. It weighs only twenty-five pounds, so it is readily transportable from one place to another.

The Duograph is illustrated in figure 16.10. It actually consists of two modules that are identical to the one used in the Unigraph. Once you have used a Unigraph, you have learned about all you need to know to operate a Duograph.

To accommodate two tracings, the chart on the Duograph is 127 mm. wide (Unigraph: 60 mm.). One difference between the Unigraph and Duograph is that it is available with ten speeds as well as two speeds. When purchased with two speeds, the speeds are identical to that of the Unigraph (2.5 and 25 mm. per second).

The Electronic Stimulator

In experiments where it is desirable to supply a precise amount of electrical stimulus to a tissue, an electronic stimulator, such as the Grass SD9 (figure 16.11), is the instrument of choice. The electronic stimulator accomplishes the following: (1) it converts AC to DC; (2) it can vary the voltage from .1 to 100 volts; (3) it can regulate the length of time that the electrical shock is delivered to the tissue; (4) it can regulate the number of electrical pulses that are delivered to the tissue in a period of time; and (5) it can regulate the length of time between electrical pulse pairs. All of these functions are performed with great precision and safety to the operator.

Control Functions

A study of the controls reveals how these various functions take place. Note that there are four control knobs with the words Frequency, Delay, Du-

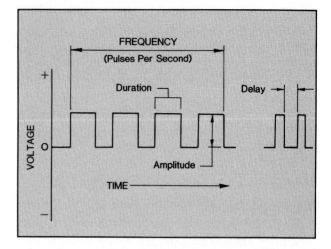

Figure 16.12 Square wave characteristics.

ration, and Volts above them. Under each control knob is a decade switch that has three or four settings (X.01, X.1, X1, X10). Figure 16.12 illustrates the wave characteristics that are produced by these controls. A discussion of each function follows.

Frequency This control, which is the upper left-hand knob, regulates the number of electrical pulses delivered per second. If the knob is set on 2 and the decade switch is set at X.1, the stimulator will deliver .2 pulses per second (or, one pulse every five seconds). If the knob is set at its maximum graduation, or 20, and the decade switch is

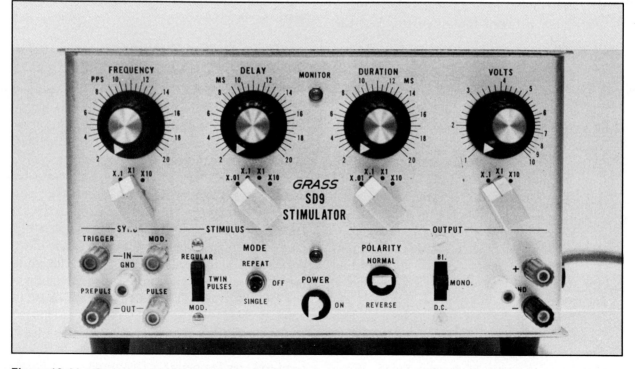

Figure 16.11 Control panel of Grass electronic stimulator.

set on its maximum, or X10, the stimulator will deliver 200 pulses per second (PPS.) Thus, we see that the range of frequency for this unit is .2 to 200 PPS. When this control is used it is necessary to place the Mode Selector Switch in the RE-PEAT position.

Delay This control is used in conjunction with the twin pulse mode. The time interval between the pairs of pulses can be varied from .02 to 200 milliseconds (msec.) with various combinations of the control knob and decade switch. Either single or repetitive pulses may be produced. When the repeat mode is used, the delay time should not exceed 50% of the period between pulses, to avoid overlap.

Duration This control determines the length of time a pulse of a given frequency and voltage is delivered to a preparation. The duration range is from .02 to 200 msec. As with the delay control, the duration setting should not exceed 50% of the interval between pulses.

Voltage The intensity of the stimulus is regulated by this control. Note in figure 16.12 that the amplitude of the wave is an expression of voltage. The voltage output of the SD9 is from .1 to 100 volts.

Operational Procedures

When setting up an electronic stimulator one should follow these steps:

1. Plug the power cord into a grounded 3-hole 110 volt outlet.
2. Insert the jacks for the electrode into the electrode posts and tighten securely. If the electrodes have designated anode and cathode prongs, be sure to insert the jack for the anode into the black negative (−) post and the cathode jack into the red positive (+) post.
3. Set the **Output Switch** on MONO (monophasic) unless instructed otherwise. The BI (biphasic) position is preferred for experiments in which stimulations are produced over

a longer period of time than we usually employ in this laboratory. There are exceptions, however.

4. Place the **Polarity Switch** in the NORMAL mode. When in this position the red post (+) acts as a positive post. If one wishes to change this output post to negative without reversing the leads on the posts, one places the polarity switch on REVERSE. This maneuver makes the red post negative.
5. Turn on the **Power Switch** to activate the unit. The red monitor lamp should light up to indicate that the circuitry is active. Although the unit is functioning, no electrical pulses will be entering the specimen as long as the mode switch is in the OFF position.
6. Set the **Voltage** controls for the desired voltage. In our first experiments the voltage will be set at minimum levels.
7. Set the **Duration** controls for the desired duration. For beginning experiments the duration is often set at 100 msec. (knob at 10, decade switch at X10).
8. If repetitive pulses are to be administered, set the **Frequency** controls.
9. If twin pulses are to be used, set the **Delay** controls. When twin pulses are administered in repetition (repeat mode), the delay should not exceed 50% of the period between each set of pulses, to avoid overlap.
10. Place the **Mode** switch in REPEAT or SINGLE as required in the experiment. When set on the repeat mode the monitor lamp will blink to indicate that the unit is in the repeat mode. The single position is only active when held in place manually; when released, it automatically goes to the OFF position.
11. When finished with the experiment, return all controls to their lowest values and turn off the instrument.

Laboratory Report

Complete the Laboratory Report by answering all the questions.

17 The Physicochemical Properties of Muscle Contraction

The microscopic changes that occur during skeletal muscle contraction will be observed here in a controlled experimental setup. The physicochemical nature of muscle contraction will be reviewed first, for clarification of this complex phenomenon.

Skeletal muscle contraction involves many components within the muscle cell. The entire process is initiated by an action potential that reaches the neuromuscular junction. As we learned in Exercise 15, acetylcholine initiates depolarization of the muscle fiber. The depolarization proceeds along the surface (sarcolemma) of the muscle fiber and deep into the sarcoplasm by way of the T-system tubules of the sarcoplasmic reticulum.

Figure 17.2 illustrates the ultrastructure of a single muscle fiber. Note that the sarcoplasm of the muscle fiber consists of many long tubular **myofibrils,** the **sarcoplasmic reticulum,** and **mitochondria. Nuclei** are seen in abundance beneath the **sarcolemma.** The abundance of **mitochondria** and their close association to the myofibrils indicates an availability of ATP for contraction.

Illustrations C and D in figure 17.2 reveal the molecular structure of the contractile **myofilaments** of the myofibrils. While myofibrils can be seen with a good compound microscope, myofilaments are visible only with electron microscopy. Note that there are two types of myofilaments: actin and myosin. **Myosin** filaments are the thickest filaments and are the principal constituents of the dark **A band.** They are slightly thicker in the middle and taper toward both ends. The thinner **actin** filaments extend about 1 μm in either direction from the **Z line** and make up the lighter **I band.**

The shortening of myofibrils to achieve muscle fiber contraction is due to the interaction of actin and myosin in the presence of calcium, mag-

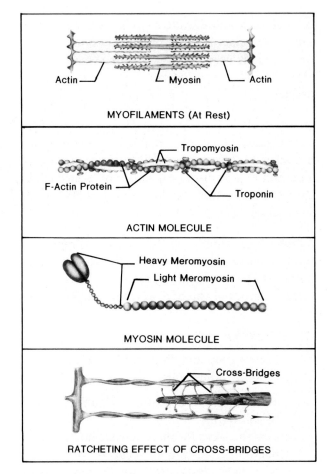

Figure 17.1 Myofilament structure.

nesium, and ATP. Two other protein molecules, troponin, and tropomyosin are also involved.

As the depolarization wave (action potential) moves along the sarcolemma and T-tubule system, large quantities of calcium ions are released by the sarcoplasmic reticulum. As soon as these ions are released, attractive forces develop between the actin and myosin filaments which causes the ends of the actin fibers to be pulled toward each other, as shown in illustration D. The exact mechanism that causes this shortening of the myofibrils is unknown, but the following explanation is the most acceptable theory we have at the present time.

It is in the nature of the actin and myosin molecules that we seek an answer to the attractive forces. The myosin molecule is made up of **heavy meromyosin** and **light meromyosin** (figure 17.1). The heavy meromyosin forms a head that is flexible and able to pivot. The actin filament consists of three components: a double stranded helix of **F-actin protein, tropomyosin,** and **troponin.** The tropomyosin is a long protein strand that lies within the groove of the F actin protein. The third protein, troponin, occurs at various active sites along the helix. The cross-bridges of heavy meromyosin are able to bridge the gap between the myosin and actin molecules.

The *ratchet theory of contraction* proposes that the following series of events takes place:

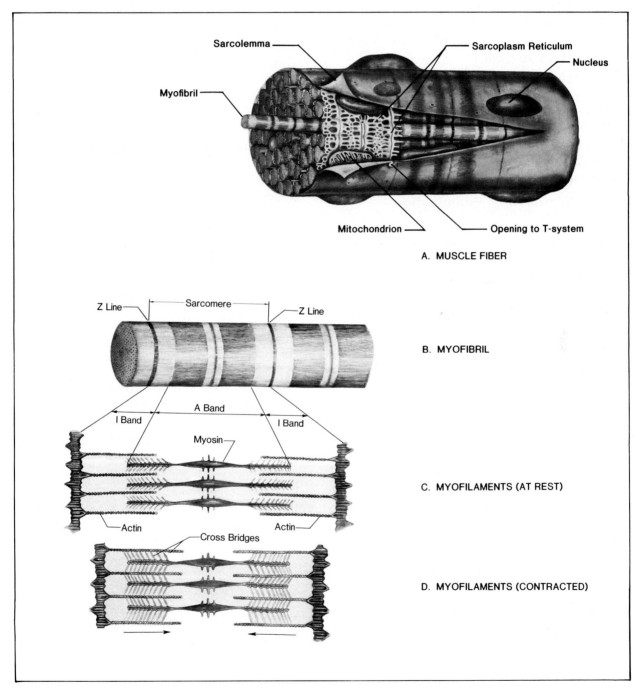

A. MUSCLE FIBER

B. MYOFIBRIL

C. MYOFILAMENTS (AT REST)

D. MYOFILAMENTS (CONTRACTED)

Figure 17.2   The ultrastructure of a muscle fiber.

1. Calcium ions combine with troponin.
2. Calcium bound troponin causes the tropomyosin to be pulled deeper into the groove of the two actin strands, exposing the active sites on the actin filament.
3. The heads of the cross-bridges form a bond with the active sites and pull on the actin filament with a ratchet-like power stroke. The energy for the first power stroke is derived from a meromyosin molecule that was previously activated by ATP.
4. After the first power stroke, ATP combines with the head of the cross-bridge, causing the head to separate from the active site to which it has been attached.
5. The ATP on the head is split by ATPase which is contained in the heavy meromyosin. The energy released by this cleavage is used for cocking the head again, attaching the head to another active site, and performing another power stroke.

Microscopic Examination

To observe the nature of muscle contraction at the microscopic level, we will induce contraction in glycerinated rabbit muscle fibers with ATP and ions of magnesium and potassium. Measurements of the muscle fibers will be made before and after contraction to determine the percentage of contraction that occurs. Microscopic comparisons will be made of fibers in both the relaxed and contracted state to determine the changes that occur.

Materials:

    test tube of glycerinated rabbit psoas muscle (muscle is tied to a small stick)
    dropping bottle of ATP in triple distilled water
    dropping bottle of 0.25% ATP, 0.05M KCl, and 0.001M $MgCl_2$
    3 microscope slides and covering glasses
    sharp scissors
    forceps
    dissecting needle
    Petri dish
    plastic ruler, metric scale
    microscopes, dissecting and compound

1. Remove the stick of glycerinated muscle tissue from the test tube and pour some of the glycerol from the tube into a Petri dish.
2. With scissors, cut the muscle bundle into segments about 2 cm. long, allowing the pieces to fall into the Petri dish of glycerol.
3. With forceps and dissecting needle, tease one of the segments into very thin groups of muscle fibers; single fibers, if obtained, will demonstrate the greatest amount of contraction. Strands of muscle fibers exceeding 0.2 mm. in cross-sectional diameter are too thick and should not be used.
4. Place one of the strands on a clean microscope slide and cover with a cover glass. Examine under a high-dry or oil immersion objective. Sketch appearance of striae on Laboratory Report.
5. Transfer three or more of the thinnest strands to a minimal amount of glycerol on a second microscope slide. Orient the strands straight and parallel to each other.
6. Measure the lengths of the fibers in millimeters by placing a plastic ruler underneath the slide. Use a dissecting microscope, if available. Record lengths on the Laboratory Report.
7. With the pipette from the dropping bottle, cover the fibers with a solution of ATP and ions of potassium and magnesium. Observe the reaction.
8. After 30 seconds or more, remeasure the fibers and calculate the percentage of contraction. Has width of the fibers changed? Record the results on the Laboratory Report.
9. Remove one of the contracted strands to another slide. Examine it under a compound microscope and compare the fibers with those seen in step 4. Record the differences on the Laboratory Report.
10. Place some fresh fibers on another clean microscope slide and cover them with a solution of ATP and distilled water (no potassium or magnesium ions). Does contraction occur?

Laboratory Report

Record data on the Laboratory Report and answer all the questions.

The Physiology of Muscle Contraction

18

In this exercise, several phenomena of skeletal muscle contraction will be studied. They are the **simple twitch, recruitment,** and **wave summation.** All these physiological activities will be observed in an exposed leg muscle of a freshly killed frog. Although the frog will be dead, the tissue will still be viable and respond to electrical stimulation. Frog tissue is used here in place of mammalian tissue because of its tolerance to temperature change and handling. The results are similar to what would be seen in more carefully controlled mammalian experiments.

Studying the physiology of skeletal muscle contraction requires the use of considerable equipment and careful handling of tissue. A frog must be properly prepared, kept viable, and hooked up to a stimulator electrode and transducer. When all hookups are made, stimuli must be administered to the skeletal muscle in a manner which will elicit the desired response. Due to the complexity of the overall procedure and the need to complete all tests within a limited laboratory period, it is necessary to perform the experiment as a team effort. The class will be divided into four or five teams of five or six students. When the teams are determined, the members of each group can decide among themselves which roles they prefer to perform on the team. The responsibilities of the various members on each team are outlined in table 18.1.

Materials:

Unigraph (or polygraph)
stimulator
event synchronization cable

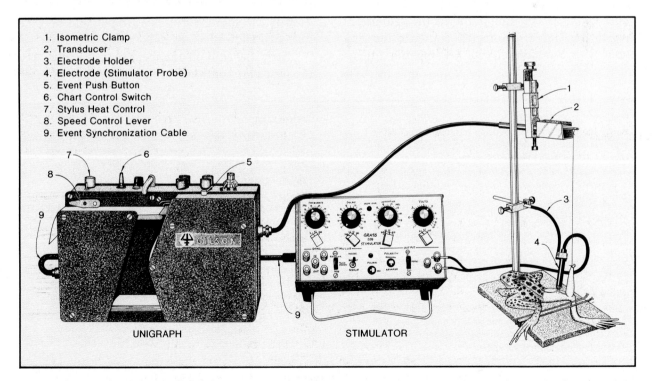

1. Isometric Clamp
2. Transducer
3. Electrode Holder
4. Electrode (Stimulator Probe)
5. Event Push Button
6. Chart Control Switch
7. Stylus Heat Control
8. Speed Control Lever
9. Event Synchronization Cable

UNIGRAPH

STIMULATOR

Figure 18.1 Instrumentation setup.

electrode
electrode holder
force transducer (Trans-Med/Biocom
    #1030)
ring stand
2 double clamps
isometric clamp (Harvard Apparatus, Inc.)
galvanized iron wire
pliers

For dissection:

small frog (one per team)
scalpel
small scissors
dissecting needle
forceps
decapitation scissors
paper towelling
frog Ringer's solution (in squeeze bottle)
spool of cotton thread

Preparations

Equipment Hookup

Set up the various pieces of equipment as shown in figure 18.1. The power cords of both the Unigraph and stimulator must be plugged into a grounded 110-volt outlet. Note that an event-sync cable connects the stimulator to the Unigraph. Attach the Harvard isometric clamp to the ring stand with a double clamp, keeping the adjustment screw of the isometric clamp upward. At-

tach the shaft of the transducer to the lower part of the isometric clamp and plug the transducer cord into the Unigraph. Put another double clamp on the ring stand shaft at the lower level and clamp the shaft of the electrode holder (label 3, figure 18.1) in place. Plug the jacks of the electrode into the stimulator at the position shown in the illustration, but do not fix the electrode to the clamp at this time since the electrode will have to be maneuvered about in the first tests.

With the completion of all the necessary connections, *balance the transducer* and *calibrate the Unigraph* according to the instructions in Appendix C.

Frog Dissection

To hold the frog for decapitation, grip it with the left hand so that the forefinger is under the neck and the thumb is over the neck. Place the frog under a water tap, allowing cool water to flow onto its head and body. This tends to calm the frog and will wash away the blood.

Insert one blade of a sharp heavy duty scissors into the mouth, well into the corner; the other scissors' blade should be poised over a line joining the tympanic membranes (eardrums). With a swift clean action, snip off the head. Immediately after this, force a probe down into the spinal canal to destroy the spinal cord. The frog should become limp and show no signs of reflexes.

Strip the skin off the right hind leg following the procedure outlined in figures 18.2 to 18.4.

EQUIPMENT ENGINEER
Hooks up the various components (Unigraph, stimulator, transducer, etc.). Balances the transducer and calibrates Unigraph. Operates the stimulator. Responsible for correct use of equipment to prevent damage. Dismantles equipment and returns all components to proper places of storage at end of period.

ENGINEER'S ASSISTANT
Works with Engineer in setting up equipment to ensure procedures are correct. Makes necessary tension adjustments before and during experiment. Applies electrode to nerve and muscle for stimulation.

RECORDER
Labels events produced on tracings. Keeps a log of the sequence of events as experiment proceeds. Sees that each member of team is provided with chart records for attachment to Laboratory Report.

SURGEON
Decapitates and piths frog. Removes skin on leg and exposes sciatic nerve. Responsible for maintaining viability of tissue. Attaches string to nerve and muscle end.

HEAD NURSE
Assists surgeon as needed during dissection. Keeps exposed tissues moist with Ringer's solution. Helps to keep work area clean.

COORDINATOR
Oversees entire operation. Communicates among team members. Reports to instructor any problems that seem to be developing that seem insurmountable. Double checks the set-up and cleanup procedures. Keeps things moving.

Table 18.1 Team member responsibilities.

Figure 18.2 The first incision through the skin is made at the base of the thigh.

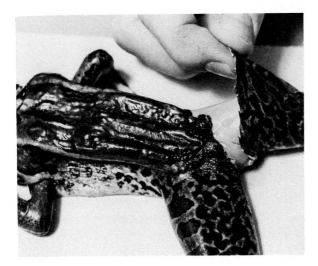

Figure 18.3 After the skin is cut around the leg, it is pulled away from the muscles.

Figure 18.4 While the body is held firmly with the left hand, the skin is stripped off the entire leg.

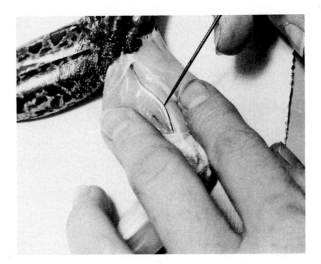

Figure 18.5 The muscles of the thigh are separated with a dissecting needle to expose the sciatic nerve.

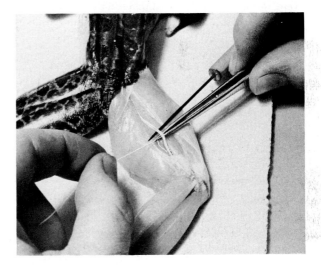

Figure 18.6 Forceps are inserted under the nerve to grasp the end of the string on the other side.

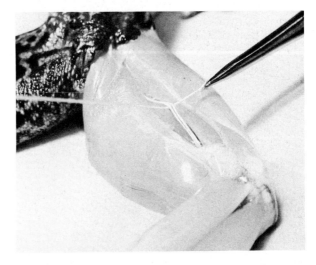

Figure 18.7 The thread used for manipulating the nerve is pulled through.

With a sharp dissecting needle, tear through the fascia and muscle tissue of the thigh to expose the sciatic nerve and insert a short length (10″ long) of cotton thread under it. This thread will be used for manipulation of the nerve. Figures 18.5 through 18.7 outline the desired procedure. *Be careful in the way you handle the nerve.* The less it is traumatized or comes in contact with metal, the better. *Keep the exposed tissues moist with Ringer's solution.*

The Muscle Twitch

When a muscle is given a single electrical shock of sufficient voltage, it gives rise to a single contraction called a *twitch.* A twitch is considered the basic unit of muscle contraction. A tracing, or *myogram,* of a single twitch is illustrated in figure 18.8. It will be noted that three distinct phases occur: a latent period, the contraction phase, and the relaxation phase. The **latent period** is a very short time lapse between the time of stimulation and the start of contraction. In most muscles of the body, it lasts about 10 milliseconds. In the **contraction phase** the muscle shortens due to the chemical changes that occur within the cells. Different muscles have different durations in this phase. After the contraction phase has reached its maximum, the muscle returns to its former relaxed state during the **relaxation phase.**

In the myogram in figure 18.8, the latent period lasts 10 milliseconds; the contraction phase, 40 milliseconds; and the relaxation phase, 50 mil-

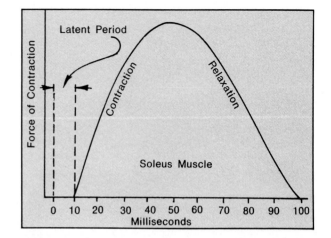

Figure 18.8 A simple twitch of the soleus muscle.

liseconds. The sum of these for the duration of the entire twitch is 100 milliseconds. These values are true only for the human soleus muscle. Other muscles may have shorter or longer contraction cycles.

The contraction of a muscle will occur only if the stimulus is of sufficient intensity and duration. The stimulus required to barely elicit a contraction is called the **threshold.** The threshold requirement is a direct consequence of the all-or-none law relating to individual muscle fibers. A stimulus of less than threshold strength is designated as being **subliminal.** Under certain conditions successive subliminal stimuli given in rapid succession can elicit a response.

Your team will now attempt to demonstrate some elements of the muscle twitch. The twitch

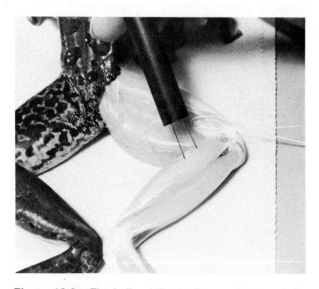

Figure 18.9 The belly of the gastrocnemius muscle is directly stimulated to produce a twitch.

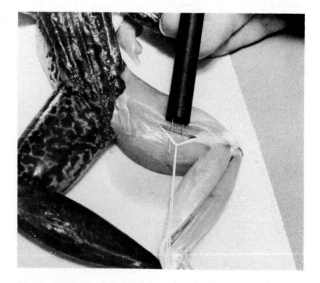

Figure 18.10 The sciatic nerve is lifted with the thread so the electrode prongs can be inserted under it.

will be observed first visually on the muscle without making a tracing via the transducer and Unigraph. After this has been observed, the muscle will be attached to the transducer and a twitch will be recorded on the Unigraph paper.

Via Muscle Stimulation

To determine the threshold stimulus for producing a visible twitch by stimulating the muscle directly, proceed as follows. *Be sure to keep the muscle and nerve moist with Ringer's solution!*

1. Set the Duration control at 15 milliseconds and the Voltage control at 0.1 volt. The Stimulus switches should be set on REGULAR and OFF. The Polarity switch should be on NORMAL and the Output switch on MONO.
2. Turn on the Power switch and depress the Mode switch to SINGLE, while holding the electrode against the belly of the gastrocnemius muscle as illustrated in figure 18.9. You are now administering a single pulse of 0.1 volt of 15 msec. duration to the muscle. This is the minimum voltage possible with the Grass stimulator. If the muscle twitches, reduce the Duration to the lowest value that will produce a twitch. In this manner we can record 0.1 volt as the **threshold stimulus** on the Laboratory Report.

   If the muscle does not twitch at 0.1 volt, 15 msec., increase the voltage at increments

of 0.1 volt until a twitch is seen. Record this voltage as the **threshold stimulus** on the Laboratory Report.

3. Now, place the foot in a flexed position and proceed to stimulate the muscle by gradually increasing the voltage until the foot extends completely, due to muscle contraction. Record this voltage on the Laboratory Report.

Via Nerve Stimulation

Gently lift the sciatic nerve with the string as shown in figure 18.10 and place the stimulator prongs under the nerve. Release the string, leaving it in place for future use. Proceed as follows to stimulate the muscle via the nerve to determine the threshold stimulus. Remember to keep the tissues moist at all times with Ringer's solution.

1. Place the foot in a flexed position.
2. Return the Voltage control to 0.1 volt and stimulate by pressing the Mode lever to SINGLE. Do you see a twitch? If not, proceed as previously, to stimulate the nerve at increasing increments of 0.1 volt until a twitch is visible. Record this voltage as the threshold stimulus by direct nerve stimulation.
3. Now, determine the minimum voltage that causes extension of the foot. Record this voltage on the Laboratory Report. (If extension occurs with a stimulus of 0.1 volt, try reducing the Duration in the same manner suggested for direct muscle stimulation.)

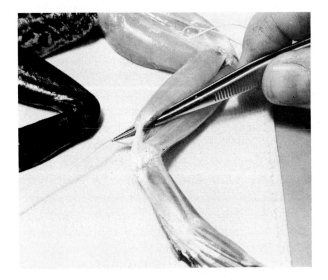

Figure 18.11 Thread is drawn through the space behind the Achilles tendon by inserting forceps through to grasp the thread.

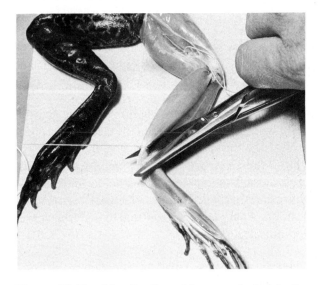

Figure 18.12 After the thread is securely tied to the Achilles tendon, the tendon is severed between the thread and joint.

Via Transducer

To produce a myogram of the twitch on Unigraph paper, it will be necessary to attach the frog's gastrocnemius muscle to the transducer with a string, as illustrated in figure 18.1. Note that the electrode is held in place by an electrode holder in such a position that the sciatic nerve rests on the prongs of the electrode. When the stimulator is activated, causing the muscle to contract, tension on the transducer leaf sends a signal to the Unigraph, activating a stylus to produce a tracing on the paper. Proceed as follows to perform each operation in this phase of the experiment.

Thread Hookup Free the connective tissue that holds the gastrocnemius muscle to adjacent muscles with a dissecting needle. Insert a pair of forceps under the muscle near the Achilles tendon (figure 18.11) to grasp hold of one end of an 18″ length of cotton thread. Pull the thread through with the forceps and tie it to the tendon a short distance away from where the tendon will be severed. Now, with a pair of scissors, cut off the tendon distal to where it is attached to the string (figure 18.12).

Tie the free end of the thread to two leaves of the transducer (the stationary one and its closest leaf). *Handle the leaves gently.* Undue stress can cause damage. Next, anchor the leg securely to the base of the ring stand. If this is not done, leg instability can create artifacts in the myogram. The easiest way to tie it down is to pass a piece of galvanized iron wire under the base and over the leg once or twice, twisting the wire securely with a pair of pliers. To get just the right tension on the thread, adjust the upper knurled knob of the isometric clamp. The tension should not be so great as to visibly deflect the leaves.

Electrode Position While holding up the sciatic nerve with the thread, position the electrode prongs so that the nerve rests on the stimulator prongs. Clamp the electrode holder to the electrode and bend the soft metal stem of the holder into a configuration that will hold the electrode securely in position. Apply Ringer's solution to both the nerve and muscle.

Unigraph Settings Turn on the power switch. Place the chart control (c.c.) lever at STBY, the red stylus heat control knob at the 2:00 position,

the speed selector lever at the slow position (opposite of position shown in figure 18.1), the red gain selector knob on 2 MV/CM, and the yellow mode selector control on TRANS (transducer).

Now, place the c.c. lever at CHART ON and note the position of the line produced on the chart. The line should be about one centimeter from the nearest margin of the paper. If it is not at this position, relocate the stylus with the centering knob. Now, stop the chart by placing the c.c. lever at STBY.

Stimulator Settings Set the Duration control at 15 msec. and the Voltage at the level that produced flexion of the leg by nerve stimulation in the previous experiment. Check the switches to make sure that the Mode switch is set at OFF, pulses are REGULAR, Polarity is on NORMAL and the last switch to the right is on MONO.

Recording Produce a simple muscle twitch record and determine the duration of each of the three phases in the contraction cycle as follows:

1. Place the c.c. lever at CHART ON. The paper should be moving at the slow rate of 2.5 mm. per second.
2. Depress the Mode switch to SINGLE and observe the tracing on the chart. The stylus travel should be approximately 1.5 centimeters. Adjust the blue sensitivity control to produce the desired stylus displacement.
3. If the stylus travel remains insufficient by adjusting the sensitivity knob, increase the sensitivity by changing the Gain control to 1 MV/CM and readjusting the sensitivity knob again.
4. Change the speed of the paper to 25 mm. per second by repositioning the speed control lever 180° to the left.
5. Administer 25 to 30 stimuli to provide enough chart material so that each member of your team will have at least five inches of chart for attachment to the Laboratory Report sheet. *If your setup does not have an event-sync cable it will be necessary to depress the event marker button on the Unigraph, simultaneously, with depression of the Mode lever on the stimulator.*
6. Calculate the duration of each of the periods, knowing that one millimeter on the chart is equivalent to .04 seconds (40 milliseconds).

Summation

The ability of whole muscles to exert different degrees of pull is achieved, primarily, by *summation.* Two types of summation are known to occur: multiple motor unit and wave summation. **Multiple motor unit summation** is an increased force caused by the recruitment of additional motor units. It is also known as recruitment or spatial summation. **Wave summation** is due to an increase in the frequency of nerve impulses, and is also referred to as temporal summation. An attempt will be made in this portion of the experiment to demonstrate both types.

Multiple Motor Unit Summation

Once the threshold stimulus has been determined for producing a twitch, it will be observed that increasing the strength of the stimulus (by voltage increase) will cause increasingly greater degrees of contraction. This phenomenon is schematically illustrated in figure 18.13. As the diagram reveals, the force of contraction is a function of the number of motor units that are stimulated.

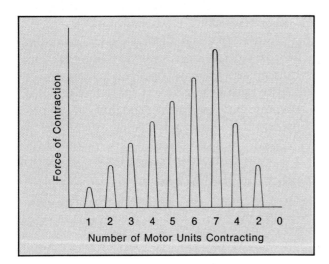

Figure 18.13 Multiple motor-unit summation.

A **motor unit** consists of a group of muscle fibers (cells) that is innervated by a single motor neuron (motoneuron). Small muscles that react quickly and precisely may have as few as two or three muscle fibers per motor unit. Large muscles that lack a fine degree of control, such as the gastrocnemius, may have as many as 1,000 muscle

fibers per motor unit. These large muscles are well adapted to a posture sustaining function.

The muscle fibers of adjacent motor units are arranged in an overlapping fashion so that each unit lends support to its neighbors. When some nerve fibers to a muscle are destroyed, as in poliomyelitis, **macromotor units** are formed by extensive arborization of the ends of surviving motoneurons to provide neural connections to all muscle fibers. These large motor units may be four or five times as large as normal units.

As voltage increases are administered, a degree of contraction is reached that cannot be exceeded by further voltage increase. This maximum contraction, called the **maximal response,** is due to the fact that all motor units are activated and no further contraction by individual stimuli can be produced. The lowest voltage that produces a maximal response is called the **maximal stimulus.** Proceed as follows to demonstrate recruitment, maximal stimulus, and maximal response.

Unigraph Settings Place the speed control lever in the slow position (lever pointing to right) and the c.c. switch at STYLUS ON. All other settings should be left as they were for the previous experiment.

Stimulator Settings Except for the Voltage control, leave all other settings as they were in the previous experiment. Return the Voltage control to 0.1 volt.

Electrode Position Leave the sciatic nerve in the same position on the electrode prongs as it was for the previous experiment.

Recording With the chart still, administer a stimulus to the nerve by depressing the Mode lever to SINGLE. Look for movement of the stylus on the chart. Advance the chart approximately 5 mm. by placing the c.c. switch at STYLUS ON temporarily.

Now raise the voltage by 0.2 volt and depress the Mode lever to SINGLE again. Continue this process of advancing the chart 5mm. and increasing the voltage by 0.2 volt until a maximal response has been achieved. Be sure to record all voltages on the chart. Repeat this process five or more times to provide all team members with chart material.

Wave Summation

If motor unit summation was the only way in which maximum contraction could be achieved by muscles, they would have to be much larger to accomplish the work they are able to do. Another phenomenon, called wave summation, plays an important role in increasing the amount of muscle contraction.

Wave summation in a skeletal muscle occurs when the muscle receives a series of stimuli in very rapid succession as illustrated in figure 18.14. If the rate of stimulation is kept very slow, only single twitches occur as indicated in figure 18.14. However, 35 pulses per second produces some summation, 70 pulses per second produces much

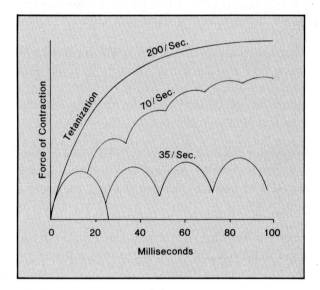

Figure 18.14 Wave summation.

more, and 200 pulses per second produces sustained or *complete tetanization.*

In this portion of the experiment, we will use the *Frequency Control* to regulate the pulses per second. Note that it is graduated from 2–20. Its Decade switch is calibrated at X.1, X1, and X10, producing a range of .2 to 200 pulses per second. To use this control, the Mode switch will be set at REPEAT.

1. Set the Voltage at the previously determined maximal stimulus and the Mode switch on OFF. Put the c.c. switch at CHART ON at slow speed.
2. Set the Frequency at 1 pps (Control at 10, decade switch at X.1), Mode switch on REPEAT, and observe that the monitor lamp is blinking at the indicated frequency. Record the tracing for 10 seconds, and return the Mode switch to OFF. Let the muscle rest for at least 2 minutes.
3. Repeat at a frequency of 2 pps for another 10 seconds. Rest the muscle for another 2 minutes.
4. Now double the frequency to 4 pps for another 10 seconds. Rest again for 2 minutes.
5. Continue to double the frequency, recording for 10 seconds and resting for 2 minutes, until the muscle goes into complete tetany.
6. After resting the muscle for a few minutes, repeat the experiment enough times to provide a record for each member of your team.

Laboratory Report

Complete the first portion of Laboratory Report 18,19, and attach the chart from the above experiment to it.

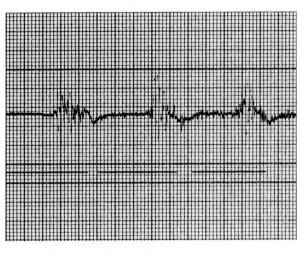

Since the early experiments of the 1920s, when muscle contraction proved to be accompanied by bioelectricity, electromyography (hereafter EMG) has become firmly established in physiology. Electrical activity associated with muscle contraction arises from nerve and muscle action potentials of the motor unit. Depolarizations initiate contraction while repolarizations accompany relaxation; together, this activity produces voltage changes that are detectable at the body surface with skin electrodes, or intramuscularly, with needle electrodes. To medicine, electromyography is valuable in the assessment of muscle dysfunction. For the student, EMG study can shed light on muscle physiology.

The type of recordings obtained in EMG depends upon the selection of electrodes. Recordings with needle electrodes inserted into muscle yield precise information concerning individual motor unit action potentials. This procedure permits the diagnostician to distinguish between myopathic and neurogenic disorders. Because of the skill required, not to mention the hazard of infection and legal implications, student applications are limited to surface electrodes. Although only a general impression of whole muscle groups are feasible with surface electrodes, single muscles may sometimes be detected.

Individual action potentials cannot be readily distinguished with skin electrodes since the region beneath the electrode often consists of several motor units. Figure 19.1 illustrates the nature of a typical EMG. The high-frequency, irregular, overlapping spikes of an EMG of this type are called an **interference pattern.** The voltage spread of these potentials may be as great as 50 millivolts.

With sufficient amplification, an interference pattern may be detectable even in muscles that are completely relaxed. This spontaneous activity is

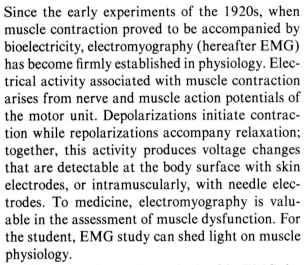

Figure 19.1 Electromyogram.

associated with the maintenance of normal muscle tone. As such a muscle is activated, however, the interference pattern becomes exaggerated.

To display an EMG, one might use an oscilloscope, audio monitor, or chart recorder. Although the Unigraph is the recorder of choice in this experiment, because of its portability a polygraph with an EMG module is excellent. A Unigraph, in which the IC-MP module has been replaced by an IC-EMG module, can also be used.

The class will be divided up into a number of teams of three or four students each. While one individual prepares the Unigraph, another will attach the electrodes to the subject, and assist in recording information on the chart.

Materials:

Unigraph, Duograph, or polygraph
handgrip dynamometer (Stoelting #19117)
3-lead patient cable for Unigraph
3 skin electrodes (Gilson self-adhering
 #E1081K)

adhesive pads for electrodes
electrode gel (EKG or other)
Scotchbrite pads (grade #7447)
alcohol or alcohol swabs

Preparations

Equipment Hookup

While the subject is being readied for EMG recording, connect the 3-lead cable to the Unigraph. Plug the power cord into an outlet. Turn on the power switch, the stylus heat control, and check out the intensity of the stylus line on the paper at slow speed.

If the Unigraph is not provided with a special EMG module, the EEG setting should be used. This mode works very well at lower gain levels for EMG monitoring. To be able to quantify the EMG in terms of millivolts, it will be necessary to calibrate the instrument according to instructions in Appendix C.

If a Gilson polygraph is used, the following modules can be used for EMG monitoring: IC-MP, IC-UM, and IC-EMG. If all modules are present, use the IC-EMG module.

Subject Preparation

Figure 19.2 illustrates the placement of three electrodes on the arm. The two recording red-wired electrodes are placed within an inch of each other on the belly of the major flexor muscle of the forearm *(flexor digitorium superficialis)*. The third electrode is a ground and should be placed some distance away over a bony area.

Prior to placing the electrodes on the arm, scrub the skin areas gently with a Scotchbrite pad and disinfect the skin with 50%–70% alcohol. The removal of some of the dead cells from the skin surface greatly facilitates conductivity.

Consult figures 19.3 through 19.6 to see how the self-adhesive pads and electrode gel are placed on the electrodes prior to attaching them to the skin. There are several other kinds of electrodes that can be used. Figure 19.7 reveals how one might substitute three ECG plate electrodes on the arm. As in the case of the other electrodes, it is necessary to scrub first with Scotchbrite and then apply electrode gel before strapping on the electrodes.

When the electrodes are firmly attached to the skin, connect the wires to the appropriate wire of the 3-lead cable. These cables usually have alli-

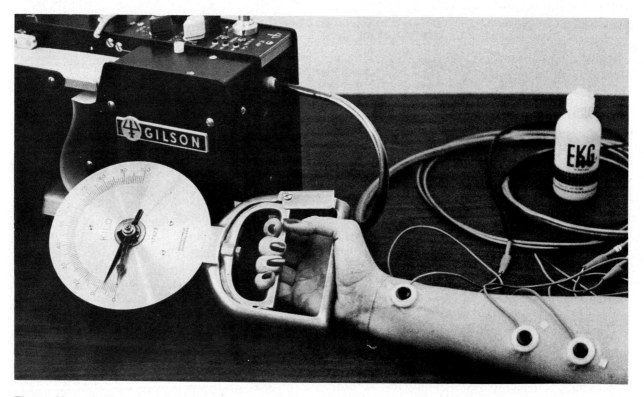

Figure 19.2 EMG monitoring setup.

gator clips and are color coded to accept the correct electrodes. Be sure that the ground electrode is attached to the black ground clip of the 3-lead cable.

Monitoring

Three phenomena of muscle activity will be studied with this setup: spontaneous activity, recruitment, and fatigue. If a Duograph, dual beam oscilloscope, or polygraph is available, it will also be possible to monitor EMGs of agonist and antagonist muscles, simultaneously.

Spontaneous Activity

With the arm completely relaxed and fully supported on the table top, turn on the Unigraph (high

speed) and observe if any evidence of motor unit activity is discernible at the lowest gain. Increase the sensitivity by changing the gain settings and turning the sensitivity control. Determine the magnitude and frequency of peaks.

Recruitment

Instruct the subject to clench his or her fingers to form a fist and note the burst of activity to form a typical EMG interference pattern on the chart. To demonstrate recruitment, have the subject grip a hand dynamometer, as in figure 19.2. Record, first, a few bursts at 5 kilograms. Then, double the force to 10 kilograms. Continue to increase the force by 5- or 10-kilogram increments to maximum, recording each force magnitude on the chart with a pen or pencil. Do you see an increase in

Figure 19.3 Skin surface is rubbed gently with a Scotchbrite pad to improve skin conductivity.

Figure 19.4 An adhesive disk is applied to the clean, dry undersurface of the skin electrode.

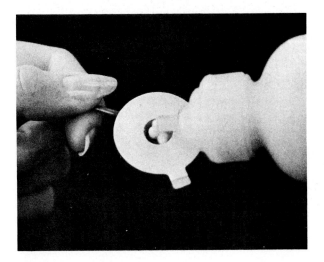

Figure 19.5 Electrode gel is applied to the depression of the skin electrode.

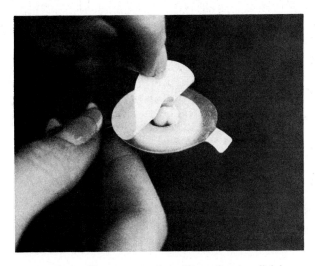

Figure 19.6 The top covering of the adhesive disk is removed and the electrode is placed on the skin.

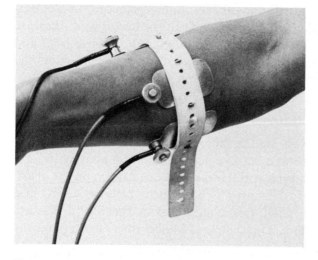

Figure 19.7 An alternate method of electrode application.

amplitude on the chart as evidence of recruitment of motor units? Produce enough tracings for all members of your team.

Fatigue

With the Unigraph turned off, direct the subject to perform some handgripping work until the hand muscles are thoroughly fatigued. When the subject cannot continue, take the dynamometer away and allow the subject's arm to recover on the table top. Start the recorder and monitor any electrical activity. How does the EMG of fatigue compare with that spontaneous activity?

Agonist vs. Antagonist Muscles

Opposite movements are achieved by muscle groups called agonists and antagonists. Flexion of the forearm against the upper arm is achieved by contraction of the *biceps brachii.* Extension occurs when the *triceps brachii* contracts. In this case, the biceps is an agonist and the triceps is the antagonist. The action of these two muscles must be coordinated so that when one contracts the other relaxes. Activity in flexors cannot be accompanied by simultaneous activity in extensors, or rigidity will occur instead of flexion or extension. Coordination may be demonstrated by recording the activity of both muscles simultaneously with a duograph, polygraph, or dual beam oscilloscope.

To demonstrate coordination, hook up three electrodes to each muscle of one arm through separate channels of whatever type of equipment that is available. Have the subject execute the following maneuvers:

1. Flexion of arm against resistance. Assistant restrains subject's fist while flexion is attempted.
2. Extension against resistance.
3. Isometric tension (contraction of both agonist and antagonist, simultaneously).

Laboratory Report

Complete the last portion of combined Laboratory Report 18,19.

Part 5 The Major Skeletal Muscles

This study of the skeletal muscles includes the principal muscles of the head, neck, thorax, arms, and legs. In no sense is this study to be considered all-inclusive, however, since it includes only seventy-five muscles. Although a few deep muscles have been included, most of these seventy-five muscles are located on the surface.

It should be kept in mind while studying these various muscles that although we study them individually, they cannot always be treated as single mechanical units. Not only do they work in conjunction with adjacent muscles, but the force of gravity also may be involved. In addition, different parts of the same muscle may have different and even antagonistic actions.

21 Body Movements

The action of muscles through diarthrotic joints results in a variety of types of movement. The nature of movement is dependent on the construction of the individual joint and the position of the muscle. Muscles working through hinge joints result in movements that are primarily in one plane. Ball-and-socket joints, on the other hand, will have many axes through which movement can occur; thus, movement through these joints is in many planes. The different types of movement are as follows:

Flexion When the angle between two parts of a limb is decreased the limb is said to be *flexed*. Flexion of the arm takes place at the elbow when the antebrachium is moved toward the brachium. The term *flexion* may also be applied to movement of the head against the chest and the thigh against the abdomen.

When the foot is flexed upward, the flexion is designated as being **dorsiflexion**. Movement of the sole of the foot downward, or flexion of the toes, is called **plantar flexion.**

Extension Increasing the angle between two portions of a limb or two parts of the body is *extension*. Straightening the arm from a flexed position is an example of extension. Extension of the foot at the ankle is essentially the same as plantar flexion. When extension goes beyond the normal posture, as in leaning backward, the term **hyperextension** is used.

Abduction and Adduction The movement of a limb away from the median line of the body is called *abduction*. When the limb moves toward the median line, *adduction* occurs. These terms may also be applied to parts of a limb, such as the fingers and toes, by using the longitudinal axes of the limbs as points of reference.

Rotation and Circumduction The movement of a bone or limb around its longitudinal axis without lateral displacement is *rotation*. Rotation of the arm occurs through the shoulder joint. The head can be rotated through movement in the cervical vertebrae. If the rotational movement of a limb through a freely movable joint describes a circle at its terminus, the movement is called *circumduction*. The arm, leg, and fingers can be circumducted.

Supination and Pronation Rotation of the antebrachium at the elbow affects the position of the palm. When the palm is raised upward from a downward facing position, *supination* occurs. When the rotation is reversed so that the palm is returned to a downward position, *pronation* takes place.

Inversion and Eversion These two terms apply to foot movements. When the sole of the foot is turned inward or toward the median line, *inversion* occurs. When the sole is turned outward, *eversion* takes place.

Sphincter Action Circular muscles such as those around the lips *(orbicularis oris)* and the eye *(orbicularis oculi)* are called sphincter muscles. When their fibers shorten they close the opening.

Muscle Grouping

For reasons of simplicity, muscles are often studied individually, as in figure 21.1. It should be kept in mind, however, that seldom, if ever, do they act singly. Rather, muscles are arranged in groups with specific functions to perform; i.e., flexion and extension, abduction and adduction, supination and pronation.

The flexors are the prime movers, or **agonists.** The opposing muscles, or **antagonists,** contribute to smooth movements by their power to maintain tone and give way to movement by the flexor group. Variance in the tension of the flexor muscles results in a reverse reaction in the extensor muscles.

Muscles that assist the agonists to reduce undesired action or unnecessary movement are called **synergists.** Other groups of muscles that hold structures in position for action are called **fixation muscles.**

Laboratory Report

Identify the types of movement in figure 21.1 and complete the last portion of Laboratory Report 20, 21.

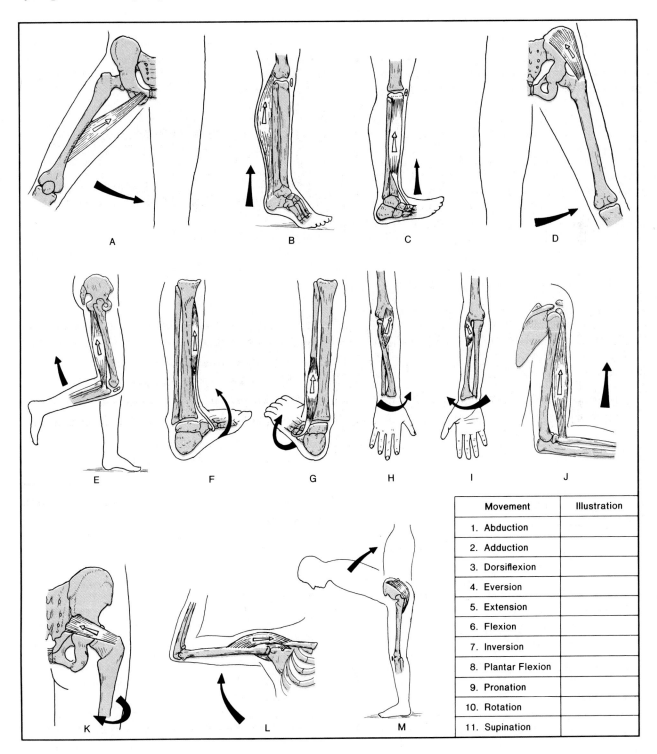

Movement	Illustration
1. Abduction	
2. Adduction	
3. Dorsiflexion	
4. Eversion	
5. Extension	
6. Flexion	
7. Inversion	
8. Plantar Flexion	
9. Pronation	
10. Rotation	
11. Supination	

Figure 21.1   Body movements.

22 Head and Trunk Muscles

The principal surface muscles of the head, neck and trunk, twenty-four in all, will be studied in this exercise. A muscle manikin, if available, would be helpful for assistance in identifying these muscles.

Head Muscles

Illustration A, figure 22.1 reveals the principal surface muscles of the head. Identify and label the following.

Epicranial The frontalis of the forehead and the occipitalis of the occiput region are joined by an aponeurosis that extends over the top of the skull. These two muscles constitute the epicranial. The **occipitalis** has its origin on the mastoid process and the occipital bone; its insertion is on the aponeurosis. The **frontalis** takes its origin on the aponeurosis and its insertion on the soft tissue of the eyebrows.

Action: The frontalis elevates the eyebrows and wrinkles the forehead. The occipitalis can pull the scalp backward.

Orbicularis Oris This muscle is a sphincter (circular) muscle that surrounds the lips of the mouth. Its *origin* is on various facial muscles, the maxilla, mandible, and septum of the nose. The *insertion* is on the lips.

Action: It closes the lips in various ways: by compression over the teeth or by pouting and pursing them; utilized in kissing.

Orbicularis Oculi This is another sphincter muscle. It surrounds the eye. It *arises* from the nasal portion of the frontal bone, the frontal process of the maxilla and the medial palpebral ligament. It *inserts* within the tissues of the eyelids (palpebrae).

Action: It causes squinting and blinking.

Zygomaticus This muscle appears as a diagonal muscle in illustration A that extends from the corner of the mouth to the zygomatic bone. Its *origin* is on the latter bone. Its *insertion* is on the orbicularis oris.

Action: Contraction of these muscles draws the

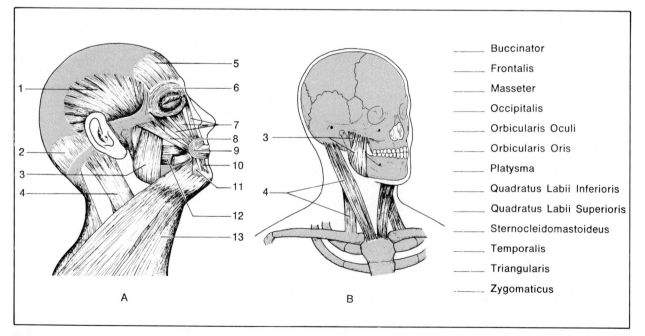

_____ Buccinator
_____ Frontalis
_____ Masseter
_____ Occipitalis
_____ Orbicularis Oculi
_____ Orbicularis Oris
_____ Platysma
_____ Quadratus Labii Inferioris
_____ Quadratus Labii Superioris
_____ Sternocleidomastoideus
_____ Temporalis
_____ Triangularis
_____ Zygomaticus

Figure 22.1 Head and neck muscles.

angles of the mouth upward and backward as in smiling and laughing.

Quadratus Labii Superioris This is a thin broad muscle with three heads that lies between the upper margin of the orbicularis oris and the region under the eye. Its *origin* is on the upper part of the maxilla and a portion of the zygomatic bone. The major portion of its *insertion* is on the superior margin of the orbicularis oris. It also inserts on the alar region of the nose.

Action: Expression of sadness results from the contraction of only the infraorbital head. Contraction of the entire muscle conveys the attitude of contempt or disdain by furrowing the upper lip.

Quadratus Labii Inferioris This muscle is a small one that extends from the lower lip and lower margin of the orbicularis oris to the mandible. Its *origin* is on the mandible.

Action: Since its *insertion* is on the lower lip it pulls the lip downward as in irony.

Triangularis This muscle is lateral to the quadratus labii inferioris. Its *origin* is on the mandible. It *inserts* on the orbicularis oris just below the point of insertion of the zygomaticus.

Action: Being an antagonist of the zygomaticus it depresses the corner of the mouth.

Platysma The broad sheet-like muscle that extends from the mandible down over the side of the neck in illustration A is the platysma. The *origin* of this muscle is primarily the fascia that covers the pectoralis and deltoideus muscles of the shoulder and chest region. Its *insertion* is on the mandible and the muscles around the mouth.

Action: It draws the outer part of the lower lip downward and backward widening the mouth as in expression of horror; assists in opening the jaws.

Sternocleidomastoideus This muscle is located on the side of the neck. It is partially covered by the platysma. Its *origin* is located on the manubrium of the sternum and the sternal end of the clavicle. Its *insertion* is on the mastoid process.

Action: When this muscle acts independently the head is drawn toward the shoulder on the same side as the muscle, rotating the head at the same time. Simultaneous contraction of both sternocleidomastoids causes the head to be flexed forward and downward on the chest.

Masseter This muscle covers the side of the mandible. It is one of four muscles of mastication. It *originates* on the zygomatic arch and is *inserted* on the lateral surface of the ramus and angle of the mandible.

Action: It raises the mandible.

Temporalis The large fan-shaped muscle on the side of the skull in illustration A is the temporalis. It *arises* on portions of the frontal, parietal and temporal bones and is *inserted* on the coronoid process of the mandible.

Action: It acts synergistically with the masseter to raise the mandible. It can also cause retraction of the mandible when only the posterior fibers of the muscle are activated.

Buccinator The horizontal muscle that obscures the teeth in illustration A of figure 22.1 is the buccinator. It is situated in the cheeks *(buccae)* on each side of the mouth. It *arises* on the maxilla, mandible and the pteromandibular raphé. It *inserts* on the orbicularis oris.

Action: It compresses the cheek to hold food between the teeth during chewing.

Assignment:

Label figure 22.1.

Upper Trunk Muscles

The surface muscles of the upper trunk and shoulder include the trapezius, pectoralis major, latissimus dorsi and deltoideus. Illustrations A and B, figure 22.2 depict these four muscles.

Trapezius The large triangular muscle that occupies the upper shoulder region of the back is the trapezius. It *arises* on the occipital bone, the ligamentum nuchae, and the spinous processes of the seventh cervical and twelfth thoracic vertebrae. The *ligamentum nuchae* is a ligament that extends from the occipital bone to the seventh cervical vertebra, uniting the spinous processes of all cervical vertebrae. The *insertion* of the muscle is on the spine and acromion of the scapula and the outer third of the clavicle.

Action: This muscle pulls the scapula toward the median line (adduction), raises the scapula as in shrugging the shoulder and draws the head backward (extension) if the shoulders are fixed.

Pectoralis Major This muscle is a thick fan-shaped muscle that occupies the upper quadrant of the chest in front. It *arises* on the clavicle, sternum, costal cartilages and aponeurosis of the external oblique. It *inserts* in the groove between the greater and lesser tubercles of the humerus.

Action: Adduction of the humerus; flexion and rotation of the humerus medially.

Deltoideus This muscle is the principal muscle of the shoulder. It *originates* on the lateral third of the clavicle, the acromion and the spine of the scapula. It *inserts* on the deltoid tuberosity of the humerus.

Action: Abduction of the humerus.

Latissimus Dorsi This large muscle of the back covers the lumbar area. It takes its *origin* in a broad aponeurosis that is attached to thoracic and lumbar vertebrae, spine of the sacrum, iliac crest, and the lower ribs. Its *insertion* is on the intertubercular groove of the humerus.

Action: Extends, adducts, and rotates the arm medially; draws the shoulder downward and backward.

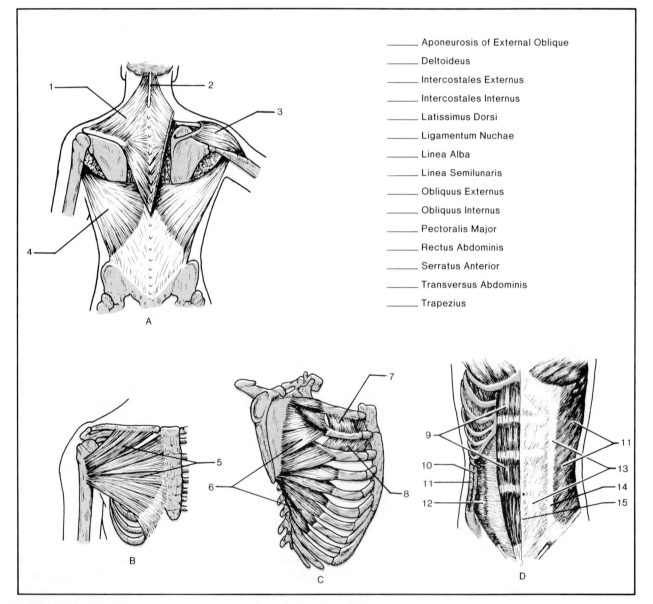

_____ Aponeurosis of External Oblique

_____ Deltoideus

_____ Intercostales Externus

_____ Intercostales Internus

_____ Latissimus Dorsi

_____ Ligamentum Nuchae

_____ Linea Alba

_____ Linea Semilunaris

_____ Obliquus Externus

_____ Obliquus Internus

_____ Pectoralis Major

_____ Rectus Abdominis

_____ Serratus Anterior

_____ Transversus Abdominis

_____ Trapezius

Figure 22.2 Trunk muscles.

Serratus Anterior *(serratus magnus)* Illustration C, figure 22.2, shows the position of this muscle. It makes its *origin* on the upper eight or nine ribs and *inserts* on the anterior surface of the scapula near the vertebral border.

Action: This muscle is the physiological antagonist of the trapezius in that it pulls the scapula forward, downward and inward toward the chest wall.

Intercostales Externi *(external intercostals)* Between the ribs on both sides are eleven pairs of short muscles, the external intercostals. The *origin* of each of these muscles is on the lower border of the upper rib; the *insertion* is the upper border of the lower rib. Their fibers are directed obliquely forward on the front of the ribs. Label 7 in figure 22.2 illustrates one of these muscles. Other external intercostals have been omitted in this illustration for clarity.

Action: They pull the ribs closer to each other resulting in their elevation. Raising of the ribs increases the volume of the thorax to cause inspiration of air in breathing.

Intercostalis Interni *(internal intercostals)* These antagonists of the external intercostals lie on the internal surface of the rib cage. Portions of two of them are shown in illustration C, figure 22.2. Each internal intercostal *arises* from the lower margin of the upper rib and *inserts* on the upper margin of the rib below. Note in the illustration that the fibers of these muscles are oriented in the opposite direction of the fibers of the external intercostals.

Action: They draw adjacent ribs together. This action has the effect of lowering the ribs and decreasing the volume of the thoracic cavity. Expiration of air from the lungs results.

Abdominal Muscles

The abdominal wall consists of four pairs of muscles. Illustration D, figure 22.2 has a portion of the right side removed to reveal its laminations. The left side shows the structures that would be seen with only the skin and fat removed. The following muscles are listed according to their positions, outermost ones being discussed first.

Obliquus Externus *(external oblique)* This muscle is the most superficial of the three layers on the side of the abdomen. It is ensheathed by the *aponeurosis of the obliquus externus,* which terminates at the linea alba and inguinal ligament. The muscle takes its origin on the external surfaces of the lower eight ribs. Although it appears to insert on the *linea semilunaris* (label 14, figure 22.2), its actual *insertion* is the *linea alba* (white line), where fibers of the left and right aponeuroses interlace on the midline of the abdomen (per *Gray's Anatomy*). The lower border of each aponeurosis forms the *inguinal ligament,* which extends from the anterior superior spine of the ilium to the pubic tubercle.

Obliquus Internus *(internal oblique)* This muscle lies immediately under the external oblique, i.e., between the external oblique and the transversus abdominis. The cut surfaces of the muscles shown in illustration D are of the external and internal obliques. Its *origin* is on the lateral half of the inguinal ligament, the anterior two thirds of the iliac crest and the thoracolumbar fascia. Its *insertion* is on the costal cartilages of the lower three ribs, the linea alba and the crest of the pubis.

Transversus Abdominis The innermost muscle of the abdominal wall is the transversus abdominis. It is the muscle revealed on the right side in illustration D which has been exposed by the absence of the external and internal obliques. It *arises* on the inguinal ligament, the iliac crest, the costal cartilages of the lower six ribs and the thoracolumbar fascia. It *inserts* into the linea alba and the crest of the pubis.

Rectus Abdominis The right rectus abdominis is the long, narrow segmented muscle running from the rib cage to the pubic bone in illustration D. It is enclosed in a fibrous sheath formed by the aponeuroses of the above three muscles. Its *origin* is on the pubic bone. Its *insertion* is on the cartilages of the fifth, sixth, and seventh ribs. The linea alba lies between these two muscles. Contraction of the rectus abdominis muscles aids in flexion of the spine and lumbar region.

Collective Action: These four abdominal muscles keep the abdominal organs compressed and assist in maintaining intra-abdominal pressure.

They act as antagonists to the diaphragm. When the latter contracts they relax. When the diaphragm relaxes they contract to effect expiration of air from the lungs. They also assist in defecation, micturition, parturition and vomiting. Flexion of the body at the lumbar region is also achieved by them.

Assignment:

Label figure 22.2.

Complete the Laboratory Report for this exercise.

23 Arm Muscles

The principal movements of the arm are controlled by the twenty-three muscles described in this exercise. They are grouped according to type of movement.

Upper Arm Movements

In the last exercise we studied the principal surface muscles of the trunk and shoulder that move the upper arm. They were the deltoideus, pectoralis major and latissimus dorsi. Partially or completely concealed by these three muscles are the five muscles shown in figure 23.1. These five muscles work together with the three major surface muscles to effect most movements of the upper arm. Descriptions follow.

Infraspinatus Illustration A, figure 23.1 shows three shoulder muscles. The infraspinatus is the largest one that lies superior to the other two. Its *origin* is on that portion of the scapula that is inferior to the scapular spine (infraspinous fossa). Its *insertion* is the middle facet of the greater tubercle of the humerus.

Action: Rotates the humerus laterally.

Teres Major This muscle is the most inferior one shown in illustration A. It *originates* in an oval area near the inferior angle of the scapula. It *inserts* on the humerus.

Action: Rotates the humerus medially and weakly adducts it.

Teres Minor The short muscle which lies inferior to the infraspinatus is the teres minor. It takes its *origin* from the lateral margin of the scapula and its *insertion* is on the lowest facet of the greater tubercle of the humerus.

Action: Assists the infraspinatus and teres major in rotating the humerus laterally.

Supraspinatus This muscle is shown in illustration B. Note that it *arises* from the fossa above the scapular spine and *inserts* on the greater tubercle of the humerus.

Action: Assists the deltoid in abduction of the humerus.

Coracobrachialis This muscle covers a portion of the upper medial surface of the humerus. Its *origin* is on the apex of the coracoid process of the scapula. Its *insertion* is near the middle of the medial surface of the humerus.

Action: Carries the arm forward in flexion and adducts the arm.

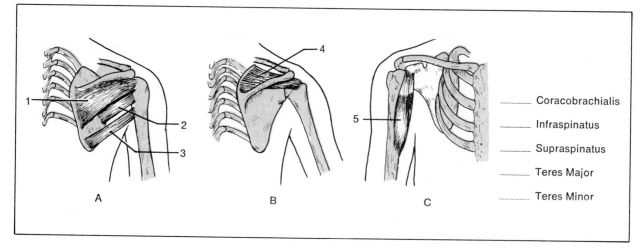

A   B   C

_____ Coracobrachialis

_____ Infraspinatus

_____ Supraspinatus

_____ Teres Major

_____ Teres Minor

Figure 23.1   Scapula to humerus muscles.

Forearm Movements

The principal movers of the forearm are the *biceps brachii, brachialis, brachioradialis,* and *triceps brachii.* These muscles are illustrated in figure 23.2.

Biceps Brachii This is the large muscle on the anterior portion of the humerus that bulges when the forearm is flexed. Its *origin* consists of two tendinous heads: a medial tendon which is attached to the coracoid process and a lateral tendon which fits into a groove (intertubercular) on the humerus. The latter tendon is attached to the supraglenoid tubercle of the scapula. At the lower end of the muscle the two heads unite to form a single tendinous *insertion* on the radial tuberosity.

Action: Flexion of the forearm; also, rolls the radius outward to supinate the hand.

Brachialis Immediately under the biceps brachii on the distal anterior portion of the humerus lies the brachialis. Its *origin* occupies the lower half of the humerus. Its *insertion* is attached to the front surface of the coronoid process of the ulna.

Action: Flexion of the forearm.

Brachioradialis This muscle is the most superficial muscle on the lateral (radial) side of the forearm. It *originates* above the lateral epicondyle of the humerus and *inserts* on the lateral surface of the radius slightly above the styloid process.

Action: Flexion of the forearm.

Triceps Brachii The entire back surface of the upper arm is covered by this muscle. It has three heads of *origin*. A long head arises from the scapula, a lateral head from the posterior surface of the humerus, and a medial head from the surface below the radial groove. The tendinous *insertion* of the muscle is attached to the olecranon process of the ulna.

Action: Extension of the forearm; antagonist of the brachialis.

Assignment:

Label figures 23.1 and 23.2.

Hand Movements

Muscles of the arm that cause hand movements are illustrated in figures 23.3 and 23.4. All illustrations are of the right arm; thus, if the thumb points to the left side of the page the anterior aspect is being viewed. Conversely, when the thumb points to the right side of the page it is the posterior aspect that is shown. These facts are pointed out to facilitate understanding the following text. Although these muscles are grouped according to function it will be seen that some have more than one type of action.

Supination of the Hand

Supination is achieved by the biceps brachii and the supinator. The **supinator** is shown in illustra-

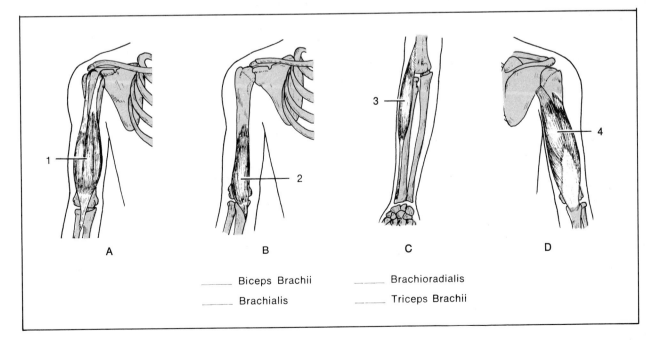

| A | B | C | D |

_____ Biceps Brachii   _____ Brachioradialis

_____ Brachialis   _____ Triceps Brachii

Figure 23.2 Upper arm muscles.

tion A. It *arises* from the lateral epicondyle of the humerus and the ridge of the ulna. It curves around the upper portion of the radius and *inserts* on the lateral edge of the radial tuberosity and the oblique line of the radius.

Pronation of the Hand

Pronation is achieved by the two muscles shown in illustration B, figure 23.3. The upper one is the *pronator teres*. The lower one is the *pronator quadratus*.

Pronator Teres This muscle *arises* on the medial epicondyle of the humerus and *inserts* on the upper lateral surface of the radius.

Pronator Quadratus This shorter pronator *originates* on the distal fourth of the ulna and *inserts* on the distal lateral portion of the radius.

> *Collective action:* Working synergistically these muscles rotate the distal end of the radius over the ulna.

Flexion of the Hand

Five muscles that flex the hand are shown in illustrations C and D of figure 23.3 and A of figure 23.4.

Flexor Carpi Radialis This muscle extends diagonally across the anterior portion of the forearm from the end of the humerus to the hand. It *arises* on the medial epicondyle of the humerus and *in-serts* on the proximal portions of the second and third metacarpals.

Action: Flexion and abduction of the hand.

Flexor Carpi Ulnaris This muscle lies medial to the flexor carpi radialis. Its *origin* is on the medial epicondyle of the humerus and the posterior surface of the ulna (olecranon process). Its *insertion* consists of a tendon that attaches to the base of the fifth metacarpal.

Action: Flexion and abduction of the hand.

Flexor Digitorum Superficialis *(flexor digitorum sublimis)* This broad muscle lies near the surface of the arm yet under the flexor carpi radialis and flexor carpi ulnaris. It extends from the distal end of the humerus to tendons of the second, third, fourth, and fifth fingers. It *arises* on three bones: the medial epicondyle of the humerus, the medial surface of the ulna and the oblique line of the radius. Its *insertion* consists of tendons that are attached to the middle phalanges of the index, middle, ring, and the little finger.

Action: Flexion of the second, third, fourth, and fifth fingers.

Flexor Digitorum Profundus This deep muscle is situated on the ulnar side of the forearm underneath the above muscle. It *arises* on the anterior proximal surface of the ulna and the interosseous membrane between the radius and ulna. It *inserts* as four tendons on the distal phalanges of the second, third, fourth, and fifth fingers.

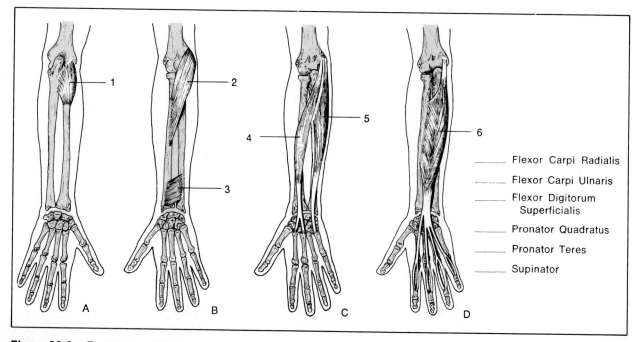

_____ Flexor Carpi Radialis

_____ Flexor Carpi Ulnaris

_____ Flexor Digitorum Superficialis

_____ Pronator Quadratus

_____ Pronator Teres

_____ Supinator

Figure 23.3 Forearm muscles.

Action: Flexes the distal phalanges of all fingers except the thumb.

Flexor Pollicis Longus (Latin: *pollex,* thumb) This muscle is the lateral one shown in illustration A of figure 23.4. Its *origin* is on portions of the radius, ulna, interosseous membrane and, occasionally, on the medial epicondyle of the humerus. Its *insertion* consists of a tendon that is attached to the distal phalanx of the thumb.

Action: Flexes the thumb.

Extension and Abduction of Hand

Illustrations B, C, and D in figure 23.4 show five extensors and one abductor of the hand and fingers. Some of the smaller muscles of the hand are not shown.

Extensor Carpi Radialis Longus and **Brevis** These two muscles extend from the lateral epicondylar region of the humerus to the bases of the second and third metacarpals on the posterior surface of the forearm. The brevis muscle *originates* on the lateral epicondyle and the longus muscle arises just superior to it on the supracondylar ridge. *Insertion* of the brevis muscle is on the proximal portion of the middle metacarpal. The longus muscle inserts on the second metacarpal near its base.

Action: These two muscles extend the wrist and assist in extension of the hand.

Extensor Carpi Ulnaris This muscle covers most of the posterior surface of the ulna. It *arises* on the lateral epicondyle of the humerus and part of the ulna. It *inserts* on the posterior surface of the fifth metacarpal.

Action: Assists the extensor carpi radialis longus and brevis in extension of the wrist.

Extensor Digitorum This muscle lies alongside the extensor carpi ulnaris. It *arises* from the lateral epicondyle of the humerus. In the wrist area its tendon divides into four tendons that *insert* on the posterior surfaces of the distal phalanges of fingers two through five.

Action: Extension of all fingers except the thumb.

Extensor Pollicis Longus This muscle is the lower one in illustration C, figure 23.4. It extends from the ulna to the end of the thumb. It *arises* on the posterior surface of both the ulna and radius. It *inserts* on the distal phalanx of the thumb. Its action is assisted by the **extensor pollicis brevis** which is not shown in this illustration. The latter muscle lies superior to the longus muscle; it inserts on the proximal phalanx of the thumb.

Action: Both of these muscles extend the thumb.

Abductor Pollicis This muscle *originates* on the interosseous membrane between the radius and ulna and *inserts* on the lateral portion of the first metacarpal and trapezium. This muscle is shown in illustration C.

Action: Abduction of the thumb.

Assignment:

Label figures 23.3 and 23.4 and complete the Laboratory Report for this exercise.

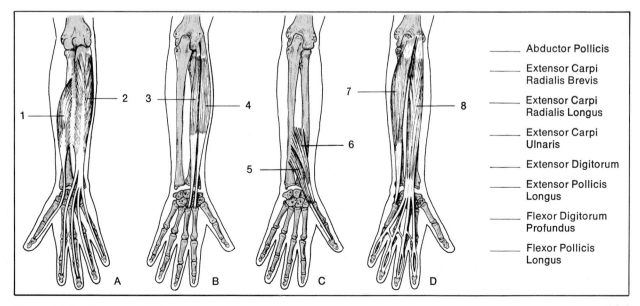

| _____ Abductor Pollicis |
| _____ Extensor Carpi Radialis Brevis |
| _____ Extensor Carpi Radialis Longus |
| _____ Extensor Carpi Ulnaris |
| _____ Extensor Digitorum |
| _____ Extensor Pollicis Longus |
| _____ Flexor Digitorum Profundus |
| _____ Flexor Pollicis Longus |

Figure 23.4 Forearm muscles.

24 Leg Muscles

This exercise describes the location and function of twenty-seven muscles of the leg. As in the case of the arm muscles they are grouped according to the type of movement.

Thigh Movements

Muscles that move the femur are anchored to some part of the pelvis and inserted on the femur. Seven such muscles are shown in figure 24.1.

Gluteus Maximus This muscle is the large superficial muscle that covers the major portion of the buttock region. It is covered by a deep fascia, the *fascia latae,* which completely invests the thigh muscles. Emerging downward from the fascia latae is a broad tendon the *iliotibial tract* (label 2), which is attached to the tibia. The gluteus maximus *arises* on the ilium, sacrum and coccyx. It *inserts* on the iliotibial tract and the posterior part of the femur.

Action: Extension and outward rotation of the femur. It is an antagonist of the iliopsoas muscle.

Gluteus Medius This muscle lies immediately under the gluteus maximus. Its *origin* is on the external surface of the ilium, covering a good portion of that bone. Its *insertion* is on the lateral part of the greater trochanter of the femur.

Action: Abduction and medial rotation of the femur.

Gluteus Minimus This is the smallest of the three gluteal muscles and lies immediately under the gluteus medius. Like the others it *arises* on the exterior surface of the ilium. It *inserts* on the anterior border of the greater trochanter.

Action: Abduction and inward rotation of the femur; slight flexion, also.

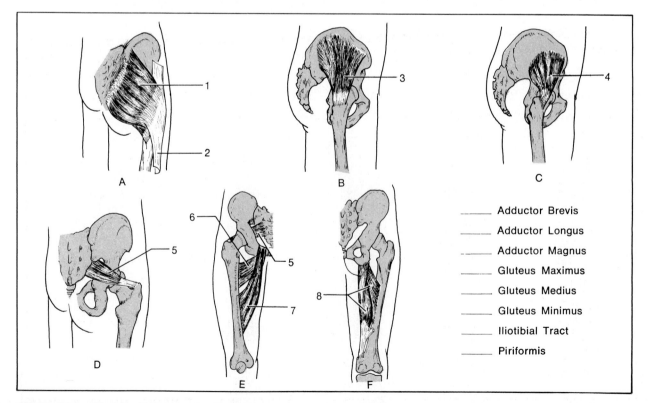

_____ Adductor Brevis

_____ Adductor Longus

_____ Adductor Magnus

_____ Gluteus Maximus

_____ Gluteus Medius

_____ Gluteus Minimus

_____ Iliotibial Tract

_____ Piriformis

Figure 24.1 Muscles that move the femur.

Piriformis This is the only muscle of this group that ties the leg to the axial skeleton. Its *origin* is on the anterior surface of the sacrum and its *insertion* is on the upper border of the greater trochanter of the femur.

Action: Outward rotation of the femur; also, some abduction and extension.

Adductor Longus and **Adductor Brevis** These two muscles are shown in illustration E. The adductor longus *originates* on the front of the pubis near the symphysis. It *inserts* on the femur (linea aspera) between the vastus medialis and adductor magnus. The adductor brevis is situated immediately behind the longus muscle and *arises* on the inferior portion of the pubis. It *inserts* on the femur at a position above the adductor longus.

Action: Although these muscles are strongest in adduction they also flex the femur and rotate it inward (medially).

Adductor Magnus This muscle is the strongest of the three adductors. It *arises* from the inferior surface of the ischium and a portion of the pubis. It is *inserted* on that portion of the femur (linea aspera) where the others insert as well as on the medial epicondyle.

Action: Works synergistically with the adductor longus and brevis.

Assignment:

Label figure 24.1.

Thigh and Lower Leg Movements

Muscles that act on the tibia and fibula are the *hamstrings, quadriceps femoris, sartorius,* and *gracilis*. The majority of these muscles are anchored to some portion of the os coxa. Although they are primarily concerned with flexion and extension of the lower part of the leg, some of them also cause rotation.

Hamstrings Three muscles, the *biceps femoris, semitendinosus,* and *semimembranosus* constitute a group of muscles on the back of the thigh known as the hamstrings. They are shown in illustration A, figure 24.2.

Biceps Femoris This muscle occupies the most lateral position of the three hamstrings. It has a long head and a short head that is obscured by the long head. The long head *arises* on the ischial tuberosity and the short head originates on the linea aspera of the femur. The muscle *inserts* on the head of the fibula and lateral condyle of the tibia.

Semitendinosus This superficial muscle lies medial to the biceps femoris. It *arises* on the ischial tuberosity and *inserts* on the upper end of the shaft of the tibia. The tendon of insertion is shown pulled away from the tibia in illustration A.

Semimembranosus This muscle occupies the most medial position of the three hamstrings and lies under the semitendinosus. Its *origin* consists of a

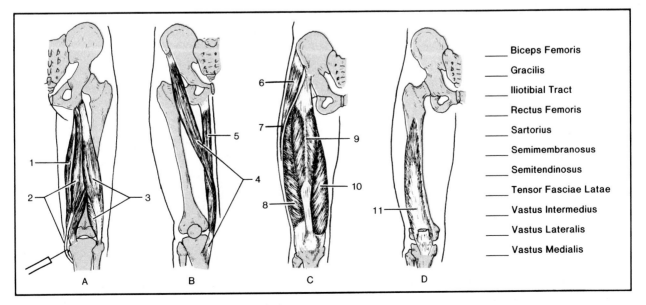

____ Biceps Femoris

____ Gracilis

____ Iliotibial Tract

____ Rectus Femoris

____ Sartorius

____ Semimembranosus

____ Semitendinosus

____ Tensor Fasciae Latae

____ Vastus Intermedius

____ Vastus Lateralis

____ Vastus Medialis

Figure 24.2   Muscles that move the tibia and fibula.

thick semimembranous tendon attached to the ischial tuberosity. Its *insertion* is primarily on the posterior medial part of the medial condyle of the tibia.

> *Action:* All of these muscles flex the calf upon the thigh. They also extend and rotate the thigh. Rotation by the biceps is outward; the other two muscles cause inward rotation.

Quadriceps Femoris The large muscle that makes up the anterior portion of the thigh is the quadriceps femoris. It consists of four parts: the *rectus femoris, vastus lateralis, vastus medialis,* and *vastus intermedius.* Its components are shown in illustrations C and D, figure 24.2. The various portions originate on the os coxa or femur. They are all united in a common tendon that passes over the patella to *insert* on the tuberosity of the tibia.

Rectus Femoris This portion of the quadriceps femoris occupies a superficial central position. It *arises* by two tendons: one from the anterior inferior iliac spine and the other from a groove just above the acetabulum. The lower portion of the muscle is a broad aponeurosis that terminates in the tendon of insertion described above.

Vastus Lateralis This is the largest and lateral portion of the quadriceps femoris. It *arises* from the lateral lip of the linea aspera.

Vastus Medialis This portion occupies the medial position on the thigh. It *arises* from the linea aspera.

Vastus Intermedius Illustration D shows the position of this portion of the quadriceps femoris. It *arises* from the front and lateral surfaces of the shaft of the femur.

> *Action:* The entire quadriceps femoris extends the leg. The rectus femoris, because of its origin on the os coxa, flexes the thigh.

Sartorius This is the longest muscle shown in illustration B. It *arises* from the anterior superior spine of the ilium and passes obliquely across the thigh and down the medial surface. The tendon at its distal end forms a broad aponeurosis that is *inserted* on the medial surface of the body of the tibia.

> *Action:* Flexes the calf on the thigh and the thigh upon the pelvis; adducts leg, permitting crossing of legs in tailor fashion; also rotates leg medially.

Tensor Fasciae Latae This muscle is shown in illustration C, figure 24.2. It *arises* from the anterior outer lip of the iliac crest; from the anterior superior iliac spine and from the deep surface of the fascia lata. It is *inserted* between the two layers of the iliotibial tract at about the junction of the middle and upper thirds of the thigh.

> *Action:* Flexes the thigh and, to some extent, rotates the thigh medially.

Gracilis This is the most superficial muscle on the inner surface of the thigh. It is the shorter one shown in illustration B. It *arises* from the lower margin of the pubic bone. It is *inserted* on the medial surface of the tibia near the insertion of the sartorius.

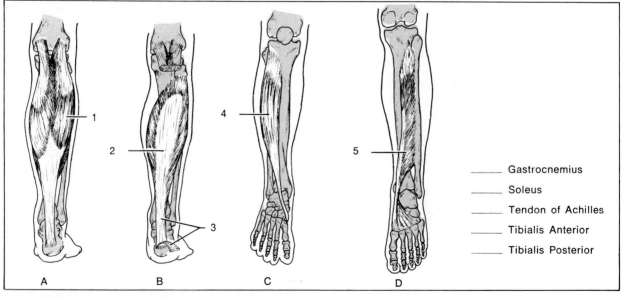

_____ Gastrocnemius

_____ Soleus

_____ Tendon of Achilles

_____ Tibialis Anterior

_____ Tibialis Posterior

Figure 24.3 Lower leg muscles.

Action: Adduction of the thigh; flexion and inward rotation of the leg.

Assignment:

Label figure 24.2.

Lower Leg and Foot Movements

Muscles of the lower part of the leg cause flexion and movements of the foot. Figures 24.3 and 24.4 include most of the muscles of this portion of the leg.

Triceps Surae The large superficial muscle that covers the calf of the leg is the triceps surae. It consists of two parts: an outer **gastrocnemius** portion and an inner **soleus.** The gastrocnemius consists of two heads that *arise* from the posterior surfaces of the medial and lateral condyles of the femur. The soleus *originates* on the head of the fibula and the tibia. Both the gastrocnemius and soleus have tendons that unite to form a common *tendon of Achilles (tendo calcaneus).* The tendon of Achilles is *inserted* on the calcaneous.

Action: Plantar flexion of foot by both muscles (standing on tiptoe). Gastrocnemius will also flex the calf on the thigh.

Tibialis Anterior The anterior lateral portion of the tibia is covered by this muscle. It *arises* from the lateral condyle and upper two-thirds of the body of the tibia. Its distal end is shaped into a long slender tendon that passes over the tarsus and *inserts* on the inferior surface of the first cuneiform and first metatarsal bones.

Action: Dorsiflexion and inversion of the foot.

Tibialis Posterior This muscle is a deep one that lies on the posterior surfaces of the tibia and fibula. It *arises* from both of these bones and the interosseous membrane that extends between them. It is *inserted* on the inferior surfaces of the navicular, the cuneiforms, the cuboid, and the second, third, and fourth metatarsals.

Action: Extension (plantar flexion) and inversion of foot; aids in maintenance of the longitudinal and transverse arches of the foot.

Peroneus Longus (Latin: *peroneus,* fibula) The three peroneus muscles are shown in illustrations A and B, figure 24.4. The peroneus longus *arises* from the head and upper two-thirds of the fibula, from deep fascia, and occasionally from the lateral condyle of the tibia. Its long slender tendon passes around the back side of the lateral malleolus under the foot to be *inserted* into the proximal portion of the first metatarsal and second cuneiform bones.

Action: Extends (plantar flexion) and everts the foot; helps to maintain the transverse arch of the foot.

Peroneus Brevis This muscle lies under the peroneus longus and is shorter and smaller. It *arises* from the distal two-thirds of the fibula and intermuscular septa. Its lower extremity consists of a tendon that passes behind the lateral malleolus along with the tendon of the previous muscle to be *inserted* on the proximal end of the fifth metatarsal bone.

Action: Extends and everts the foot.

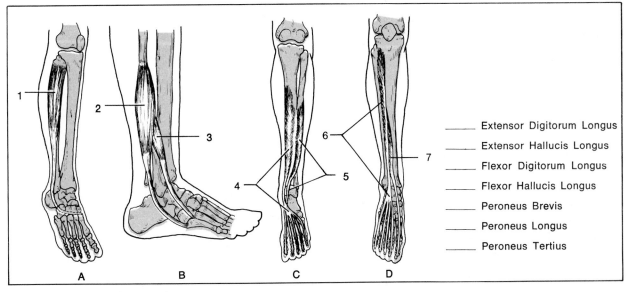

____ Extensor Digitorum Longus

____ Extensor Hallucis Longus

____ Flexor Digitorum Longus

____ Flexor Hallucis Longus

____ Peroneus Brevis

____ Peroneus Longus

____ Peroneus Tertius

Figure 24.4 Lower leg muscles.

Peroneus Tertius This is the small muscle shown in illustration B. It *arises* from the anterior surface of the lower portion of the fibula. It *inserts* on the base of the proximal portion of the fifth metatarsal bone.

Action: Flexes and everts the foot.

Flexor Hallucis Longus (Latin: *hallux*, big toe) This muscle lies on the posterior lateral surface of the leg. It *originates* from the lower two-thirds of the fibula and intermuscular septa. Its long distal tendon runs obliquely under the foot to *insert* at the base of the distal phalanx of the great toe.

Action: Flexes the great toe.

Flexor Digitorum Longus This muscle is situated on the medial (tibial) side of the leg. It *originates* on the posterior surface of the tibia and the fascia that covers the tibialis posterior. Its distal end divides into four tendons that pass along the bottom of the foot to *insert* into the bases of the last phalanges of the second, third, fourth, and fifth toes.

Action: Flexes the distal phalanges of the four small toes.

Extensor Digitorum Longus This muscle is situated on the lateral portion of the leg posterior to the tibialis anterior. It is the longer muscle shown in illustration D, figure 24.4. It *originates* on the lateral condyle of the tibia, the anterior surface of the fibula, and a part of the interosseous membrane. Its tendon of insertion divides into four parts that *insert* on the superior surfaces of the second and third phalanges of the four smaller toes.

Action: Extends the proximal phalanges of the four small toes; also flexes and pronates the foot.

Extensor Hallucis Longus This muscle lies between the tibialis anterior and the extensor digitorum longus. It is seen in illustration D, figure 24.4. It makes its *origin* on the anterior surface of the fibula and the interosseous membrane. Its tendon of insertion passes over the first metatarsal to *insert* on the superior surface at the base of the distal phalanx of the great toe.

Action: Extends the proximal phalanx of the great toe and aids in dorsiflexion of the foot.

Assignment:

Label figures 24.3 and 24.4.

Surface Muscles Review

Figures 24.5 and 24.6 have been included in this exercise to summarize our study of the muscles of the body as a whole. Only surface muscles are shown. To determine your present understanding of the human musculature, attempt to label these diagrams first by not referring back to previous illustrations. This type of self-testing will determine what additional study is needed.

Assignment:

Complete the Laboratory Report for this exercise.

Part 6 The Nervous System

The nervous system consists of an intricate maze of neurons and receptors that functions to coordinate various activities of the body. It includes the brain and spinal cord of the *central nervous system* and all the cranial and spinal nerves of the *peripheral nervous system*. It is divided functionally into the *somatic* (voluntary) and *autonomic* (involuntary) systems.

This unit consists of five exercises. Since much of the activity of the nervous system pertains to reflexes, the first exercise relates to the structure of the spinal cord, the peripheral nerves, and the two types of reflex arcs. A thorough study of reflex arcs appears in Exercise 25, followed by an indepth study of specific somatic reflexes in Exercise 26. The next two exercises pertain to the structure and functions of the human brain. Preserved human brains, as well as brain models, will be used for study in the laboratory. Because dissections of human material will not be permitted, preserved sheep brains will be available for dissection. In Exercise 29 the electrical activity of the brain is studied by recording an electroencephalogram (EEG) in the laboratory. Students will work in teams of four students to record the brain waves of members within their team.

25 The Spinal Cord and Reflex Arcs

Automatic stereotyped responses to various types of stimuli enable animals to adjust quickly to adverse environmental changes. These automatic responses, generated by the nervous system, are called **reflexes.** They play a major role in most physiological activities and are essential to survival.

Reflexes that result in automatic regulation of body function can be either somatic or visceral. **Somatic reflexes** involve skeletal muscle responses and **visceral reflexes** involve the adjustments of smooth muscle, cardiac muscle, and glands to stimuli.

The neural pathway utilized in performing a reflex is called a **reflex arc.** This pathway involves receptors, spinal nerves, the spinal cord and effectors. It is the purpose of this exercise to study each of the components that are involved in both somatic and visceral reflexes. The Laboratory Report for this study should be completed prior to performing the reflex experiments in Exercise 26.

The Spinal Cord

The spinal cord is a downward extension of the medulla oblongata of the brain. Figure 25.1 reveals its gross structure, as seen posteriorly. To expose it, the posterior portions of the vertebrae and sacrum have been removed.

The spinal cord starts at the upper border of the atlas and terminates as the **conus medullaris** at the lower border of the first lumbar vertebra. In fetal life the spinal cord occupies the entire length of the vertebral canal (spinal cavity), but as the vertebral column continues to elongate, the spinal cord fails to lengthen with it; thus, the vertebral canal extends downward beyond the end of the spinal cord.

Extending downward from the conus medullaris is an aggregate of fibers called the **cauda equina** (horse's tail), which fills the lower vertebral canal. The innermost fiber of the cauda equina, which is located on the median line, is called the **filum terminale interna.** This delicate median prolongation of the conus medullaris becomes the **filum terminale externa** after it passes through the dural sac. The **dural sac** (dura mater) is continuous with the dura mater that surrounds the brain. Only a portion of it (label 25) is shown in figure 25.1. The remainder of the dura mater has been left out of this illustration for clarity.

The Spinal Nerves

The spinal nerves emerge from the spinal cord in pairs through the intervertebral foramina on each side of the spinal cavity. There are eight cervical, twelve thoracic, five lumbar, and five sacral pairs. These thirty pairs of nerves plus one pair of coccygeal nerves makes a total of thirty-one pairs of spinal nerves.

Reference to the sectional view of the spinal cord in figure 25.1 reveals that each spinal nerve is connected to the spinal cord by structures called **anterior** and **posterior roots.** Note that the posterior root (label 21) has an enlargement, the **spinal ganglion,** which contains cell bodies of sensory neurons. Neuronal fibers from these roots pass through the outer **white matter** of the spinal cord into the inner **gray matter** where neuronal connections *(synapses)* are made.

Cervical Nerves The *first cervical nerve* (C_1) emerges from the spinal cord in the space between the base of the skull and the atlas. The *eighth cervical nerve* (C_8) emerges between the seventh cervical and first thoracic vertebrae. Note that the first four cervical nerves are united in the neck region to form a network called the **cervical plexus.** The remaining cervical nerves (C_5, C_6, C_7, and C_8) and the first thoracic nerve unite to form the **brachial plexus** in the shoulder region.

Thoracic (Intercostal) **Nerves** The *first thoracic nerve* (T_1) emerges through the intervertebral foramen between the first and second thoracic vertebrae. The *twelfth thoracic nerve* (T_{12}) emerges through the foramen between the twelfth thoracic and first lumbar vertebrae. Each of these nerves lies adjacent to the lower margin of the rib above it.

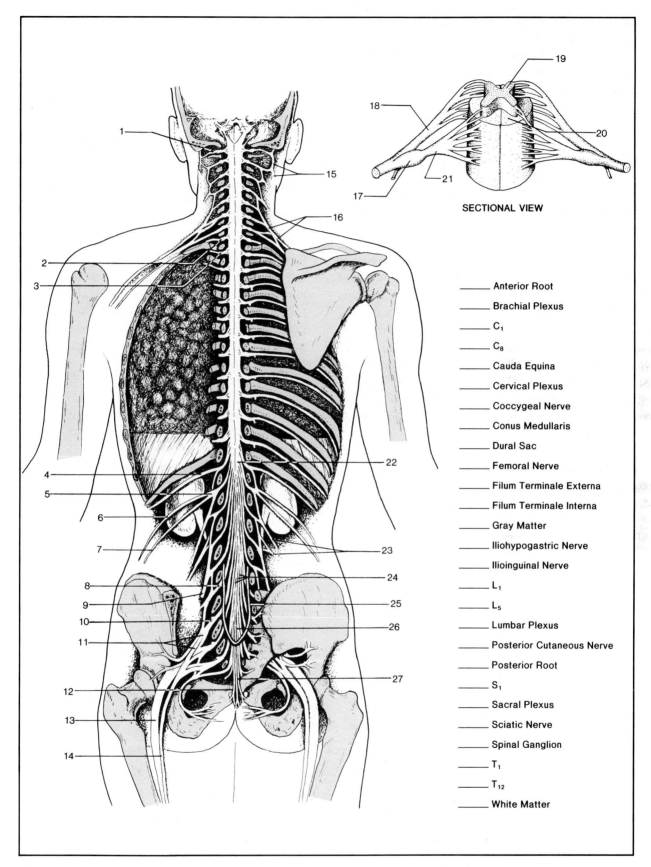

SECTIONAL VIEW

_____ Anterior Root

_____ Brachial Plexus

_____ C$_1$

_____ C$_8$

_____ Cauda Equina

_____ Cervical Plexus

_____ Coccygeal Nerve

_____ Conus Medullaris

_____ Dural Sac

_____ Femoral Nerve

_____ Filum Terminale Externa

_____ Filum Terminale Interna

_____ Gray Matter

_____ Iliohypogastric Nerve

_____ Ilioinguinal Nerve

_____ L$_1$

_____ L$_5$

_____ Lumbar Plexus

_____ Posterior Cutaneous Nerve

_____ Posterior Root

_____ S$_1$

_____ Sacral Plexus

_____ Sciatic Nerve

_____ Spinal Ganglion

_____ T$_1$

_____ T$_{12}$

_____ White Matter

Figure 25.1   The spinal cord and spinal nerves.

Lumbar Nerves The *first lumbar nerve* (L_1) emerges from the intervertebral foramen between the first and second lumbar vertebrae. Close to where the nerve emerges from the spinal cavity, it divides to form an upper **iliohypogastric** nerve and a lower **ilioinguinal** nerve.

Note in figure 25.1 that L_1, L_2, L_3, and most of L_4 are united to form the **lumbar plexus** (label 23). The largest trunk emanating from this plexus is the **femoral** nerve, which innervates part of the leg. In reality, the femoral is formed by the union L_2, L_3, and L_4.

Sacral and **Coccygeal Nerves** The remaining nerves of the cauda equina include five pairs of sacral and one pair of coccygeal nerves. A **sacral plexus** (label 11) is formed by the union of roots of L_4, L_5, and the first three sacral nerves (S_1, S_2, and S_3). The largest nerve that emerges from this plexus is the **sciatic** nerve which passes down into the leg. The smaller nerve that parallels the sciatic is the **posterior cutaneous** nerve of the leg. The **coccygeal** nerve (label 12) contains nerve fibers from the fourth and fifth sacral nerves (S_4 and S_5).

Assignment:

Label figure 25.1.

Spinal Cord and Nerve Protection

Injury to the spinal cord is life-threatening in that once the spinal cord is traumatized or severed the damage is permanent. Fortunately, the bony protection of the spinal cavity and the tissues around the spinal cord afford maximum protection. The presence of areolar connective tissue, meninges, and cerebrospinal fluid between the bony canal and the cord act as a combined shock absorber to protect the vulnerable cord. The cross-sectional view shown in figure 25.2 illustrates how the cord and spinal nerves are surrounded by these protective layers.

Cord Structure Before examining the surrounding tissues, identify the following structures on the cord in figure 25.2. The most noticeable characteristic of the spinal cord is the pattern of the **gray matter,** which resembles the configuration of the outstretched wings of a swallow-tailed butterfly. Surrounding this darker material is the **white matter,** which consists primarily of myelinated nerve fibers.

Along the median line of the spinal cord are two fissures and a small canal. The **posterior median sulcus** is the upper fissure in figure 25.2. The **anterior median fissure** is the wider fissure on the anterior surface of the spinal cord. In the gray

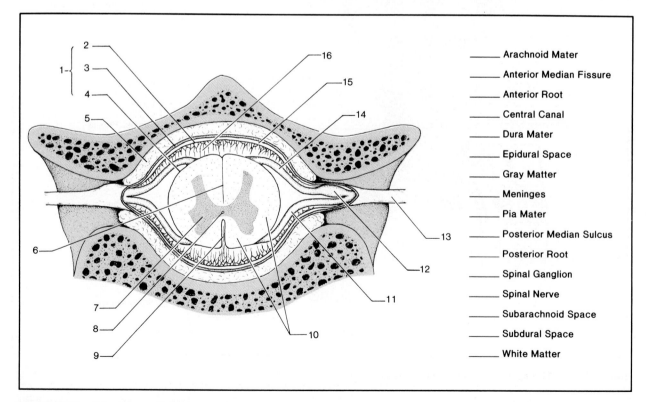

_____ Arachnoid Mater
_____ Anterior Median Fissure
_____ Anterior Root
_____ Central Canal
_____ Dura Mater
_____ Epidural Space
_____ Gray Matter
_____ Meninges
_____ Pia Mater
_____ Posterior Median Sulcus
_____ Posterior Root
_____ Spinal Ganglion
_____ Spinal Nerve
_____ Subarachnoid Space
_____ Subdural Space
_____ White Matter

Figure 25.2 The spinal cord.

matter on the median line lies a tiny **central canal,** evidence of the tubular nature of the spinal cord. This canal is continuous with the ventricles of the brain and extends into the filum terminale. It is lined with ciliated ependymal cells.

Spinal Nerves Note how the **spinal nerves** (label 13) emerge from the vertebrae on each side of the spinal column. Identify the **posterior root** with its **spinal gangion** and the **anterior root.**

The Meninges Surrounding the spinal cord (and brain) are three meninges (*meninx,* singular). The outermost membrane is the **dura mater** (label 3), which also forms a covering for the spinal nerves. A cutaway portion of the dura on each side in figure 25.2 reveals the inner spinal nerve roots. The dura mater is the toughest of the three meninges, consisting of fibrous connective tissue. Between the dura and the bony vertebrae is an **epidural space,** which contains areolar connective tissue and blood vessels.

The innermost meninx is the **pia mater.** It is a thin delicate membrane that is actually the outer surface of the spinal cord. Between the pia mater

and dura mater is the third meninx, the **arachnoid mater.** Note that the outer surface of the arachnoid mater lies adjacent to the dura mater. The space between the dura and arachnoid mater is actually a potential cavity that is called the **subdural space.** The inner surface of the arachnoid mater has a delicate fibrous texture that forms a net-like support around the spinal cord. The space between the arachnoid mater and the pia mater is the **subarachnoid space.** This space is filled with cerebrospinal fluid that provides a protective fluid cushion around the spinal cord.

Assignment:

Label figure 25.2.

The Somatic Reflex Arc

A brief mention of the somatic reflex arc was made in Exercise 13. In that exercise our prime concern was with the structure of neuronal elements that make up this circuit. Figure 25.3 is very similar to figure 13.1, although figure 25.3 is simpler.

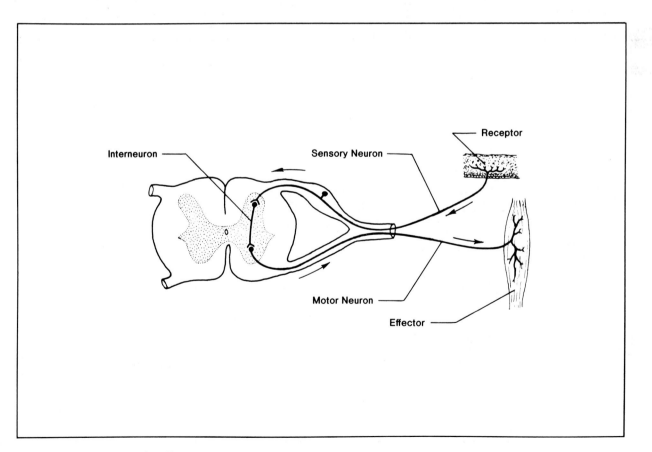

Figure 25.3 The somatic reflex arc.

The principal components of this type of reflex arc are: (1) a **receptor,** such as a neuromuscular spindle or a cutaneous endorgan that receives the stimulus; (2) a **sensory** (afferent) **neuron,** which carries impulses through a peripheral nerve and posterior root to the spinal cord; (3) an **interneuron** (association neuron), which forms synaptic connections between the sensory and motor neurons in the gray matter of the spinal cord; (4) a **motor** (efferent) **neuron,** which carries nerve impulses from the central nervous system through the ventral root to the effector via a peripheral nerve; and (5) an **effector** (either a muscle or a gland), which responds to the stimulus by contracting or secreting.

Reflex arcs may have more than one interneuron or may be completely lacking in interneurons. If no interneuron is present, as is true of the "knee-jerk" reflex, the reflex arc is said to be *monosynaptic.* When one or more interneurons exist, the reflex arc is classified as being *polysynaptic.*

The Visceral Reflex Arc

Muscular and glandular responses of the viscera to internal environmental changes are controlled by the *autonomic nervous system.* The reflexes of this system follow a different pathway in the nervous system. Figure 25.4 illustrates the components of a visceral reflex arc.

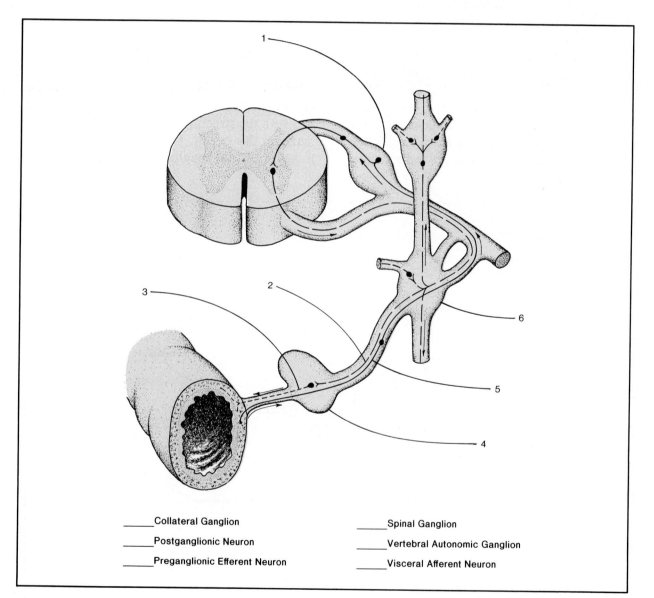

_____Collateral Ganglion

_____Postganglionic Neuron

_____Preganglionic Efferent Neuron

_____Spinal Ganglion

_____Vertebral Autonomic Ganglion

_____Visceral Afferent Neuron

Figure 25.4 The visceral reflex arc.

The principal anatomical difference between somatic and visceral reflex arcs is that the latter type has two efferent neurons instead of one. These two neurons synapse are outside of the central nervous system in an autonomic ganglion. Two types of autonomic ganglia are shown in figure 25.4: vertebral and collateral. The **vertebral autonomic ganglia** are united to form a chain that ties along the vertebral column. The **collateral ganglia** are located further away from the central nervous system.

A visceral reflex originates with a stimulus acting on a receptor in the viscera. Impulses pass along the dendrite of a **visceral afferent neuron.** Note that the cell bodies of these neurons are located in the **spinal ganglion.** The axon of the visceral afferent neuron forms a synapse with the **preganglionic efferent neuron** in the spinal cord. Impulses in this neuron are conveyed to the **postganglionic efferent neuron** at synaptic connections in either type of autonomic ganglion (note the short portions of cut-off postganglionic efferent neurons in the vertebral autonomic ganglia). From these ganglia the postganglionic efferent neuron carries the impulses to the organ innervated.

In this brief mention of the autonomic nervous system, the student is reminded that the system consists of two parts: the sympathetic and parasympathetic divisions. The *sympathetic (thoracolumbar) division* involves the spinal nerves of the thoracic and lumbar regions. The *parasympathetic (craniosacral) division* incorporates the cranial and sacral nerves. Most viscera of the body are innervated by nerves from both of these systems. This form of double innervation enables an organ to be stimulated by one system and inhibited by the other.

Assignment:

Label figure 25.4.

Laboratory Report

Complete the Laboratory Report for this exercise.

26 Somatic Reflexes

We learned in Exercise 25 that somatic reflexes are automatic responses that occur in skeletal muscles when the proper stimulus is applied to the appropriate receptor. In this laboratory period we will study somatic reflexes from two aspects: first, the characteristics of reflexes in general will be studied using a frog; and, secondly, human reflexes will be studied from a diagnostic standpoint.

Frog Experiments

The following five characteristics of somatic reflexes will be studied here: (1) functional nature; (2) reaction time; (3) reflex radiation; (4) reflex inhibition; and (5) synaptic fatigue. Since it is desirable to study spinal reflex activity that is independent of cerebral control, it will be necessary to use a frog in which the brain has been destroyed; the spinal cord, however, will be left intact.

Frog Preparation

To destroy the brain of the frog in preparation for spinal reflex tests, proceed as follows: Rinse the frog under cool tap water and grip it in your left hand, as illustrated in figure 26.1. Use your left index finger to force the frog's nose downward so that the head makes a sharp angle with the trunk. Locate the transverse groove between the skull and the vertebral column by pressing down on the skin with your fingernail to mark its location. Now, force a sharp dissecting needle into the crevice, forcing it forward into the center of the skull through the foramen magnum. Twist the needle from side to side to destroy the brain. This process of destroying the brain is called *single pithing* and the animal is referred to as a *spinal frog.*

After five or six minutes, test the effectiveness of the pithing by touching the cornea of each eye with the dissecting needle. If the lower eyelid on each eye does not rise to cover the eyeball, the pithing is complete. This test is valid only after approximately five minutes to allow recovery from a temporary state of neural shock to the entire spinal cord. Neural or *spinal shock* caused by pithing has ended when a pinch of the toes with forceps causes a withdrawal reflex. As soon as you are certain that spinal shock has disappeared from your specimen, observe how the specimen positions its limbs. Proceed to perform the following five reflex tests.

Functional Nature

Muscle tone, movement, and coordinated action in a spinal frog would indicate that these physiological activities occur independently of cognition and are the result of reflex activity. To see if any of these phenomena are present, proceed as follows:

Materials:

> spinal frog
> ring stand and clamp
> galvanized iron wire
> pliers
> 28% acetic acid
> filter paper squares (3 mm.)
> 10% NaHCO₃ in squeeze bottle
> beaker (250 ml. size)

With the frog in a squatting position, gently draw out one of the hind legs. Is any pull exerted by the animal? When the foot is released, does it return to its previous position? Squeeze the muscles of the hind legs. Do they feel flaccid or firm? Does muscle tone seem to exist? Can the animal be induced to jump by prodding its posterior? Place the animal in a sink basin filled with water. Does it attempt to swim? Does it float or sink? Answer all these questions on the Laboratory Report.

Suspend the frog with a wire hook through the lower jaw as illustrated in figure 26.2. Dip a small piece of filter paper (3 mm. square) into 28% acetic acid, shake off excess, and apply to the ventral surface of the thigh for about **10 seconds.** Is the frog able to remove the irritant? Do you think that the frog feels the burning sensation caused by the acid? Explain what is happening.

Wash the acid off the leg with 10% $NaHCO_3$, spraying it on the affected area with a squeeze bottle. Allow the liquid to drain from the leg into an empty beaker. Let stand for 1 or 2 minutes and rinse with water from squeeze bottle. Place another piece of filter paper and acid on another portion of the body (leg or abdomen). Observe reaction. What happens if you restrain the limb that attempts removal? Does any other adaptive behavior take place? Record these observations on the Laboratory Report.

Reaction Time

To determine whether or not the reaction time to a stimulus is influenced by the strength of the stimulus, proceed as follows:

Materials:

7 test tubes (20 mm. diam.) in test tube rack
3 beakers (150 ml. size)
Kimwipes
1 graduate (10 ml. size)
1 wash bottle of tap water
1 squeeze bottle of 1% HCl
1 squeeze bottle of 1% $NaHCO_3$

1. Dispense 30 ml. of 1% $NaHCO_3$ into a beaker.
2. Label seven test tubes: 1.0, .5, .4, .3, .2, .1, and .05%.
3. With a graduate, measure 10 ml. of 1% HCl from the squeeze bottle and pour into the 1% tube.
4. Into the other six tubes, measure out the correct amounts of 1% HCl and tap water to make 10 ml. each of the correct percentages of HCl. (Examples: 5 ml. of 1% HCl and 5 ml. of water in the .5% tube; 4 ml. of 1% HCl and 6 ml. of water in the .4% tube; etc.) All liquids are dispensed from the two squeeze bottles.
5. Pour the contents from the .05% tube into a

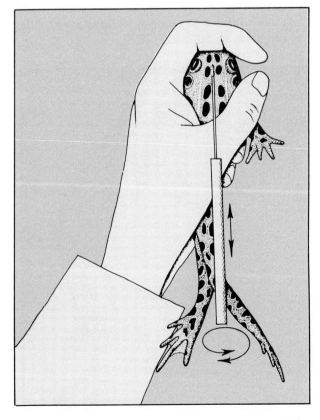

Figure 26.1 The nose of frog is depressed to find the pithing spot.

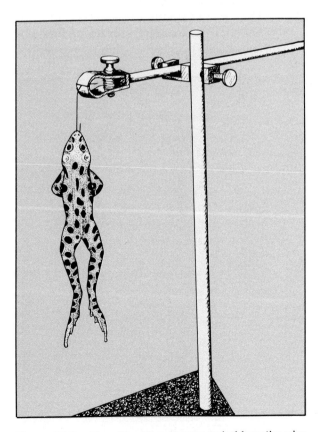

Figure 26.2 The pithed frog is suspended from the wire for reflex testing.

clean beaker and allow the long toe of one of the hind legs to be immersed in it for **90 seconds.** Be sure to prevent other parts of the foot from touching the sides of the beaker. If the foot is withdrawn, record on the Laboratory Report the number of seconds that elapsed from the time of immersion to withdrawal. If there is no withdrawal, remove the beaker of 0.05 HCl at 90 seconds.

6. Bathe the foot in the beaker of 1% $NaHCO_3$ for 15 seconds and rinse the foot by spraying with tap water from wash bottle. Use the third, unused beaker to catch any water sprayed on the foot. Dry the foot off with Kimwipes and let rest for 2 or 3 minutes. Make two more tests with .05 HCl and record an average for three tests. Be sure to neutralize, wash and dry between tests.

7. Pour the .05% HCl back into its test tube and empty the tube of 0.1% HCl into the beaker. Place the long toe into this solution, as previously, and record the withdrawal time in seconds again. After neutralization, bathing, drying, and resting the foot, repeat the test two times, as above, with this solution. Be sure not to exceed 90 seconds.

8. Repeat these procedures for all other tubes of diluted HCl, progressively increasing the acid concentration from low to high.

9. Plot the reaction times against the concentration on the graph provided on the Laboratory Report.

Reflex Radiation

If the foot of a spinal frog is subjected to increasing intensity of electrical stimulation, a phenomenon called *reflex radiation* occurs. To demonstrate it, one applies a tetanic stimulus, first at a low level and then gradually increases it. Proceed as follows to observe what takes place.

Materials:

> stimulator
> 2-wire leads with small fish hooks soldered to the leads

1. Connect the jacks of the 2-wire leads to the output posts of the stimulator. (These leads have very small fish hooks soldered to one end to enable easy attachment to the skin. Handle them gingerly. They are sharp!)
2. Attach the hooks to the skin of the left foot as illustrated in figure 26.3. They should be positioned about ¼" apart.
3. Set the stimulator to produce a minimum tetanic stimulus as follows: Voltage at 0.1 volt, Duration at 50 msec., Frequency at 50 pps.,

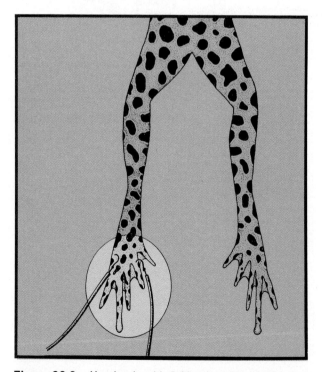

Figure 26.3 Use leads with fishhooks when testing for reflex radiation and inhibition to prevent separation during sudden leg movements.

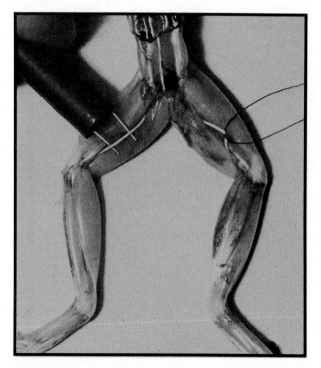

Figure 26.4 The setup required for inducing synaptic fatigue in the left leg. Note the thread on the right sciatic nerve.

Stimulus switch at REGULAR, Polarity at NORMAL, and Output switch on BIphasic.

4. Press the Mode switch to REPEAT and look for a response. If no response occurs, sequentially increase the voltage by doubling each time (.2, .4, .8, etc.) until a response occurs. Describe the response on the Laboratory Report.

5. Increase the voltage in steps of 2 or 3 volts at a time and note any changes that occur in the reflex pattern.

Reflex Inhibition

Just as it is possible to stifle a sneeze or prevent a knee jerk, it should be possible to override the reflex to acid with electrical stimulation. With the stimulator still hooked up to the left foot from the previous experiment, lower the toes of the right foot into a beaker of HCl of a concentration that produced a moderately fast reflex response. As the right foot is lowered into the acid, a moderate tetanizing stimulus is administered to the left foot. Adjust the voltage until the foot in the acid is inhibited. Record on the Laboratory Report the acid concentration and voltage that inhibited the reflex action.

Synaptic Fatigue

If your specimen is still responding well to stimuli, it may be used in this final experiment. If it does not prove viable, replace it with a freshly pithed specimen for this last test. In this experiment, it will be necessary to remove the skin from both hind legs and expose the sciatic nerves of both thighs (see figure 26.4). Stimulate the sciatic nerve of the left leg and note that muscle contraction occurs in both legs. Continue this stimulation until the muscles in the right leg fail to respond. Wait thirty seconds and stimulate again to make sure that no right leg muscle contraction occurs. Now, stimulate the sciatic nerve in the right leg. Does this cause muscles in the right leg to contract? Where did the fatigue take place when the left sciatic nerve was over-stimulated?

Reflex Diagnostics

Reflex testing is a standard, useful clinical procedure employed by physicians in search of neurological pathology. Damage to intervertebral disks, tumors, polyneuritis, apoplexy, and many other conditions can be better understood with the aid of reflex studies. The diagnostic reflexes studied here are the ones most frequently employed by physicians.

Interpretation of reflex responses is often subjective and requires considerable experience on the part of the diagnostician. Our purpose in performing the tests here, obviously, is not to diagnose, but, rather, to test and observe and to understand why a particular test is performed.

Two Types of Reflexes

Clinically, reflexes are categorized as being either deep or superficial. The **deep reflexes** include all reflexes that are elicited by a sharp tap on an appropriate tendon or muscle. They are also called *jerk, stretch,* or *myotatic* reflexes. The receptors (spindles) for these reflexes are located in the muscle, not in the tendon. When the tendon is tapped, the muscle is stretched; stretching the muscle, in turn, activates the muscle spindle, which triggers the reflex response. **Superficial reflexes** are withdrawal reflexes elicited by noxious or tactile stimulation. They are also called *cutaneous reflexes.* Instead of percussion initiating these reflexes, the skin is stroked or scratched to induce a response.

Reflex Aberrations

Abnormal responses to stimuli may be diminished (hyporeflexia), exaggerated (hyperreflexia), or pathological. *Hyporeflexia* may be due to malnutrition, neuronal lesions, aging, deliberate relaxation, etc. *Hyperreflexia* is often accompanied by marked muscle tone due to the loss of inhibitory control by the motor cortex; among other things, it may also be caused by strychnine poisoning. *Pathological reflexes* are reflex responses that occur in one or more muscles other than the muscle where the stimulus originates. It is these three deviations that we will look for in performing these tests. The following scale should be used for evaluating each response:

+ + + + very brisk, hyperactive; often indicative of pathology; may be associated with clonus.

+ + + brisker than average; may or may not be indicative of pathology.

+ + average; normal.

+ somewhat diminished.

0 no response.

Reflexes of Arm and Hand

Biceps Reflex This reflex is a deep reflex which causes flexion of the arm. It is elicited by holding the subject's elbow with the thumb pressed over the tendon of the *biceps brachii,* as shown in figure 26.5. To produce the desired response, strike a sharp blow to the first digit of the thumb with the reflex hammer.

If the reflex is absent or diminished (0 or +), apply reinforcement by asking the subject to clench his/her teeth or squeeze his/her thigh with the other hand. Test both arms and record the degree of response on the Laboratory Report.

This reflex functions through C_5 and C_6 spinal nerves.

Triceps Reflex This deep reflex causes extension of the arm in normal individuals. To demonstrate this reflex, flex the arm at the elbow, holding the wrist as shown in figure 26.6 with the palm facing the body. Strike the *triceps brachii* tendon above the elbow with the pointed end of the reflex hammer. Use the same reinforcement techniques as above, if necessary. Test both arms and record the degree of response on the Laboratory Report.

This reflex functions through C_7 and C_8 spinal nerves.

Brachioradialis Reflex In normal individuals this reflex manifests itself in flexion and pronation of the forearm and flexion of the fingers. To demonstrate this reflex, direct the subject to rest the hand on the thigh in the position shown in figure 26.7. Strike the forearm with the wide end of the reflex hammer about 1″ above the end of the radius. The illustration reveals the approximate spot. Use reinforcement techniques, if necessary. The tendon that is struck here is for the *brachioradialis* muscle of the arm.

This reflex functions through the same spinal nerves (C_5 and C_6) as the *biceps brachii.*

Hoffmann's Reflex This reflex is an example of a deep reflex in which the response does not occur only in the muscle that is stretched. As stated above, a positive reflex is classified as pathological. Hoffmann's reflex also exhibits characteristics of hyperreflexia.

This reflex is produced by flicking the terminal phalanx of the index finger upward, as illustrated in figure 26.8. In individuals who have

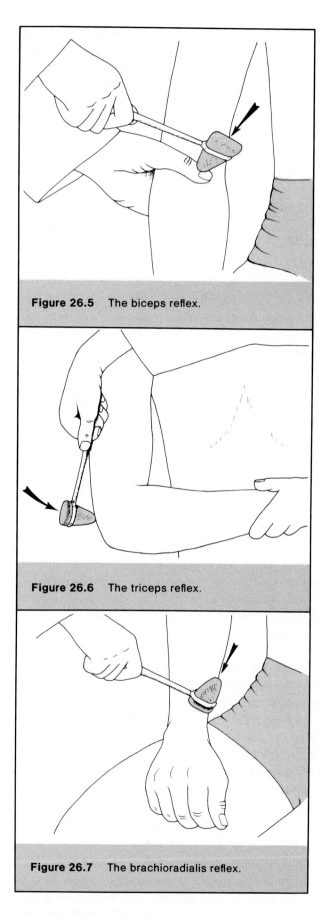

Figure 26.5 The biceps reflex.

Figure 26.6 The triceps reflex.

Figure 26.7 The brachioradialis reflex.

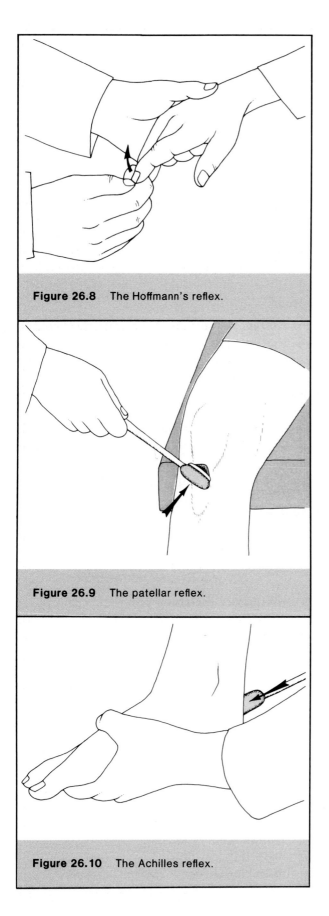

Figure 26.8 The Hoffmann's reflex.

Figure 26.9 The patellar reflex.

Figure 26.10 The Achilles reflex.

pyramidal tract lesions, the thumb of the hand is adducted and flexed, and other fingers exhibit twitch-like flexion. Such a broad field of reflex action resulting from a brief stretch of the flexor muscles of one finger manifests deep reflex hyperactivity and pathological characteristics.

Test both hands of the subject and record results on the Laboratory Report.

Reflexes of the Leg

Patellar Reflex This monosynaptic reflex is variously referred to as the *knee reflex, knee jerk,* or *quadriceps reflex.* To perform this test, the subject should be seated on the edge of a table with the leg suspended and somewhat flexed over the edge.

To elicit the typical response, strike the patellar tendon, which is just below the knee cap. The correct spot is shown in figure 26.9.

If the response is negative, utilize the *Jendrassic maneuver* for facilitation. This technique involves the subject locking the fingers in front of the body and pulling each hand against each other, isometrically.

A phenomenon called *clonus* is sometimes seen when inducing this reflex. Clonus is characterized by a succession of jerk-like contractions that follow the normal response and persist for a period of time. This condition is a manifestation of hyperreflexia and indicates damage within the central nervous system.

Test both legs, recording the results on the Laboratory Report. This reflex functions through L_2, L_3, and L_4 spinal nerves.

Achilles Reflex This reflex is also referred to as the *ankle jerk.* It is characterized by plantar flexion when the Achilles tendon is struck a sharp blow. Stretching this tendon affects the muscle spindles in the *triceps surae,* causing it to contract. To perform this test, grip the foot with the left hand as shown in figure 26.10, forcing it upward somewhat. With subject relaxed, strike the Achilles tendon as shown. This reflex functions through S_1 and S_2 spinal nerves. Hyporeflexia here is often a sign of hypothyroidism.

Laboratory Report

Complete the Laboratory Report for this exercise.

27 Brain Anatomy: External

This study of the human brain will be done in conjunction with dissection of the sheep brain. The ample size and availability of sheep brains make them preferable to the brains of most other animals. The anatomical differences that exist between the human and sheep brain are insignificant compared to the similarities. Practically all the major components of the human brain have their counterparts in the sheep brain. Although preserved human brains will be available for study in the laboratory, they should not be dissected. All dissection will be performed **only** on the sheep brains. A discussion of the meninges will precede the sheep brain dissection.

The Meninges

It was observed in Exercise 25 that the spinal cord, as well as the brain, is surrounded by three membranes called **meninges.** Due to the importance of these protective structures it will be necessary to reemphasize certain anatomical characteristics of these membranes.

Dura Mater The outermost meninx is the **dura mater,** which is a thick, tough membrane made up of fibrous connective tissue. The lateral view of the opened skull in figure 27.1 reveals the dura mater lifted away from the exposed brain. On the median line of the skull it forms a large **sagittal sinus** (label 11) which collects blood from the surface of the brain. Note in the lateral view that a large number of cerebral veins empty into this sinus. Extending downward between the two halves of the cerebrum is an extension of the dura mater, the **falx cerebri.** This structure is shown in the frontal section. At its anterior end, the falx cerebri is attached to the crista galli of the ethmoid bone.

Arachnoid Mater Inferior to the dura mater lies the second meninx, the **arachnoid mater** (label 8, figure 27.1). It is a delicate net-like membrane. Between this membrane and the surface of the

brain is the **subarachnoid space,** which contains cerebrospinal fluid. Note that the arachnoid mater has small projections of tissue, the **arachnoid granulations,** which extend up into the sagittal sinus. These granulations allow cerebrospinal fluid to diffuse from the subarachnoid space into the blood.

Pia Mater The surface of the brain is covered by the third meninx, the **pia mater.** This membrane is very thin. Note that the surface of the brain has grooves, or **sulci,** which increase the surface area of the **gray matter** (label 4). As in the case of the spinal cord, the gray matter is darker than the inner **white matter** because it contains cell bodies of neurons. The white matter consists of neuron fibers that extend from the gray matter to other parts of the brain and spinal cord.

Assignment:

Label figure 27.1.

Sheep Brain Dissection

As you dissect the sheep brain and identify the various structures on it, compare the sheep brain with the human brain. Note that illustrations of comparable views of both brain types are on the same page or opposite pages. Once you have located a structure on the sheep brain, try to find the same structure on the human brain. Performing the dissection in this way will help you to learn human brain anatomy.

Materials:

preserved sheep brains
preserved human brains
human brain models
dissecting instruments
dissecting trays

Place a sheep brain on a dissecting tray and, by comparing it to figures 27.3 and 27.4, identify the **cerebrum, cerebellum, pons Varoli, midbrain,** and **medulla oblongata** on the specimen. Study each of these major divisions of the brain to identify those fissures, grooves, ridges, and other structures as follows:

Cerebral Hemispheres In both the sheep and man, the cerebrum is the predominant portion of the brain. Note that when observed from the dorsal or ventral aspect, the cerebrum is divided along

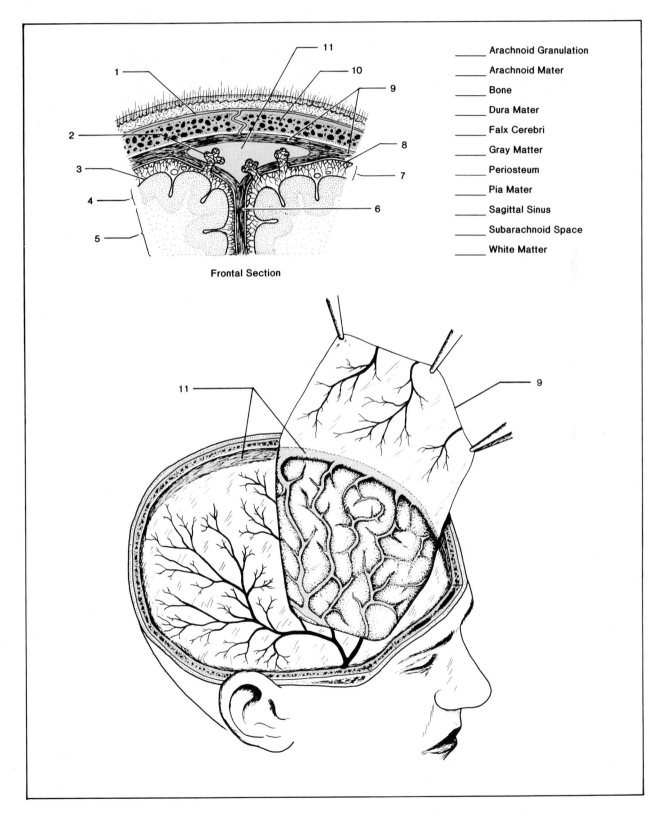

_____ Arachnoid Granulation

_____ Arachnoid Mater

_____ Bone

_____ Dura Mater

_____ Falx Cerebri

_____ Gray Matter

_____ Periosteum

_____ Pia Mater

_____ Sagittal Sinus

_____ Subarachnoid Space

_____ White Matter

Frontal Section

Figure 27.1 The meninges.

the median line to form two **cerebral hemispheres** by the **longitudinal cerebral fissure.**

Observe that the surface of the cerebrum is covered with ridges and furrows of variable depth. The deeper furrows are called **fissures,** and the shallow ones, **sulci.** The ridges or convolutions are called **gyri.** Figure 27.1 reveals how this infolding of the cerebral surface greatly increases the amount of gray matter of the brain.

The principal fissures of the human cerebrum are the **central sulcus (fissure of Rolando), lateral cerebral fissure (fissure of Sylvius),** and the **parieto-occipital fissure.** In figure 27.3 the central sulcus is label 14 and the lateral cerebral fissure is label 12. The greater portion of the parieto-occipital fissure is seen on the medial surface of the cerebrum (label 1, figure 28.2). If the lateral cerebral fissure is retracted as shown in the lower illustration of figure 27.3, an area called the **island of Reil** is exposed.

Each half of the human cerebrum is divided into four lobes. The **frontal lobe** is the most anterior portion. Its posterior margin falls on the central sulcus and its inferior border consists of the lateral cerebral fissure. The **occipital lobe** is the most posterior lobe. The anterior margin of this lobe falls on the parieto-occipital fissure. Since most of this fissure is seen on the medial face of the cerebrum, an imaginary dotted line in figure 27.3 reveals its position. The **temporal lobe** lies inferior to the lateral cerebral fissure and extends back to the parieto-occipital fissure. An imaginary dotted line provides a border between it and part of the **parietal lobe.**

Cerebellum Examine the surface of the sheep cerebellum. Note that its surface is furrowed with sulci. How do these sulci differ from those of the cerebrum? The human cerebellum is constricted in the middle to form right and left hemispheres. Can the same be said for the sheep's cerebellum? This part of the brain plays an important part in the maintenance of posture and the coordination of complex muscular movements.

Midbrain Since the midbrain lies concealed under the cerebellum and cerebrum, expose its dorsal surface by forcing the cerebellum downward with the thumb as illustrated in figure 27.2. Observe that you have exposed five roundish bodies in this area: four bodies of the corpora quadrigemina and a single pineal gland on the median line.

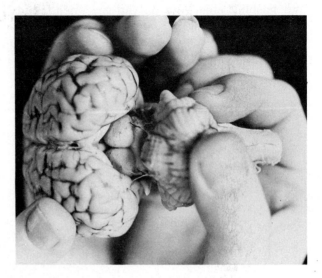

Figure 27.2 Method of exposing midbrain.

The two pairs of rounded eminences that dominate this area constitute the **corpora quadrigemina.** The larger pair are called the **superior colliculi;** the smaller ones, the **inferior colliculi.** In some animals, the superior colliculi are called **optic lobes** because of their close association with the optic tracts. In animals, the colliculi are important analytical centers concerned with brightness and sound discrimination. Their functions in man are still obscure.

The **pineal gland** lies on the median line of the midbrain, anterior to the corpora quadrigemina. From an evolutionary point of view, there are indications that the pineal gland is a remnant of a third eye that exists in certain reptiles. This gland produces a hormone called **melatonin,** which in lower animals inhibits the estrus cycle. The exact role of melatonin in humans is not completely understood at this time; it may have something to do with the estrus cycle. It is probably significant that this structure is, histologically, glandular until puberty, and then becomes fibrous in nature after puberty.

Pons Varoli Locate this portion of the sheep's brain by examining its ventral surface and comparing it with figures 27.4 and 27.5. This part of the brain stem contains fibers that connect parts of the cerebellum and the medulla with the cerebrum. It also contains nuclei of the fifth, sixth, seventh, and eighth cranial nerves.

Medulla Oblongata This portion of the brain stem is also known as the **spinal bulb.** It contains centers of the gray matter that control the heart, respiration, and vasomotor reactions. The last four cranial nerves emerge from the medulla.

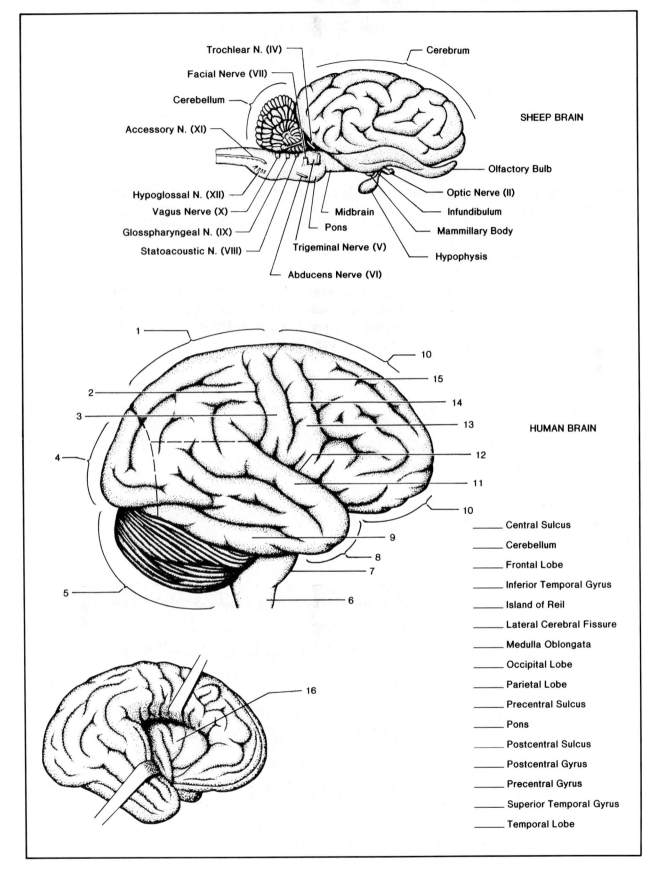

Figure 27.3   Lateral aspects of sheep and human brains.

_____ Central Sulcus

_____ Cerebellum

_____ Frontal Lobe

_____ Inferior Temporal Gyrus

_____ Island of Reil

_____ Lateral Cerebral Fissure

_____ Medulla Oblongata

_____ Occipital Lobe

_____ Parietal Lobe

_____ Precentral Sulcus

_____ Pons

_____ Postcentral Sulcus

_____ Postcentral Gyrus

_____ Precentral Gyrus

_____ Superior Temporal Gyrus

_____ Temporal Lobe

Assignment:

Label figure 27.3.

The Cranial Nerves

There are twelve pairs of nerves that emerge from various parts of the brain and pass through foramina of the skull to innervate parts of the head and trunk. Each pair has a name as well as a position number. Although most of these nerves contain both motor and sensory fibers (**mixed nerves**), a few contain only sensory fibers (**sensory nerves**). The sensory fibers have their cell bodies in ganglia outside of the brain; the cell bodies of the motor neurons, on the other hand, are situated within **nuclei** of the brain.

Compare your sheep brain with figures 27.3 and 27.4 to assist you in identifying each pair of nerves. As you proceed with the following study, avoid damaging the brain tissue with the dissecting needle. Once a nerve is broken off it is difficult to determine where it was.

I Olfactory Nerve This cranial nerve contains sensory fibers for the sense of smell. Since it extends from the mucous membranes in the nose through the ethmoid bone to the olfactory bulb, it cannot be seen on your specimen. In the **olfactory bulb,** synapses are made with fibers that extend inward to olfactory areas in the cerebrum. Note that on the human brain an **olfactory tract** (label 2) extends between the bulb and the brain.

II Optic Nerve This sensory nerve functions in vision. It contains axon fibers from ganglion cells of the retina of the eye. Part of the fibers from each optic nerve cross over to the other side of the brain as they pass through the **optic chiasma.** From the optic chiasma the fibers pass through the **optic tracts** to the thalamus and finally to the visual areas of the cerebrum. Identify these structures on the sheep brain.

III Oculomotor Nerve This nerve emerges from the midbrain and supplies somatic nerve fibers to the **levator palpebrae** (eyelid muscle) and four of the extrinsic ocular muscles: **superior rectus, medial rectus, inferior rectus,** and **inferior oblique.** (The superior oblique and lateral rectus are innervated by IV and VI, respectively).

The oculomotor nerve also supplies parasympathetic fibers to the **constrictor pupillaris** of the iris and the **ciliaris muscle** of the ciliary body. These fibers constrict the iris and change the lens shape as needed in accommodation (focusing).

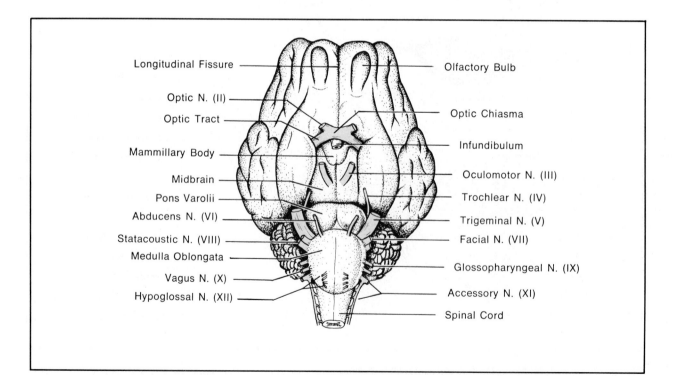

Figure 27.4 Ventral surface of sheep brain.

If the oculomotor nerves are not visible on your sheep brain, it may be that they are obscured by the hypophysis (pituitary gland). The sheep brain illustrated in figure 27.4 lacks this structure. To remove the hypophysis without damaging other structures, lift it up with a pair of forceps and cut the infundibulum with a scalpel or a pair of scissors.

IV Trochlear Nerve This nerve provides muscle sense and motor stimulation of the **superior oblique** muscle of the eye. As in the case of the oculo-

motor, this nerve also emerges from the midbrain. Because of its small diameter it is one of the harder nerves to locate.

V Trigeminal Nerve Just posterior to the trochlear nerve lies the largest cranial nerve, the trigeminal. Although it is a mixed nerve, its sensory functions are much more extensive than its motor functions. Because of its extensive innervation of parts of the mouth and face, it is described in detail on pages 156 to 157.

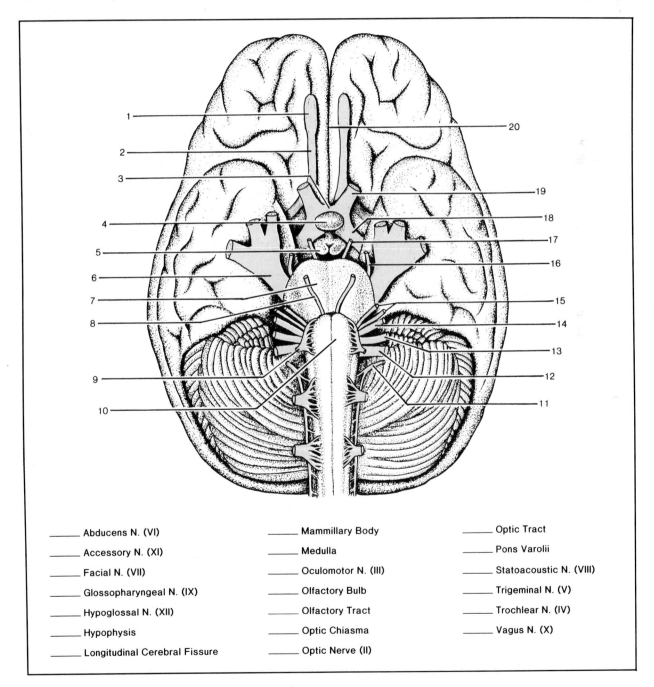

_____ Abducens N. (VI)

_____ Accessory N. (XI)

_____ Facial N. (VII)

_____ Glossopharyngeal N. (IX)

_____ Hypoglossal N. (XII)

_____ Hypophysis

_____ Longitudinal Cerebral Fissure

_____ Mammillary Body

_____ Medulla

_____ Oculomotor N. (III)

_____ Olfactory Bulb

_____ Olfactory Tract

_____ Optic Chiasma

_____ Optic Nerve (II)

_____ Optic Tract

_____ Pons Varolii

_____ Statoacoustic N. (VIII)

_____ Trigeminal N. (V)

_____ Trochlear N. (IV)

_____ Vagus N. (X)

Figure 27.5   Inferior aspect of human brain.

VI Abducens Nerve This small nerve provides innervation of the **lateral rectus** muscle of the eye. It is a mixed nerve in that it provides muscle sense as well as muscular contraction. On the human brain it emerges from the lower part of the pons near the medulla.

VII Facial Nerve This nerve consists of motor and sensory divisions. Label 15, figure 27.5 reveals the double nature of this nerve on the human brain. It innervates muscles of the face, salivary glands, and taste buds of the anterior two-thirds of the tongue.

VIII Statoacoustic (Auditory) **Nerve** This nerve goes to the inner ear. It has two branches: the **vestibular division,** which innervates the semicircular canals, and the **cochlear division,** which innervates the cochlea. The former branch functions in maintaining equilibrium; the cochlear portion is auditory in function. The statoacoustic nerve emerges just posterior to the facial nerve on the human brain.

IX Glossopharyngeal Nerve This mixed nerve emerges from the medulla posterior to the statoacoustic. It functions in reflex control of the heart, taste, and swallowing reflexes. Taste buds on the back of the tongue are innervated by some of its fibers. Efferent fibers innervate muscles of the pharynx (swallowing) and the parotid salivary glands (secretion).

X Vagus Nerve This cranial nerve, which originates on the medulla, exceeds all of the other cranial nerves in its extensive ramifications. In addition to supplying parts of the head and neck with nerves, it has branches that extend down into the chest and abdomen. It is a mixed nerve. Sensory fibers in this nerve go to the heart, external acoustic meatus, pharynx, larynx, and thoracic and abdominal viscera. Motor fibers pass to the pharynx, base of the tongue, larynx, and to the autonomic ganglia of thoracic and abdominal viscera. Many references will be made to this nerve in subsequent discussions of the physiology of the lungs, heart, and digestive organs.

XI Accessory Nerve Since this nerve emerges from both the brain and spinal cord it is sometimes called the **spinal accessory** nerve. Note that on both the sheep and the human brain, it parallels the lower part of the medulla and the spinal cord

with branches going into both structures. On the human brain it appears to occupy a more posterior portion than the twelfth nerve. Afferent and efferent spinal components innervate the **sternocleidomastoid** and **trapezius** muscles. The cranial portion of the nerve innervates the pharynx, upper larynx, uvula, and palate.

XII Hypoglossal Nerve This nerve emerges from the medulla to innervate several muscles of the tongue. It contains both afferent and efferent fibers. On the human brain in figure 27.5 it has the appearance of a cut-off tree trunk with roots extending into the medulla.

The Hypothalamus

The ventral portion of the brain that includes the mammillary bodies, infundibulum, and part of the hypophysis is collectively called the *hypothalamus.* The deeper portions of the hypothalamus that surround the third ventricle are seen in figure 28.5 (label 6). This portion of the brain is a part of the diencephalon which lies between the cerebral hemispheres and the mid-brain. It will be studied more thoroughly in the next exercise when the internal parts are revealed by dissection.

Mammillary Bodies This part of the hypothalamus lies superior to the hypophysis. Although it appears as a single body on the sheep, it consists of a pair of rounded eminences just posterior to the infundibulum on the human. The mammillary bodies receive fibers from the olfactory areas of the brain and ascending pathways and send fibers to the thalamus and other brain nuclei.

Hypophysis This structure, which is also called the pituitary gland, is attached to the base of the brain by a stalk, the *infundibulum.* It consists of two distinctly different parts—an anterior adenohypophysis and a posterior neurohypophysis. Only the neurohypophysis and infundibulum are considered to be part of the hypothalamus because they both have the same embryonic origin as the brain. The adenohypophysis originates as an outpouching of the pharynx and is quite different histologically. A close functional relationship exists between the hypothalamus, hypophysis, and the autonomic nervous system.

Assignment:

Label figure 27.5.

Functional Localization of Cerebrum

Extensive experimental studies on monkeys, apes, and humans have resulted in considerable knowledge of the functional areas of the cerebrum. Figure 27.6 shows the positions of these various centers. Before attempting to identify each of these areas from the following description be sure you know the boundaries (sulci and fissures) of the cerebral lobes (figure 27.3).

Somatomotor Area This area occupies the surface of the *anterior central gyrus* of the frontal lobe. This gyrus lies just anterior to the central sulcus. Electrical stimulation of this portion of the cerebral cortex in a conscious human results in movement of specific muscular groups.

Premotor Area The large area anterior to the somatomotor area is the premotor area. It exerts control over the motor area.

Somatosensory Area This area occupies the *posterior central gyrus,* which is the most anterior portion of the parietal lobe adjacent to the central sulcus. This area functions to localize very precisely those points on the body where sensations of light, touch, and pressure originate. It also assists in determining organ position. Other sensations, such as aching pain, crude touch, warmth, and cold, are localized by the thalamus rather than this area.

Motor Speech Area This area is located in the frontal lobe just above the lateral cerebral fissure, anterior to the somatomotor area. It exerts control over the muscles of the larynx and tongue that produce speech.

Visual Area This area is located on the occipital lobe. Note that only a small portion of it is seen on the lateral aspect; the greater portion of this area is seen on the medial surface of the cerebrum. On this surface it extends anteriorly to the parieto-occipital fissure, becoming narrower as it approaches the fissure. This area receives impulses from the retina via the thalamus. Destruction of this region causes blindness, although light and dark are still discernable.

Auditory Area This area receives nerve impulses from the cochlea of the inner ear via the thalamus.

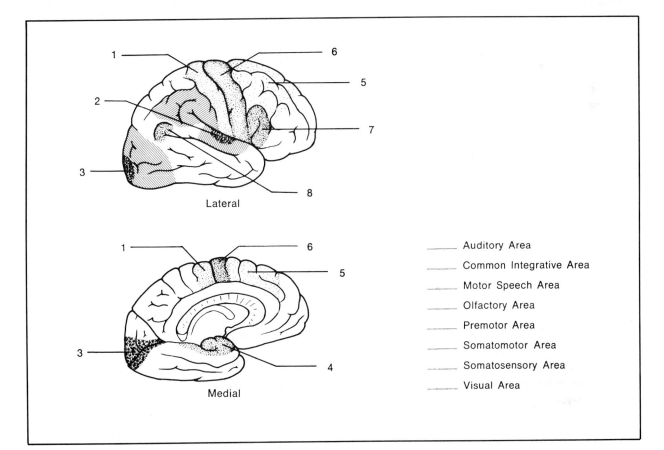

_____ Auditory Area
_____ Common Integrative Area
_____ Motor Speech Area
_____ Olfactory Area
_____ Premotor Area
_____ Somatomotor Area
_____ Somatosensory Area
_____ Visual Area

Lateral

Medial

Figure 27.6   Brain areas.

It is responsible for hearing and speech understanding. It is located in the gyrus of the temporal lobe that borders on the lateral cerebral fissure *(superior temporal gyrus).*

Olfactory Area This area is located on the medial surface of the temporal lobe (see lower illustration, figure 27.6). The recognition of various odors occurs here. Tumors in this area cause individuals to experience nonexistent odors of various kinds, both pleasant and unpleasant.

Association Areas Adjacent to the somatosensory, visual, and auditory areas are association areas that lend meaning to what is felt, seen, or heard. These areas are lined regions in figure 27.6.

Common Integrative Area The integration of information from the above three association areas and the olfactory and taste centers is achieved by a small area which is called the common integrative or *gnostic* area. This area is located on the *angular gyrus,* which is positioned approximately midway between the three association areas.

Assignment:

Label figure 27.6.

The Trigeminal Nerve

Although a detailed study of all cranial nerves is precluded in an elementary anatomy course, the trigeminal nerve has been singled out for a thorough study here. This fifth cranial nerve is the largest one that innervates the head and is of particular medical-dental significance.

Figure 27.7 illustrates the distribution of this nerve. You will note that it has branches that supply the teeth, tongue, gums, forehead, eyes, nose, and lips. The following description identifies the various ganglia and nerves.

On the left margin of the diagram is the severed end of the nerve at a point where it emerges from the brain. Between this severed end and the three main branches is an enlarged portion, the **Gasserian ganglion.** This ganglion contains the nerve cell bodies of the sensory fibers in the nerve. The three branches that extend from this ganglion are the *ophthalmic, maxillary,* and *mandibular nerves.* Locate the bony portion of the skull in figure 27.7 through which these three branches pass.

Ophthalmic Nerve The ophthalmic nerve is the upper branch which passes out of the cranium through the *superior orbital fissure.* It has three branches which innervate the lacrimal gland, the upper eyelid and the skin of the nose, forehead, and scalp. One of the branches also supplies parts of the eye such as the *cornea* (window of the eye); the *ciliary body* (muscle attached to the lens of the eye that is concerned with focusing the eye); and the *iris* (color band of the eye that controls the amount of light that enters the eye). Superior to the lower branch of the ophthalmic is the **ciliary ganglion,** which contains nerve cell bodies that are concerned with the control of the ciliary body of the eye. The ophthalmic nerve and its branches are not involved in dental anesthesia.

Maxillary Nerve The maxillary nerve, or *second division* of the trigeminal nerve, is a sensory nerve that provides innervation to the nose, upper lip, palate, maxillary sinus, and upper teeth. It is the branch that comes off just inferior to the ophthalmic nerve.

Note that the maxillary nerve has two short branches that extend downward to an oval body, the **sphenopalatine ganglion.** The major portion of the maxillary nerve becomes the **infraorbital nerve,** which passes through the **infraorbital nerve canal** of the maxilla. This canal is shown with dotted lines in figure 27.7. This latter nerve emerges from the maxilla through the **infraorbital foramen** to innervate the tissues of the nose and upper lip.

The upper teeth are innervated by three nerves. The **posterior superior alveolar nerve** is a branch of the maxillary nerve that enters the posterior surface of the upper jaw and innervates the molars. The **anterior superior alveolar nerve** is an anterior branch of the infraorbital nerve that supplies the anterior teeth and bicuspids with nerve fibers. The anterior superior alveolar nerve forms a loop with the **middle superior alveolar nerve.** This latter nerve is shown clearly where a portion of the maxilla has been cut away. It contains fibers that pass to the bicuspids and first molar.

To desensitize all of the upper teeth on one side of the maxilla, a dentist can inject anesthetic near either the maxillary nerve or the sphenopalatine ganglion. Desensitization of the maxillary nerve is called a *second division nerve block.* This type of nerve block will affect the palate as well as the teeth because it involves fibers of the **anterior pal-**

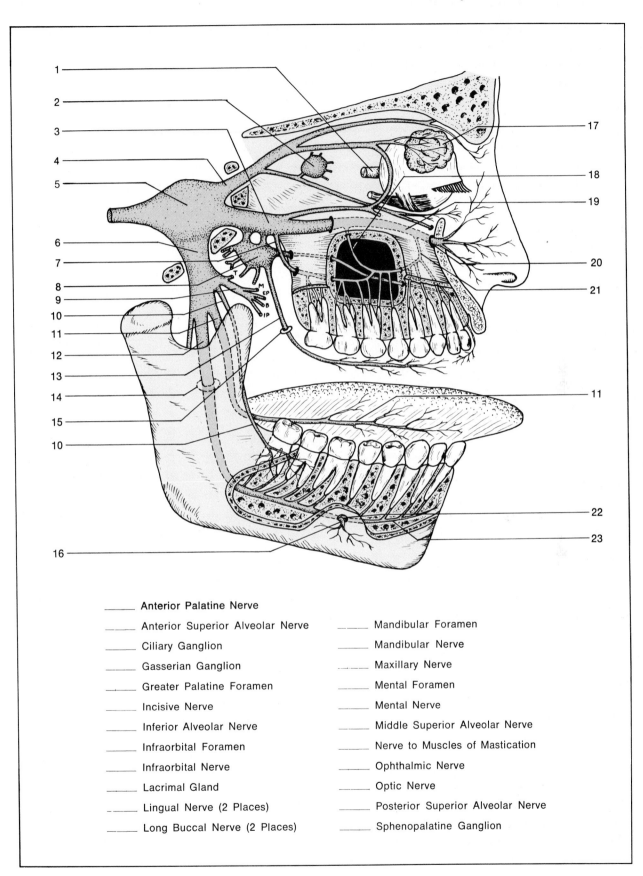

Figure 27.7 The trigeminal nerve.

_____ Anterior Palatine Nerve

_____ Anterior Superior Alveolar Nerve

_____ Ciliary Ganglion

_____ Gasserian Ganglion

_____ Greater Palatine Foramen

_____ Incisive Nerve

_____ Infraorbital Foramen

_____ Infraorbital Nerve

_____ Lacrimal Gland

_____ Lingual Nerve (2 Places)

_____ Long Buccal Nerve (2 Places)

_____ Mandibular Foramen

_____ Mandibular Nerve

_____ Maxillary Nerve

_____ Mental Foramen

_____ Mental Nerve

_____ Middle Superior Alveolar Nerve

_____ Nerve to Muscles of Mastication

_____ Ophthalmic Nerve

_____ Optic Nerve

_____ Posterior Superior Alveolar Nerve

_____ Sphenopalatine Ganglion

atine nerve. This latter nerve extends downward from the sphenopalatine ganglion to the soft tissues of the palate through the *greater palatine foramen.*

Mandibular Nerve The mandibular nerve, or *third division* of the trigeminal nerve, is the lowest branch of this nerve. It innervates the teeth and gums of the mandible, the muscles of mastication, the anterior part of the tongue, the lower part of the face, and some skin areas on the side of the head.

Moving downward from the Gasserian ganglion, the first branch of the mandibular nerve that we encounter is the **nerve to the muscles of mastication.** This nerve has five branches with small identifying letters near their cut ends (T—temporalis, M—masseter, EP—external pterygoid, B—buccinator, and IP—internal pterygoid). The fibers in these nerves control the motor activities of these five muscles.

The largest branch of the mandibular nerve is the **inferior alveolar nerve,** which enters the ramus of the mandible through the **mandibular foramen** and supplies branches to all the teeth on one side of the mandible. At the mental foramen the inferior alveolar nerve becomes the **incisive nerve,** which innervates the anterior teeth. Emerging from the **mental foramen** is the **mental nerve,** which supplies the soft tissues of the chin and lower lip.

Desensitization of the mandibular teeth is generally achieved by injecting the anesthetic near the mandibular foramen. This type of injection, called a *lower nerve block,* or *third division nerve block,* is widely used because the external and internal alveolar plates of the mandibular alveolar process are too dense to facilitate infiltration types of injection. The lower nerve block of one side of the mandible desensitizes all of the teeth on one side except for the anterior teeth. The fact that the anterior teeth receive nerve fibers from both the left and right incisive nerves prevents complete desensitization.

A branch of the mandibular nerve that emerges just above the inferior alveolar nerve and innervates the tongue is the **lingual nerve.** As in the case of the inferior alveolar nerve, it is sensory in function.

The **long buccal nerve** is another sensory branch of the mandibular nerve that innervates the buccal gum tissues of the molars and bicuspids. This nerve arises from a point which is just superior to the lingual nerve.

Assignment:

Label figure 27.7.

Laboratory Report

Complete the Laboratory Report for this exercise.

28 Brain Anatomy: Internal

The interrelations of the various divisions of the brain can be seen only by studying sections such as those in figures 28.1 and 28.2. Bundles of interconnecting fibers unite the cerebral hemispheres, cerebellum, pons, and medulla to each other in a manner that enables the brain to function as an integrated whole. In this part of our brain study we will identify these important fiber tracts as well as brain cavities, meningeal spaces, and other related parts.

Materials:

    preserved sheep brains
    preserved human brains
    human brain model
    long sharp knife
    dissecting instruments
    dissecting tray

Midsagittal Section

With the preserved sheep brain resting ventral side down on a dissecting tray, place a long sharp meat cutting knife (butcher knife) in the longitudinal cerebral fissure. Carefully slice through the tissue, cutting the brain into right and left halves on the midline. If only a scalpel is available, attempt to cut the tissue as smoothly as possible. Proceed to identify the structures on the sheep brain and their counterparts on the human brain as follows.

If the brain has been cut exactly on the midline, the only portion of the cerebrum that is cut is the **corpus callosum.** Locate this long band of white matter on the sheep brain by referring to figure 28.1. Also, compare your specimen with the human brain in figure 28.2. The corpus callosum is a commissural tract of fibers that connects the right and left cerebral hemispheres, correlating their functions. Note that the corpus callosum on the human brain is significantly thicker.

The Diencephalon

Between the corpus callosum of the cerebral hemispheres and the midbrain exists a section called the *diencephalon* or *interbrain.* This region

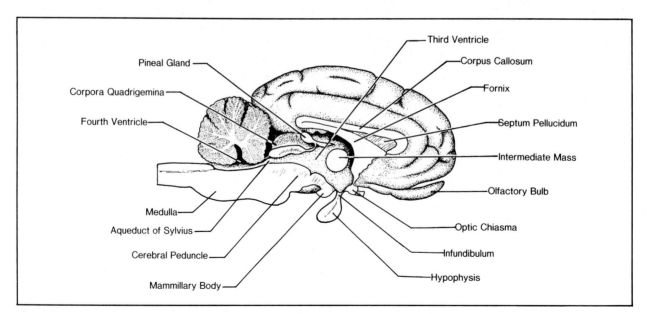

Figure 28.1 Midsagittal section of sheep brain.

includes the fornix, third ventricle, thalamus, intermediate mass, and hypothalamus.

Fornix Locate this structure on the sheep brain. This body contains fibers that are an integral part of the olfactory mechanism of the brain, the *rhinencephalon*. Note how much larger, proportionately, it is on the sheep than in the human brain.

Intermediate Mass This oval area in the midsection of the diencephalon is the only part of the thalamus that can be seen in the midsagittal section. The *thalamus* is a large ovoid structure on the lateral walls of the third ventricle. It is an important relay station in which sensory pathways of the spinal cord and brain form synapses on their way to the cerebral cortex. The *intermediate mass* contains fibers that pass through the third ventricle, uniting both sides of the thalamus.

Hypothalamus This portion of the diencephalon extends about two centimeters up into the brain from the ventral surface. It has many nuclei that control various functions of the body (temperature regulation, hypophysis control, coordination

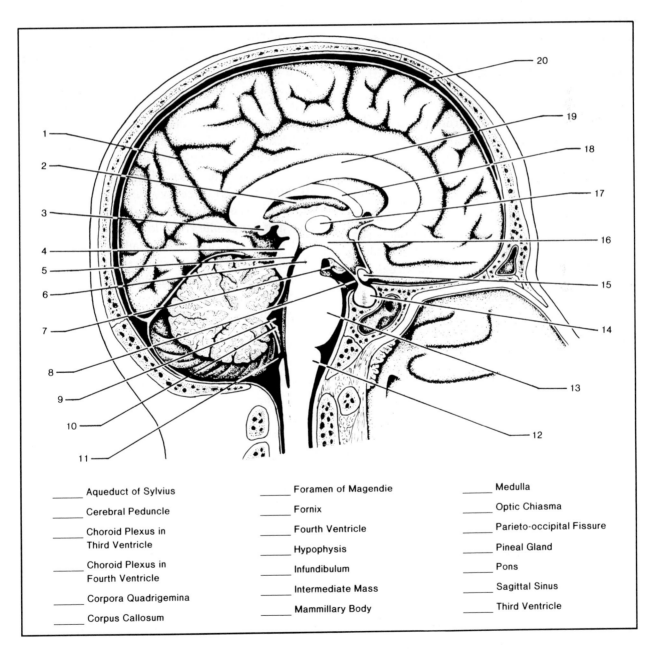

_____ Aqueduct of Sylvius	_____ Foramen of Magendie	_____ Medulla
_____ Cerebral Peduncle	_____ Fornix	_____ Optic Chiasma
_____ Choroid Plexus in Third Ventricle	_____ Fourth Ventricle	_____ Parieto-occipital Fissure
_____ Choroid Plexus in Fourth Ventricle	_____ Hypophysis	_____ Pineal Gland
_____ Corpora Quadrigemina	_____ Infundibulum	_____ Pons
_____ Corpus Callosum	_____ Intermediate Mass	_____ Sagittal Sinus
	_____ Mammillary Body	_____ Third Ventricle

Figure 28.2 Midsagittal section of human brain.

of the autonomic nervous system). The only part of the hypothalamus that is labeled on the human brain is the left **mammillary body** (label 7).

Inferior and anterior to the mammillary body in figure 28.2 lies the **infundibulum** and **hypophysis.** Note how the latter structure lies encased in the bony recess of the *sella turcica.* The **optic chiasma** is the small round body just above the hypophysis, anterior to the infundibulum.

The Midbrain

Locate the following structures of the sheep's midbrain: *corpora quadrigemina, pineal body, cerebral peduncle,* and *aqueduct of Sylvius.*

Cerebral Peduncles This part of the midbrain consists of a pair of cylindrical bodies made up largely of ascending and descending fiber tracts that connect the cerebrum with the other three brain divisions.

Aqueduct of Sylvius This duct runs longitudinally through the midbrain, connecting the third and fourth ventricles.

The Cerebellum

Examine the cut surface of the cerebellum and note that gray matter exists near the outer surface. The cerebellum receives nerve impulses along cerebellar peduncles from motor and visual centers of the brain, the semicircular canals of the inner ears, and muscles of the body. It transmits nerve impulses to all the motor centers of the body wall, maintaining posture, equilibrium, and muscle tonus. None of the activities of the cerebellum are of a conscious nature. It is significant that each hemisphere controls the muscles on its side of the body.

The Pons and Medulla

These two parts make up the greater portion of the brain stem. The pons, medulla, and midbrain constitute the entire *brain stem.* The pons is that portion of the brain stem between the midbrain and medulla. Note that it consists primarily of white matter (fiber tracts). Locate the pons and medulla on both the sheep and human brains.

The Ventricles and Cerebrospinal Fluid

The brain has four cavities, or *ventricles,* which contain cerebrospinal fluid. There are two **lateral ventricles** situated in the lower medial portions of each cerebral hemisphere. Although they are not visible in a midsagittal section, the **septum pellucidum,** a thin membrane may be seen which separates these two cavities. A frontal section of the brain, such as figure 28.5, shows the position of the lateral ventricles (label 2). The **third** and **fourth ventricles,** however, are readily visible in figure 28.1. Note that the intermediate mass passes through the third ventricle. Within this ventricle is seen a **choroid plexus** (label 2, figure 28.2), which secretes cerebrospinal fluid. Each ventricle has its own choroid plexus.

The path of cerebrospinal fluid is indicated in figure 28.3. From the lateral ventricles the fluid enters the third ventricle through the **foramen of Monro.** Note in figure 28.3 that this foramen lies anterior to the fornix and choroid plexus of the third ventricle. From the third ventricle the cerebrospinal fluid passes into the **aqueduct of Sylvius.** The duct conveys the cerebrospinal fluid to the fourth ventricle. The **choroid plexus of the fourth ventricle** lies on the posterior wall of this cavity. The cerebrospinal fluid exits from the fourth ventricle into the subarachnoid space around the cerebellum through three foramina: one foramen of Magendie and two foramina of Luschka. The **foramen of Magendie** is the lower opening in figure 28.3 which has two arrows leading from it. That part of the subarachnoid space that it empties into is the **cisterna cerebello-medullaris.** Since the **foramina of Luschka** are located on the lateral walls of the fourth ventricle, only one is shown in figure 28.3.

From the cisterna cerebello-medullaris the cerebrospinal fluid moves up into the **cisterna superior** (above the cerebellum) and finally into the subarachnoid space around the cerebral hemispheres. From the subarachnoid space this fluid escapes into the blood of the sagittal sinus through the **arachnoid granulations.** The cerebrospinal fluid passes down the subarachnoid space of the spinal cord on the posterior side and up along its anterior surface.

Assignment:

After identifying all the structures of the sheep brain, label figure 28.2.

Label figure 28.3.

Frontal Sections

To get a better understanding of the special relationships of the thalamus, ventricles, and other parts of the brain it is necessary to study frontal sections of the sheep brain. To accomplish this we will study two frontal sections, identifying previously mentioned structures.

Infundibular Section

Place a whole sheep brain on a dissecting tray with the dorsal side down. Cut a frontal section with a long knife by starting at the point of the infundibulum. Cut through perpendicularly to the tray and to the longitudinal axis of the brain. Cut an-

other slice through the posterior portion about ¼ inch further back. This latter cut should be through the center of the hypophysis. Place this slice, with the first cut upward, on a piece of paper towelling for study. Illustration A, figure 28.4 is of this surface. Proceed to identify the following structures. You may need to look back to figure 28.1 for reference.

Note the pattern of the **gray matter** of the cerebral cortex. Force a probe down into a sulcus. What meninx do you break through as the probe

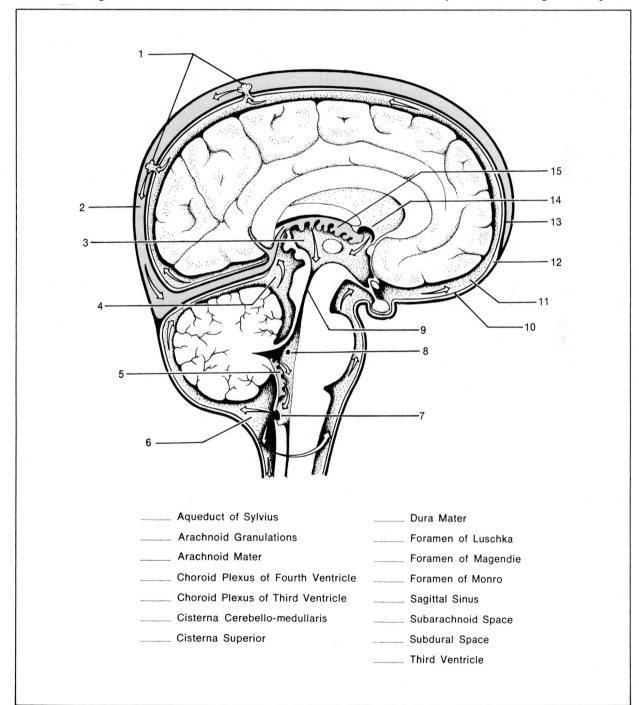

_____ Aqueduct of Sylvius	_____ Dura Mater
_____ Arachnoid Granulations	_____ Foramen of Luschka
_____ Arachnoid Mater	_____ Foramen of Magendie
_____ Choroid Plexus of Fourth Ventricle	_____ Foramen of Monro
_____ Choroid Plexus of Third Ventricle	_____ Sagittal Sinus
_____ Cisterna Cerebello-medullaris	_____ Subarachnoid Space
_____ Cisterna Superior	_____ Subdural Space
	_____ Third Ventricle

Figure 28.3 Origin and circulation of cerebrospinal fluid.

moves inward? What is the average thickness of the gray matter in the sulci of the dorsal part of the brain?

Locate the two triangular **lateral ventricles.** Probe into one of the ventricles to see if you can locate the **choroid plexus.** What is its function? Identify the **corpus callosum** and **fornix.**

Identify the small portion of the **third ventricle,** which is situated under the fornix. Also identify the **thalamus** and **intermediate mass.** Does the thalamus appear to consist of white or gray matter? What is the function of all the white matter between the thalamus and cerebral cortex?

Hypophyseal Section

Cut another ¼ inch slice off the posterior portion of the brain. Place this slice, with its anterior face upward, on a piece of paper towelling for study. This section through the hypophysis should look like the lower illustration in figure 28.4. Identify the structures that are labeled.

Assignment:
Label figure 28.4.

Basal Ganglia

In the walls of the ventricles are situated masses of gray matter called the *basal ganglia.* They include the caudate nuclei, putamen, and globus pallidus. Figure 28.5 is a frontal section of the human brain which shows the position of these various centers.

The **caudate nuclei** are masses of gray matter that are situated in the walls of the lateral ventricles superior to the thalamus. These nuclei have something to do with muscular coordination since surgically produced lesions in this area can correct certain kinds of palsy.

The **putamen** and **globus pallidus** are below and lateral to the thalamus. The two combined form a triangular mass in each cerebral hemisphere called the **lentiform nucleus.** The globus pallidus

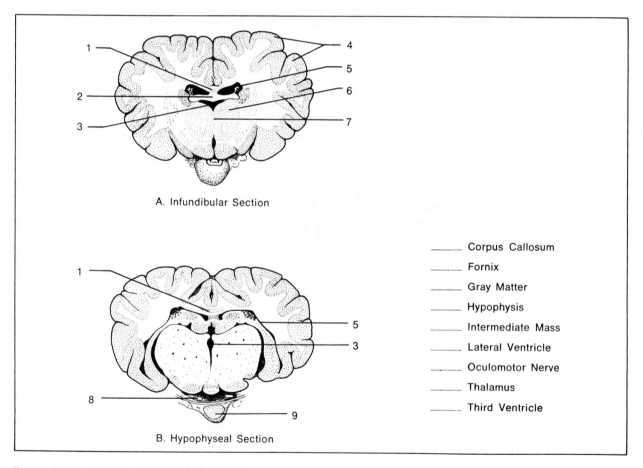

A. Infundibular Section

B. Hypophyseal Section

_____ Corpus Callosum
_____ Fornix
_____ Gray Matter
_____ Hypophysis
_____ Intermediate Mass
_____ Lateral Ventricle
_____ Oculomotor Nerve
_____ Thalamus
_____ Third Ventricle

Figure 28.4 Frontal sections of sheep brain.

is medial to the putamen. The main effect of these ganglia on the body is to exert a steadying effect on voluntary movement.

The hypothalamus and thalamus are sometimes included as basal ganglia even though, functionally, they are quite different.

Assignment:

Label figure 28.5.

Laboratory Report

Complete the Laboratory Report for this exercise.

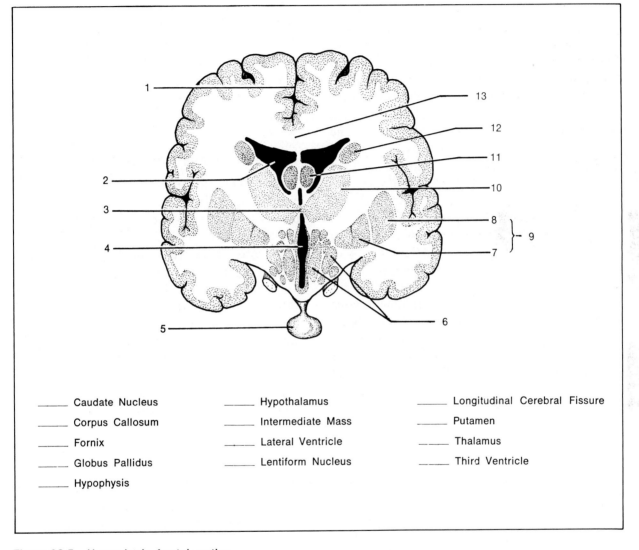

_____ Caudate Nucleus   _____ Hypothalamus   _____ Longitudinal Cerebral Fissure

_____ Corpus Callosum   _____ Intermediate Mass   _____ Putamen

_____ Fornix   _____ Lateral Ventricle   _____ Thalamus

_____ Globus Pallidus   _____ Lentiform Nucleus   _____ Third Ventricle

_____ Hypophysis

Figure 28.5   Human brain, frontal section.

29 Electroencephalography

The movement of nerve impulses from one neuron to another throughout the brain produces background electrical activity that is observable with electronic equipment. The study of this electrical activity is called *electroencephalography*. The monitoring of this phenomenon may be achieved with needle electrodes inserted directly into the scalp, or by placing the electrodes on the skin, externally. The record produced by either method is called an **electroencephalogram, or EEG.**

Although considerable variation exists in the electric activity of the brains of different individuals, normal patterns have been established. Electroencephalograms have many medical applications. They can be used for diagnosing organic malfunctions such as epilepsy, brain tumor, brain abscess, cerebral trauma, meningitis, encephalitis, and certain congenital conditions. They are particularly useful in establishing the existence of cerebral thrombosis and embolism. In spite of its diagnostic value, however, a completely normal EEG is frequently seen where considerable brain damage is present, and in most nonorganic disorders, one encounters normal electroencephalograms.

Normal Wave Patterns

Most of the time, brain waves are irregular and exhibit no general pattern. At other times, distinct patterns are seen. While some of these patterns are characteristic of specific abnormalities, such as epilepsy, there are distinctly normal patterns such as alpha, beta, theta, and delta waves.

Alpha waves are the most dominant pattern seen in normal adults. This pattern is seen when the subject is in a relaxed state of mind with eyes closed and not concentrating on any particular thought. The waves occur at a rate of 8–13 cycles per second (c.p.s.). When the eyes are opened and the subject's attention is drawn to a thought or visual object, the alpha rhythm changes to fast,

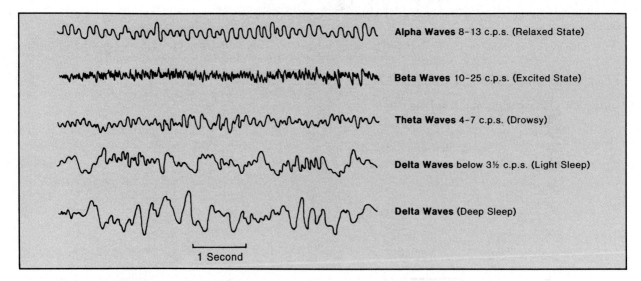

Alpha Waves 8–13 c.p.s. (Relaxed State)

Beta Waves 10–25 c.p.s. (Excited State)

Theta Waves 4–7 c.p.s. (Drowsy)

Delta Waves below 3½ c.p.s. (Light Sleep)

Delta Waves (Deep Sleep)

1 Second

Figure 29.1 Normal electroencephalograms.

irregular, shallow waves. This change is called *alpha block*.

Beta waves occur at a frequency of 14–25 c.p.s. There appears to be two types of beta waves: beta I waves that disappear with increased mental activity, and beta II waves that are elicited by intense mental activity.

Theta waves have frequencies of 4–7 c.p.s. They are normal for children, but usually represent emotional stress in adults. Disappointment and frustration are frequently manifested as theta waves in adults. These waves are also present in certain brain disorders.

Delta waves occur at frequencies below 3½ c.p.s. They are normal in infancy and occur during sleep in adults.

The brain wave pattern of infants and children exhibit the fast beta rhythms as well as the slower theta and delta rhythms. As children mature into adolescence, the rhythms develop into the characteristic alpha pattern of adulthood. Physiological conditions, such as low body temperatures, low blood sugar levels, and high levels of carbon dioxide in the blood can slow down the alpha rhythm; the opposite conditions speed it up.

Pathological Evidence

The five electroencephalograms shown in figure 29.2 are representative of what might be expected of individuals with brain damage. Note the absence of stability that is typical for the alpha rhythm of figure 29.1.

In evaluating electroencephalograms, the neurologist looks for patterns that are considered *slow* or *fast*, as well as those that are paroxysmal. Records that are moderately slow or moderately fast are considered *mildly abnormal*. Those that are very slow or very fast are *definitely abnormal*. A very fast one would be the fourth record in figure 29.2 of an individual that suffered motor seizures and amnesia from barbiturate intoxication. Considerable extremes in voltage are also seen here. Epileptic patterns vary with the types of seizures that occur. They characteristically have large spikes, indicating considerable voltage differential.

Procedure

The clarity of wave patterns produced in EEG measurements is a function of the type of equipment and technique. A good oscilloscope produces excellent results. The procedures outlined here are for the Unigraph.

The electrical potentials emitted by the brain are of such low magnitude that the Unigraph will be operated at its maximum sensitivity of 0.1 MV/CM. Ideally, one would use needle electrodes that are inserted ino the scalp to pick up these signals. For obvious legal reasons, such electrodes are impractical here; instead, we will use small disk electrodes that will be pressed tightly against the skin. To overcome skin resistance, the skin will be gently scrubbed with a Scotchbrite pad before applying the electrodes. Electrode jelly will also be used to

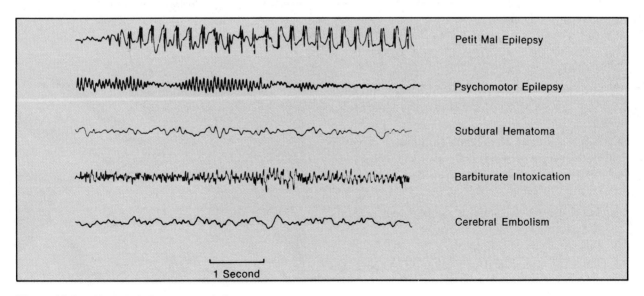

Figure 29.2 Abnormal electroencephalograms.

enhance skin conductivity. An elastic headband will exert constant pressure to the electrodes.

To record an EEG, students will work in groups of four. One member of the group will be the subject, another member will prepare the subject, and the remaining two will work with the equipment.

Three electrodes will be attached to the subject: two on the forehead and one at the base of the skull in the occiput region. The electrodes will be attached to the three leads of a patient cable. The latter is plugged into the Unigraph.

The success of this experiment hinges considerably on environmental conditions as well as electrode placement. Ideally, the room should be darkened and quiet. A relaxed subject is essential, particularly, when trying to record alpha waves. Subjects who are trained in Yoga or are capable of self-hypnosis can produce EEG patterns that are quite striking.

While the subject is being prepared, it will be necessary for the other two members to calibrate the Unigraph. Proceed as follows:

Materials:

  Unigraph or polygraph
  patient cable (3 leads)
  3 EEG disk-type electrodes
  3 circular Band-Aids
  electrode jelly
  alcohol swabs
  Scotchbrite pad (grade #7447)
  elastic wrapping

Calibration

For the EEG tracing to have any significant meaning, it is necessary to calibrate the Unigraph so that a one-centimeter deflection of the pen equals 0.1 millivolt; thus, each millimeter of vertical deflection will equal 0.01 millivolt. Proceed as follows to make this calibration:

1. Turn the Mode selector control to CC/Cal (Capacitor Coupled Calibrate).
2. Turn the power switch on.
3. Turn the stylus heat control knob to the 2:00 position.
4. Check the speed selector level to see that it is in the slow position. The free end of the lever should be pointed toward the styluses.

5. Move the c.c. switch to Chart On and observe the width of the tracing that appears on the paper. Adjust the stylus heat control to produce the desired width of tracing.
6. Set the Gain control knob on .1 MV/CM.
7. Push down the .1 MV button to determine the amount of deflection. Hold the button down a full 2 seconds before releasing. If it is released too quickly the tracing will not be perfect. Adjust the deflection so that it deflects exactly 1 centimeter in each direction by turning the blue sensitivity knob. The unit is now calibrated so that you get 1 cm. deflection for .1 millivolt.
8. Set the Mode selector control on EEG and return the c.c. switch to STBY. Place the speed selector lever in the fast position. The instrument is now ready for use.

Preparation of Subject

Two electrodes will be fixed to the forehead near the hair line and one will be positioned in the occiput region of the head on the right side. The electrodes that are attached to the back of the head and right forehead will pick up the brain waves. The electrode on the left forehead acts as a ground. Proceed as follows:

1. With a Scotchbrite pad, *lightly* scrub the right and left forehead areas four or fives times to remove the dead surface cells. Refer to figure 29.3 to note the general area. Examine the

Figure 29.3 The skin surface is prepared by rubbing lightly with a Scotchbrite pad.

surface of the pad for evidence of cleansing (white powder on pad). Do not scrub so much that you break through to the capillaries. It is not necessary to draw blood.

2. Disinfect each area with an alcohol swab.
3. Affix an electrode to a round Band-Aid or strip of adhesive tape, add a drop of electrode jelly, and attach to the forehead, as shown in figures 29.4, 29.5, and 29.6. Be sure to position the electrode so that its flat surface will be against the skin. Follow this procedure for both electrodes that are attached to the forehead.
4. Position the third electrode on the back of the head on the right side in the occiput region, using electrode jelly. Do not use a Band-Aid. To hold it in place, put an elastic wrapping over it and the other two electrodes. See figure 29.7.
5. Clamp the leads of the patient cable to the wires of the electrodes as follows:

Lead #1 to the occiput electrode.
Lead #2 to the electrode on right forehead.
Unnumbered Lead to electrode on left forehead.

6. Have the patient lie down on a cot or laboratory table. Make sure that all leads are free and not binding under the head or shoulders. Leads should not be disturbed by body movements. Subject should lie perfectly still.
7. Plug the end of the 3-lead cable into the Unigraph. You are now ready to monitor the tracing.

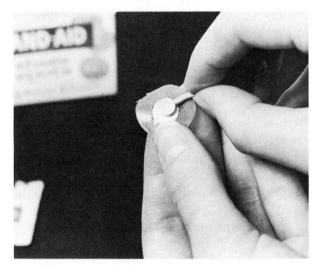

Figure 29.4 With flat surface upward, the electrode is placed on an adhesive patch.

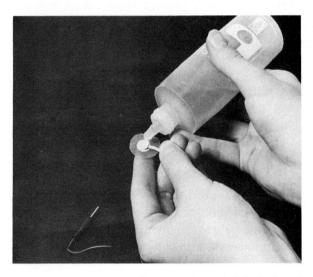

Figure 29.5 A very small drop of electrode jelly is placed on the electrode.

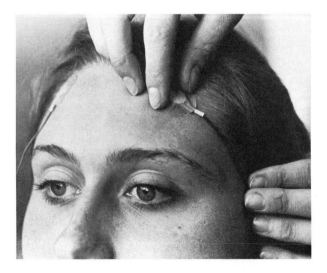

Figure 29.6 The forehead electrodes are positioned near the hairline.

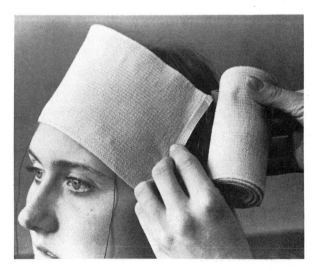

Figure 29.7 An elastic band is wrapped tightly over all three electrodes.

Monitoring

With the Unigraph set at .1 MV/CM in the EEG Mode, place the c.c. switch at CHART ON and observe the EEG pattern produced on the chart. The chart should be moving at the fast speed (25 mm. per second). In most cases, the typical wave pattern is irregular. Be sure to keep the subject as quiet as possible. Even eye movements will alter the pattern. After about 2 minutes of recording, attempt to induce alpha waves.

Alpha Rhythm Although alpha waves are the most common wave form in normal adults, they are not always easy to demonstrate in a laboratory situation. Producing complete relaxation in the subject is very important. Extraneous disturbances, such as conversation, slamming doors, furniture moving, etc., will prevent them. With the c.c. switch at CHART ON, make the following type of statement to the subject in an attempt to induce relaxation:

> "With your eyes closed, relax all the muscles of your body as much as you can. Try not to concentrate on any one thought. Let your thoughts flit lightly from one thought to another. You have absolutely no problems. Life is beautiful. Everybody is kind, good, and your friend."

Allow the tracing to run intermittently for about five minutes; then, place the c.c. switch in the OFF position. Count the waves per second. Is it an alpha rhythm pattern? What is the maximum number of microvolts seen in a specific wave?

Alpha Block To demonstrate alpha block, place the c.c. switch on Chart On again and get the subject in the same mental state as previously. When the alpha rhythm is present, have the subject open his or her eyes (as you press the event marker button) and ask some questions that require concentration. Observe the pattern for evidence of alpha block. Continue to present the subject with questions and problems for a period of 4 or 5 minutes. Mark the paper with a pencil where the questioning began.

Hyperventilation Start the chart moving again. Have the subject relax with closed eyes to reestablish the alpha rhythm again. Depress the event marker and ask the subject to take deep breaths at a rate of 40–50 per minute for a period of 2 or 3 minutes. This will hyperventilate the lungs, causing a lowering of the level of carbon dioxide in the blood. Observe the tracing and note the changes that occur.

Laboratory Report

Remove the chart from the Unigraph and review the results with the subject. Attach samples of the EEG to the Laboratory Report sheet in correct places, using tape or adhesive.

If time permits, repeat the experiment with other members of the group acting as subjects.

Answer the questions on the Laboratory Report.

Part 7 The Sense Organs

This unit is comprised of four exercises. The first two pertain to the eye and the last two are concerned with the ear. In Exercise 30, the anatomy of the eye will be studied through dissection and visual examination of its interior. A fresh beef eye will be used for dissection. To observe the inside of the eyeball, an ophthalmoscope will be used. Exercise 31 pertains entirely to eye tests. Some of these tests will be used to reveal normal physiological phenomena; other tests will be used to determine deviations from normal vision.

Exercises 32 and 33 are concerned with different functions of the ear. In Exercise 32 that part of the ear that functions in hearing will be studied. Different kinds of hearing tests will be employed here to detect deviations from normal. In Exercise 33 the part of the ear that functions in the maintenance of equilibrium will be studied. A variety of tests pertaining to equilibrium will be performed here.

30 Anatomy of the Eye

The eyes, functioning somewhat as a pair of cameras, record images that can be interpreted by the brain. The perception of clear images requires that the eye be able to focus equally well on far and near objects *(accommodation to distance)* and adjust to variabilities in light intensity *(accommodation to light)*. To avoid double images the eyes must be coordinated for *convergence* and *divergence*. A mechanism must exist that enables the receptors to achieve differentiation of various wave lengths of light *(color perception)*. Needless to say, the eyes, as receptor organs of vision, incorporate amazingly intricate functional mechanisms to achieve these physiological activities.

In this exercise we will study the anatomy of the eye through dissection and ophthalmoscopy.

Before these laboratory experiments are performed, however, the illustrations in figures 30.1, 30.2 and 30.3 should be labeled. A fresh beef eye will be used for dissection. Its large size makes it an ideal dissection specimen. Ophthalmoscopy will be used to study the interior of the living eye to identify structures on the retina.

External Anatomy

Each eye lies protected in a bony recess of the skull and covered externally by the **eyelids.** Lining the eyelids and covering the exposed surface of the eyeball is a thin membrane, the **conjunctiva.** The surfaces of the conjunctiva are kept constantly moist by secretions of the **lacrimal gland,** which lies between the upper eyelid and the eyeball.

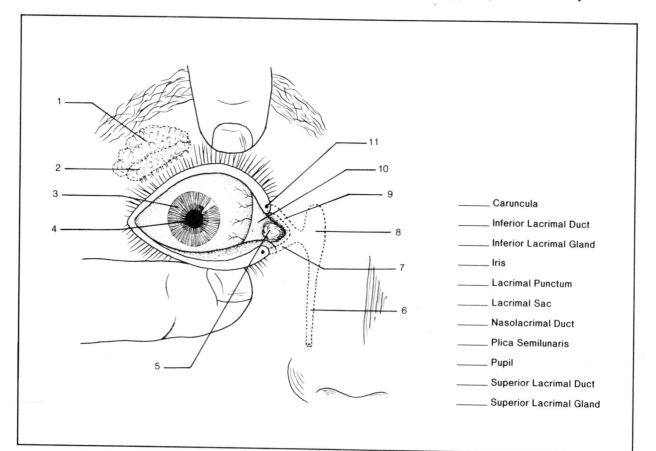

_____ Caruncula

_____ Inferior Lacrimal Duct

_____ Inferior Lacrimal Gland

_____ Iris

_____ Lacrimal Punctum

_____ Lacrimal Sac

_____ Nasolacrimal Duct

_____ Plica Semilunaris

_____ Pupil

_____ Superior Lacrimal Duct

_____ Superior Lacrimal Gland

Figure 30.1   The lacrimal apparatus of the eye.

The Lacrimal Apparatus

Figure 30.1 reveals the external anatomy of the right eye, with particular emphasis on the lacrimal apparatus. Note that the lacrimal gland consists, actually, of two portions: a large **superior lacrimal gland** and a smaller **inferior lacrimal gland.** Secretions from these glands empty through the conjunctiva of the eyelid via six to twelve small ducts. The flow of fluid from these ducts moves across the eyeball to the medial canthus (angle) of the eye, where the fluid enters two small orifices, the **lacrimal puncta.** These openings lead into the **superior** and **inferior lacrimal ducts:** the superior one is in the upper eyelid and the inferior one is in the lower eyelid. These two ducts empty directly into an enlarged cavity, the **lacrimal sac,** which, in turn, is drained by the **nasolacrimal duct.** The tears finally exit into the nasal cavity through the end of the nasolacrimal duct.

Two other structures, the caruncula and the plica semilunaris are seen near the puncta in the medial canthus of the eye. The **caruncula** is a small red conical body which consists of a mound of skin containing sebaceous and sweat glands and a few small hairs. It is this structure that produces a whitish secretion which constantly collects in this region. Lateral to the caruncula is a curved fold of conjunctiva, the **plica semilunaris.** This structure contains some smooth muscle fibers; in the cat and many other animals it is more highly developed and is often referred to as the "third eyelid."

Assignment:

Label figure 30.1 and examine your laboratory partner's eyes to identify as many of these structures as possible.

The Extrinsic Muscles

Figure 30.2 illustrates the six extrinsic muscles and the superior levator palpebrae. The muscle on the side of the eye, which has a portion removed to reveal the optic nerve is the **lateral rectus.** On the other side of the eye is its opposing muscle, the **medial rectus** which is only partially visible. On the superior surface of the eye is attached the **superior rectus,** and opposing it on the inferior surface is the **inferior rectus.** Just above the stub of the lateral rectus is seen the insertion of the **superior oblique** muscle which passes through a cartilaginous loop, the **trochlea.** Opposing the superior oblique is the **inferior oblique,** of which only the insertion is seen below the stub of the lateral rectus. The muscle which extends into the upper eyelid is the **superior levator palpebrae.** It raises the eyelid.

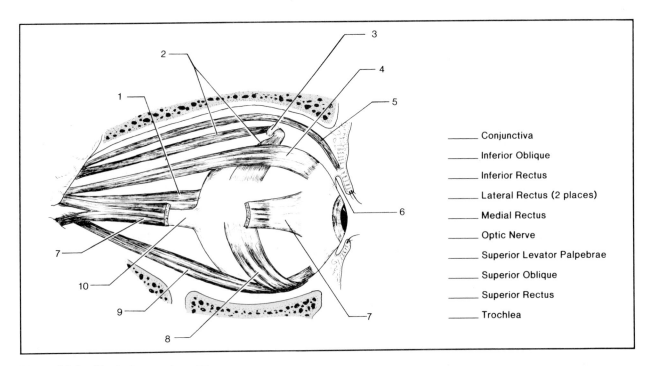

_____ Conjunctiva

_____ Inferior Oblique

_____ Inferior Rectus

_____ Lateral Rectus (2 places)

_____ Medial Rectus

_____ Optic Nerve

_____ Superior Levator Palpebrae

_____ Superior Oblique

_____ Superior Rectus

_____ Trochlea

Figure 30.2 Extrinsic muscles of the eye.

Assignment:

Label figure 30.2.

Internal Anatomy

Figure 30.3 is a horizontal section of the right eye as seen looking down upon it. Note that the wall of the eyeball consists of three layers: an outer **scleroid coat,** a middle **choroid coat,** and an inner **retina.** The continuity of the surface of the retina is interrupted only by two structures, the optic nerve and the fovea centralis. The **optic nerve** (label 3) contains fibers leading from the rods and cones of the retina to the brain. That point on the retina where the optic nerve makes its entrance is lacking in rods and cones. The absence of these receptors of light at this spot makes it insensitive to light; thus, it is called the **blind spot.** It is also called the **optic disk.** Note that the optic nerve is wrapped in a sheath, the **dura mater,** which is an extension of the dura that surrounds the brain. To the right of the blind spot is a pit, the **fovea centralis,** which is the center of a round yellow spot,

the **macula lutea.** The macula lutea, which is only a half a millimeter in diameter, is composed entirely of cones and is that part of the eye where all critical vision occurs.

Light entering the eye is focused on the retina by the **lens,** an elliptical crystalline clear structure suspended in the eye by a **suspensory ligament.** Attached to the suspensory ligament is the **ciliary body,** which is concerned with changing the shape of the lens for focusing at different distances. In front of the lens is a chamber that contains a watery fluid, the **aqueous humor.** Between the lens and the retina is a larger chamber which contains a more viscous, jelly-like substance, the **vitreous body.** Immediately in front of the lens lies the circular colored portion of the eye, the **iris.** It consists of circular and radiating muscle fibers that can change the size of the **pupil** of the eye. This structure regulates the amount of light that enters the eye through the pupil. Covering the anterior portion of the eye is the **cornea,** a clear, transparent structure that is an extension of the scleroid coat. It acts as a window to the eye, allowing light to enter.

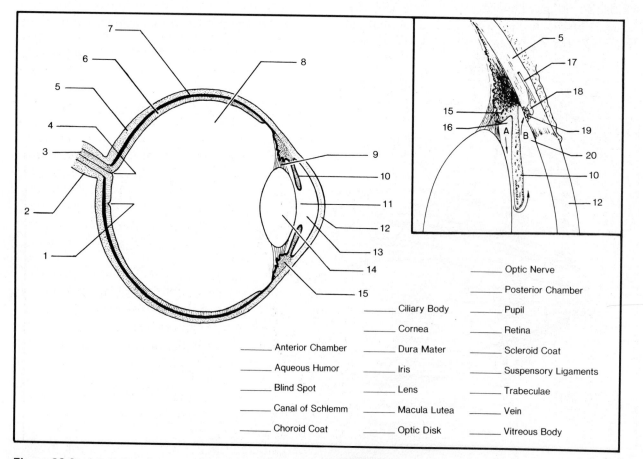

	_____ Ciliary Body	_____ Optic Nerve
	_____ Cornea	_____ Posterior Chamber
_____ Anterior Chamber	_____ Dura Mater	_____ Pupil
_____ Aqueous Humor	_____ Iris	_____ Retina
_____ Blind Spot	_____ Lens	_____ Scleroid Coat
_____ Canal of Schlemm	_____ Macula Lutea	_____ Suspensory Ligaments
_____ Choroid Coat	_____ Optic Disk	_____ Trabeculae
		_____ Vein
		_____ Vitreous Body

Figure 30.3   Internal anatomy of the eye.

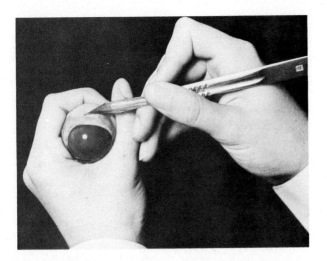

Figure 30.4 The wall of the eyeball is pierced through the sclera with a sharp scalpel.

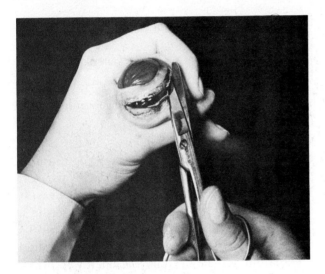

Figure 30.5 The wall is cut all the way around with sharp scissors.

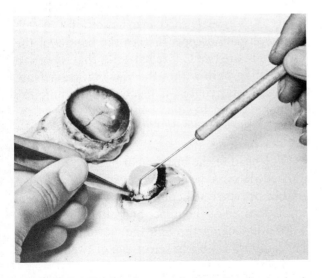

Figure 30.6 The lens is separated from the vitreous body with a dissecting needle.

Aqueous humor is constantly being renewed in the eye. The enlarged inset of the lens-iris area in figure 30.3 illustrates where the aqueous humor is produced (point A) and where it is reabsorbed (point B). At point A, which is in the **posterior chamber,** aqueous humor is produced by capillaries of the ciliary body. As this fluid is produced it passes through the pupil into the **anterior chamber** and is reabsorbed into minute spaces called **trabeculae** at the juncture of the cornea and the iris (point B). From the trabeculae the fluid passes to the **canal of Schlemm,** which empties into the venous system.

Assignment:

Label figure 30.3.

Beef Eye Dissection

The nature of the tissues of the eye can best be studied by actual dissection of an eye. A cow's eye will be used for this study.

Materials:

  beef eye (preferably fresh)
  dissecting instruments
  tray

1. Examine the back and side surfaces of the eye. Identify the **optic nerve.** Look for the **extrinsic muscles.** Only traces of these muscles will be present.
2. Note the shape of the **pupil.** Is it round or elliptical? Identify the **cornea.**
3. Holding the eyeball as shown in figure 30.4, make an incision into the scleroid coat with a sharp scalpel about one-quarter of an inch away from the cornea. Note how difficult it is to penetrate. Insert scissors into the incision and carefully cut all the way around the cornea, *taking care not to squeeze the fluid out of the eye.*
4. Gently lift the front portion off the eye and place it on the tray with the inner surface upward. The lens usually remains attached to the vitreous body, as shown in figure 30.6. If the eye is preserved (not fresh), the lens may remain in the anterior part of the eye.
5. Looking at the inner surface of the anterior portion identify the thickened, black circular **ciliary body.** What function does the black pigment perform?
6. Examine the **iris** carefully. Can you distinguish **circular** and **radial** muscle fibers?

7. Is there any **aqueous humor** left between the cornea and the iris? Compare its consistency with the **vitreous body.**

8. With a dissecting needle separate the lens from the vitreous body by working the needle around its perimeter as shown in figure 30.6. Hold the lens up by its edge with a forceps and look through it at some distant object. What is unusual about the image seen?

9. Place the lens on printed matter. Is it magnified?

10. Compare the consistency of the lens at its center and near its circumference. Use a probe or forceps. Do you detect any differences?

11. Locate the **retina** at the back of the eye. It is a thin colorless inner coat which separates easily from the pigmented **choroid coat.** Now, locate the **blind spot.** This is the area where the retina is attached to the back of the eyeball.

12. Note the iridescent nature of a portion of the choroid coat. This reflective surface is the **tapetum lucidum.** It causes the eyes of animals to reflect light at night and appears to enhance night vision by reflecting some light back into the retina.

13. Answer all questions on the Laboratory Report that pertain to this dissection.

Ophthalmoscopy

Routine physical examinations invariably involve an examination of the *fundus,* or interior, of the eyeball. A careful examination of the fundus provides valid information about the general health of the patient. The retina of the eye is the only part of the body where relatively large blood vessels may be inspected without surgical intervention. Figure 30.7 illustrates what one might expect to see in a normal eye (A); the eye of a diabetic (B); and the eye of one who has hypertension, a brain tumor, or meningitis (C). In *diabetic retinopathy* the peripheral blood vessels are generally irregular, fewer in number, and often show small hemorrhages. In *papilledema,* or choked disk, the optic disk is blurred, enlarged, and often hemorrhagic; the veins are also considerably enlarged. These are only a few examples of pathology showing up in the fundus.

The instrument that one uses for examining the fundus is called an *ophthalmoscope.* In this

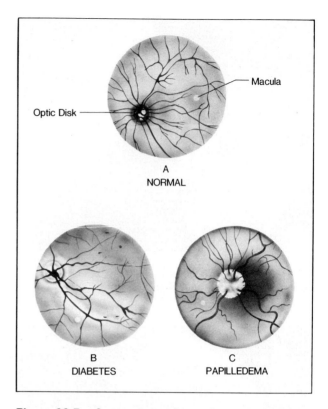

Figure 30.7 One normal and two abnormal retinas, as seen through an ophthalmoscope.

exercise you will have an opportunity to examine the interior of the eyes of your laboratory partner. Your partner, in turn, will examine your eyes. It will be necessary to make this examination in a darkened room. Before this examination can be made, however, it is essential that you understand the anatomy of the eye and the operation of the ophthalmoscope.

The Ophthalmoscope

Figure 30.8 reveals the construction of a Welch Allyn ophthalmoscope. Within the handle of the instrument are two C-size batteries that power an illuminating bulb within the viewing head. Note that a rheostat control is located at the juncture of the neck and handle. This control can be rotated to activate the light bulb after depressing the red lock button.

The head of the instrument has two rotatable notched disks: a large lens selection disk and a smaller aperture selection disk. A viewing aperture is located at the upper end of the head.

While looking through the viewing aperture, the examiner is able to select one of twenty-three

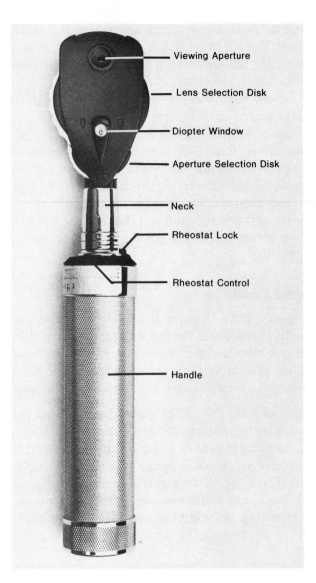

Figure 30.8 The ophthalmoscope.

Figure 30.9 Viewing lenses of the ophthalmoscope are rotated into place by moving the lens selection disk with the index finger.

small lenses by rotating the **lens selection disk** with the index finger held as shown in figure 30.9. Twelve of the lenses on this disk are positive (convex) and eleven of them are negative (concave). Numbers that appear in the **diopter window** indicate which lens is in place. The positive lenses are represented as black numbers; negative lens numbers are red. A black "0" in the window indicates that no lens is in place. If the eyes of both the subject and examiner are *emmetropic* (normal), no lens is needed. If the eye of the subject or examiner is *hypermetropic* (far-sighted), the positive lenses will be selected. *Myopic* (near-sighted) eyes require the use of negative lenses. The degree of myopia or hypermetropia is indicated by the magnitude of the number. Since the eyes of both the examiner and subject affect the lens selection, the selection becomes entirely empirical. The diopter value of a lens is the reciprocal of its focal length *(f)* in meters, or $D = \dfrac{1}{f}$.

A lens of one diopter (1D) has a focal length of 1 meter, $(D = \dfrac{1}{f} = \dfrac{1}{1} = 1)$; a 2D lens has a focal length of .5 meter, or $\dfrac{1}{f} = \dfrac{1}{.5} = 2$; a 4D lens is $\dfrac{1}{.25} = 4$; etc.

The **aperture selection disk** enables the examiner to change the character of the light beam that is projected into the eye. The light may be projected as a grid, straight line, white spot, or green spot. The green spot is most frequently used because it is less irritating to the eye of the subject, and it causes the blood vessels to show up more clearly.

Using an ophthalmoscope for the first time seems difficult for some and relatively easy for others. The examiner with emmetropic eyes has a definite advantage over one with myopia, astigmatism, or hypermetropia. Even for individuals with normal eyes, there is the troublesome reflection of light from the retina back into the examiner's eyes. The best way to minimize this problem is to *direct the light beam toward the edge of the pupil* rather than through the center. Practicing the entire procedure prior to entering a darkened room can be helpful. An important thing to keep in mind with respect to the subject's well-being is not to overly expose the eyes to the intense light of the ophthalmoscope. **A safe time limit of exposure is one minute.** After one minute's exposure allow several minutes of rest for recovery of the retina.

Procedure

Prior to using the ophthalmoscope in the darkened room, it will be necessary for you to become completely familiar with its mechanics.

Materials:

    ophthalmoscope
    metric tape or ruler

Desk Study of Ophthalmoscope

Turn on the light source of the ophthalmoscope by depressing the lock button and rotating the rheostat control. Observe how the light intensity can be varied from very dim to very bright. Rotate the aperture selection disk as you hold the ophthalmoscope about two inches away from your desk top. Keep the ophthalmoscope parallel to the desk surface. How many different light patterns are there? Select the large white spot for the next phase of this study.

Rotate the lens selection disk until a black *5* appears in the diopter window. While peering through the viewing window at the print on this page, bring the letters into sharp focus. Have your lab partner measure with a metric tape the distance in millimeters between the ophthalmoscope

and the printed page when the lettering is in sharp focus. Record this distance on the Laboratory Report. Do the same thing for the 10D, 20D, and 40D lenses, recording all measurements.

Now, set the lens selection disk on "0" and the light source on the green spot. The instrument is properly set now for the beginning of an eye examination. Read over the following procedure in its entirety and practice preliminary examinations in the lab before entering the darkened room. Due to the small size of the pupil in the bright room, examination will be difficult, but at least you will have an opportunity to get the feel of the instrument.

The Examination

Figures 30.10 through 30.13 illustrate how the ophthalmoscope is held in different viewing situations. Note that the examiner uses her right hand when examining the right eye and the left hand when examining the left one. Proceed as follows:

1. With the "0" in the diopter window and the light turned on, grasp the ophthalmoscope, resting the index finger on the lens selection disk, as shown in figure 30.9.

2. Place the viewing aperture in front of your right eye and steady the ophthalmoscope by resting the top of it against your eyebrow. Start viewing the subject's right eye at a dis-

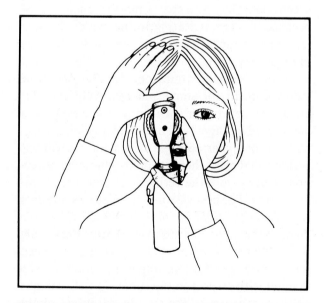

Figure 30.10 When examining the right eye, the ophthalmoscope is held with the right hand.

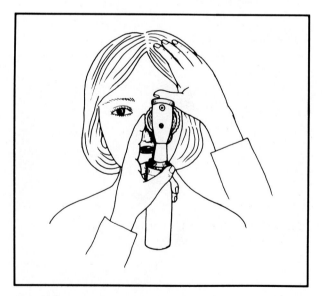

Figure 30.11 When examining the left eye, the ophthalmoscope is held with the left hand.

tance of about 12 inches, as illustrated in figure 30.12. Keep your subject on your right and instruct your subject to look straight ahead at a fixed object at eye level.

3. Direct the beam of light into the pupil and examine the lens and vitreous body. As you look through the pupil, a red *reflex* will be seen. If the image is not sharp, adjust the focus with the lens selection disk.

4. While keeping the pupil in focus, move in to within two inches of the subject, as illustrated in figure 30.13. The red reflex should become more pronounced. Try to *direct the light beam toward the edge of the pupil* rather than in the center to minimize reflection from the retina.

5. Search out the optic disk and adjust the focus, as necessary, to produce a sharp image. Observe how blood vessels radiate out from the optic disk. Follow one or more of them to the periphery.

   Locate the macula. It is situated about two disk diameters *laterally* from the optic disk. Note that it lacks blood vessels. It is easiest to observe when the subject is asked to look directly into the light, a position that must be limited to **one second only.**

6. Examine the entire retinal surface. Look for any irregularities, depressions, or protrusions from its surface. It is not unusual to see a roundish elevation, not unlike a pimple, on its surface. To examine the periphery, instruct the subject to: (1) *look up* for examination of the superior retina, (2) *look down* for examination of the inferior retina, and (3) *look laterally and medially* for those respective areas.

7. **Caution:** Remember to limit the examinations to **one minute!** After this time limit is reached, examine the left eye using the left hand to hold the ophthalmoscope and reversing the entire procedure.

8. Switch roles with your laboratory partner. When you have examined each other's eyes, examine the eyes of other students in the examination room. Report any unusual conditions to your instructor.

Laboratory Report

Answer the questions on the first part of combined Laboratory Report 30, 31.

Figure 30.12 The proper position for viewing the lens and vitreous body of the right eye is shown.

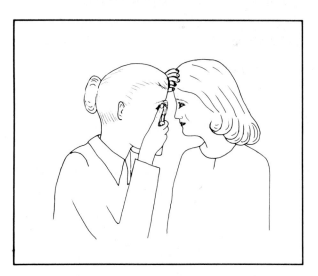

Figure 30.13 The proper position for examining the retinal surfaces of the right eye is shown.

31 Visual Tests

The various tests outlined in this exercise are of two types. Some of them are concerned with the observation of normal conditions in the eye, others relate to the detection of abnormalities. Through the performance of all these tests, much can be learned about the physiology of vision. Note in the materials listed below that equipment is listed according to the specific test or experiment of this exercise.

Materials:

> Penlite (image suppression)
> 12" ruler (blind spot)
> 3" × 5" card and pins (near point)
> Snellen eye charts (visual acuity)
> White-tipped pointer, 4 ft. long (glaucoma)
> Ishihara or Stilling charts (color blindness)
> Colored yarn kits (color blindness)
> Laboratory lamp (reflexes)

Image Suppression
(The Purkinje Tree)

When you examined the fundus of the eye with an ophthalmoscope, it was observed that blood vessels and capillaries were very much in evidence. These branching vessels lie very close to receptor cells and cast sharp shadows on most regions of the retina. You may wonder why these shadows do not show up in one's field of vision. The answer is that the brain has the capacity to suppress disturbing images of this sort that are always in the normal direct line of vision.

To illustrate that these blood vessels do cast an image on the retina and that they can be brought into one's field of vision, one need only illuminate the interior of the eye with rays of light

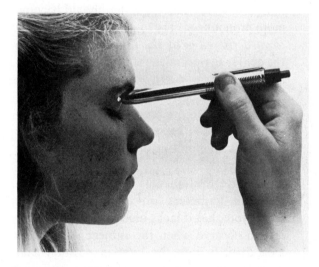

Figure 31.1 The proper position of penlite when observing the Purkinje tree is shown.

that enter through the eyelid and sclera instead of through the pupil. Since light rays do not normally enter the eye this way, the brain will not suppress any shadows cast by these blood vessels on the retina. The branching image that one sees is called the **Purkinje tree.**

To perform this experiment on your own eye, proceed as follows: Hold a penlite with your right hand against the eyelid of your right eye at an angle of approximately 45° to the right, as shown in figure 31.1. The eyelid may be open or closed. If the eyelid is open, look at a dimly lighted wall. If the eyelid is closed, do not face a brightly lighted window. While exposing the eye to the light, move the penlite from side to side to produce the Purkinje image. The amount of movement should be very slight—only a few millimeters from side to side. A most striking image can be produced if this experiment is performed in a dark closet. Describe the image on Laboratory Report 30, 31.

Figure 31.2 The blind spot test.

The Blind Spot

The examination of the fundus in Exercise 30 clearly revealed the nature of the optic disk. It was noted that where the optic nerve enters the eyeball, there is an absence of rods and cones, which renders that part of the retina insensitive to light. The illustration in figure 31.2 can be used to detect the presence of this blind spot in each eye.

To test the right eye, close the left eye and stare at the plus sign with the right eye as the book is moved from about 18 inches toward the face. At first, both the plus sign and dot are seen simultaneously. Then, at a certain distance from the eye, the dot will disappear as it comes into focus on the optic disk of the retina.

Perform this test on both of your eyes. To test the left eye, it is necessary to look at the dot instead of the plus sign. Have your laboratory partner measure the distance from your eye to the test chart with a ruler. Record the measurements on the chart on Laboratory Report 30, 31.

Near Point Determination

The ability of the lens of the eye to produce a sharp image on the retina is, partially, a function of its elasticity. When focusing on close objects, the lens must be considerably more spherical, or convex, than when focusing on distant objects. To become more convex, the tension on the lens periphery is relaxed as the ciliary muscle is stimulated.

In an infant, the ability of the eye to accommodate to distance is at its maximum. On the average, an infant's eye lens has a range of 17 diopters from distance sighting to near objects. As an individual gets older, the lens gradually becomes less elastic and the degree of accommodation diminishes. Between the ages of 45 and 50, one's accommodation is reduced from 17 to 2 diopters. Beyond 60 years of age, accommodation is nearly nonexistent. This condition is called **presbyopia.** This loss of lens elasticity is probably due to protein denaturation in the lens. It is for this reason that most individuals over 45 years of age find it necessary to acquire bi- or tri-focal lens glasses.

A measure of the elasticity of the lens in the eye can be made by determining the near point of the eye. The *near point* is the closest distance at which one can see an object in sharp focus. At 20 years of age the near point is approximately 3½″; at 30, 4½″; at 40, 6½″; at 50, 20½″; and at 60 it may be 33″.

To determine the near point in each eye, use the letter "T" at the beginning of this paragraph. Close one eye and move the page up to your eye until the letter becomes blurred; then move it away until you get a clear undistorted image. Have your laboratory partner measure the distance from your eye to the page with a ruler. The closest distance at which the image is clear will be the near point. Test the other eye also and record your results on the Laboratory Report.

Scheiner's Experiment When an object, such as a common pin, is viewed through two pinholes in a 3″ × 5″ card, an interesting optical phenomenon is observed. The two pinholes should be pierced in the center of the card and be no more than 2 mm. apart, center to center.

The card is placed in front of the eye as close as possible to the face. See figure 31.3. While peering through the two holes, it will be noted that the holes overlap so that it seems as if you are looking through a single opening; however, if the

Figure 31.3 Scheiner's experiment. An object is viewed through two pinholes in a card.

two holes are distinct with an opaque bridge between them, the holes are too far apart. Now, if a pin is viewed through the holes at arm's length while looking at some distant object, it should appear as a double pin. The double image of the pin indicates that the eye is accommodating for distant objects. If a single pin is seen while trying to focus on a distant object, the subject is actually focusing on the pin. *It is important that the pin be viewed through the space where the two luminous circles overlap.*

If the eye is focused on the pin to produce a single image and the pin is gradually moved from arm's length toward the eye and kept in the overlapping area of the two holes, a point is reached where the image changes from single to double. *The distance from the eye at which the image changes from single to double is the near point.*

Repeat Scheiner's experiment several times to establish an average near point. How does this measurement compare with the technique used previously? Record the measurements on Laboratory Report 30, 31.

Visual Acuity

A completely normal human eye, lacking any form of deviation, is able to differentiate at a distance of ten meters, two points that are only one millimeter apart. Points that are less than a millimeter apart at this distance will be seen as a single spot. The size and proximity of cones in the fovea determine this degree of visual acuity.

If pinpoints of light from two different objects strike adjacent cones, only a single image will be seen. This is due to the fact that the brain lacks the ability to record the stimuli that it receives as separate entities. However, if the images fall on two cones separated by an unexcited cone, the brain recognizes two separate points. Because the diameter of a cone is approximately two micrometers (1/500 mm.), the images on the fovea must be at least two micrometers apart. On the basis of this information it can be calculated that the brain can differentiate two pinpoints that enter the eye at an angle of 26 seconds. This is somewhat less than one-half of a degree!

The Snellen eye chart (figure 31.4) has been developed with this information in mind. It is printed with letters of various sizes. When you stand at a certain distance from it, usually 20 feet, and are able to read the letters on a line desig-

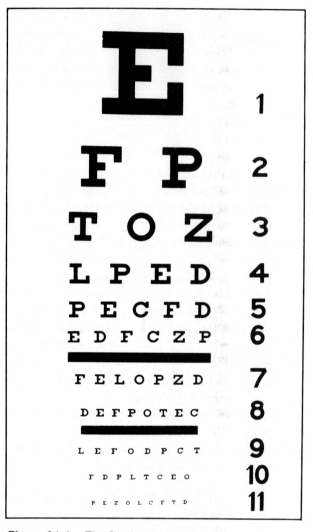

Figure 31.4 The Snellen eye chart.

nated to be read at 20 feet, you are said to have 20/20 vision. The ability to read these letters indicates that there are no aberrations of the lens or cornea to interfere with the angle of pinpoints of light reaching the retina of the eye. If you are only able to read the larger letters, such as those that should be read at 200 feet, you are said to have 20/200 vision.

At some place in the laboratory, a Snellen eye chart will be posted on the wall. A twenty-foot mark will also be designated on the floor. Working with your laboratory partner, test each other's eyes. Test one eye at a time, with the other eye covered. Record your test results on the Laboratory Report.

Test for Astigmatism

If the lens of an eye has an uneven curvature on one of its surfaces, a condition called *astigmatism* exists. Figure 31.5 reveals the difference between normal and astigmatic lenses. Note that the astigmatic lens has a greater curvature on its left upper surface than on the lower left quadrant. This type of lens will cause greater bending of light rays as they pass through one axis of the lens than when passing through another axis. The image that one sees with this type of lens will be blurred in one axis and sharp in other axes.

To determine the presence of astigmatism, look at the center of the diagram in figure 31.6 and note if all radiating lines are in focus and have the same intensity of blackness. If all lines are sharp and equally black, no astigmatism exists. Of course, the presence of other refractive abnormalities can make this test impractical.

Look at the wheel-like chart with each eye while closing the other and record your results on the Laboratory Report.

Test for Glaucoma

Glaucoma is a very prevalent disease in which the intraocular pressure of the eye becomes excessive. Normally, the intraocular pressure is approximately 17 mm. Hg. When this pressure reaches heights of 25 to 30 mm. Hg., blindness can result from damage to fibers of the optic nerve. Extreme pressures of 35 to 50 mm. Hg. can cause complete blindness within a few days. Although glaucoma

can be caused by infection, it is usually due to hereditary impairment of aqueous humor reabsorption into the canal of Schlemm.

We learned in Exercise 30 that the aqueous humor is constantly being produced in the posterior chamber and reabsorbed by the trabeculae in the anterior chamber. Glaucoma appears to be due to blockage of the trabeculae, causing an increase of intraocular pressure. Stimulation of the constrictor pupillae of the iris with drugs, such as neostigmine, usually opens up these spaces so that reabsorption is facilitated and the pressure is reduced. Surgical procedures have also been devised for patients in which drugs are ineffective.

The most reliable test for detection of glaucoma is to measure the intraocular pressure with a special gauge that is placed on the surface of the cornea. Since this is a medical procedure beyond the scope of this course, we must utilize a test for another characteristic of glaucoma: the *loss of peripheral vision.*

One of the early symptoms of glaucoma is the loss of peripheral vision. As the disease progresses, an increasing amount of peripheral vision is lost until the field is finally reduced to a very narrow angle (tunnel vision). At this late stage, the only light receptors that are functional are the cones of the macula lutea.

To determine if peripheral vision has been impaired, the subject is seated in front of a dark wall (or blackboard) that is approximately 6′–8′

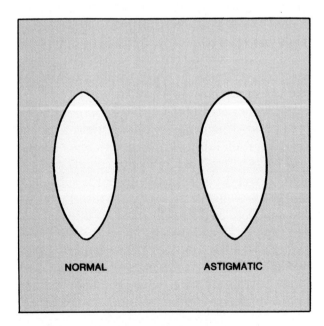

Figure 31.5 Normal and astigmatic lenses.

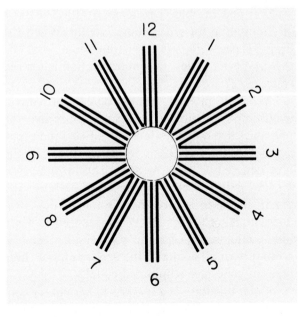

Figure 31.6 Astigmatism test chart.

square. In the center of the area is a round white dot, about ½″ diameter. The subject covers one eye and stares straight ahead at the white dot with the other eye. The examiner, using a pointer with a white tip, moves the pointer on the black area from the periphery toward the white dot, continually asking the subject if the pointer can be seen. If the subject lacks side vision, he or she will not see the pointer until it gets closer to the white dot.

If facilities are available for this test, work with your laboratory partner to test each other's eyes. Have the subject sit about 4′ from the targeted area. Test all four quadrants of the field for loss of peripheral vision. Record your results on the Laboratory Report.

Tests for Color Blindness

The perception of color is a sensation achieved partly by the retina and partly by the brain. The receptors of the retina that are sensitive to color are the cones. According to the *Young-Helmholtz Theory of Color Perception,* there are three different types of cones, each of which responds, maximally, to a different color. The three colors which cause maximum response in these cones are red, blue, and green. The degree of stimulation that each type of cone gets from a particular wavelength of light determines what color is perceived by the brain.

When the retina is exposed to red monochromatic light (wavelength at 610 nanometers), the red cones are stimulated at 75%, the green cones at 13%, and the blue cones, not at all. The ratio of stimulation for red is thus, 75:13:0. When the brain receives this ratio of stimulation from the three types of cones, the interpretation is for red color.

When a monochromatic blue light (wavelength of 450 nanometers) strikes the retina, the red cones are not stimulated at all (0), the green cones are stimulated to a value of 14%, and the blue cones to a value of 86%. The ratio here of 0:14:86 is interpreted by the brain as blue. For green, the ratio is 50:85:15. Orange-yellow produces a ratio of 100:50:0. When exposed to white light, which has no specific wavelength since it is a mixture of all colors of the spectrum, the three types of cones are stimulated equally.

Color blindness is a sex-linked hereditary con-

dition which affects 8% of the male population and 0.5% of females. The most common type is red-green color blindness, in which either the red or green cones are lacking. If red cones are lacking, a condition called *protanopia* exists. Individuals that have this condition see blue-greens and purplish tinted reds as greys. Individuals that lack green cones see greens and purple-reds as grey. A lack of green cones is designated as *deuteranopia.* Although both protanopes and deuteranopes have difficulty differentiating reds and greens, their visual spectrums differ enough so that they can be diagnosed with color test charts. Illustration 4, figure 31.7 is a test plate for differentiating protanopes and deuteranopes. While a normal person would see the number 96 on this plate, a protanope would see only the number 6 and a deuteranope would see only the number 9. Other than protanopia and deuteranopia, there are rarer forms of color vision deficiencies, such as blue-color weakness, yellow-color weakness, and total-color blindness. These forms of color vision deficiencies are not as well understood.

There are two simple ways to determine whether an individual has color blindness or color weakness. One method is to use test charts such as the Ishihara system (figure 31.7). The other method is to use a collection of pieces of variously colored yarn. Work with your laboratory partner to test each other's eyes, using both methods.

Ishihara Plates These test charts are arranged in book form with instructions for interpretation. Figure 31.7 is a sampling of four of these plates. Note that each plate is a profusion of differently colored spots. Within the arrangement of spots is concealed a number or other configuration which the subject attempts to identify.

Procure a book of these plates (*Ishihara's Test for Colour-Blindness,* Concise Edition) from the demonstration table. Working with your laboratory partner, test each other by holding the test plates about 30 inches away from the subject. If sunlight is available, use it. Start with Plate No. 1 and proceed consecutively through all 14 plates. The subject should respond with an answer for each plate within 3 seconds. The examiner will record the responses on the chart on the Laboratory Report. After all plates have been observed and recorded, compare the responses with the correct

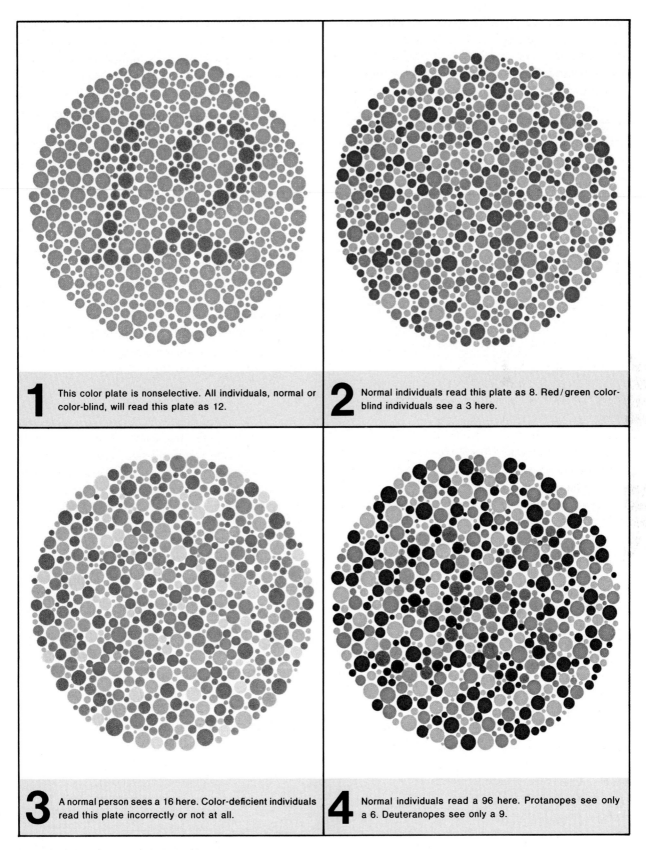

1 This color plate is nonselective. All individuals, normal or color-blind, will read this plate as 12.

2 Normal individuals read this plate as 8. Red/green color-blind individuals see a 3 here.

3 A normal person sees a 16 here. Color-deficient individuals read this plate incorrectly or not at all.

4 Normal individuals read a 96 here. Protanopes see only a 6. Deuteranopes see only a 9.

Figure 31.7 Ishihara color test plates.

The above has been reproduced from *Ishihara's Tests for Colour Blindness* published by Kanehara & Co. Ltd., Tokyo, Japan, but tests for color blindness cannot be conducted with this material. For accurate testing, the original plates should be used.

answers and make a statement at the bottom of the chart (conclusion).

Yarn Method Procure a bundle of yarn samples from the demonstration table and sort them into groups of like-colored pieces: yellows in one pile, greens in another, blue-greens in another, etc. When you are satisfied that you have them sorted properly, ask your laboratory partner to evaluate the results. If you and your laboratory partner disagree, call in a third party to determine who is color-blind.

Pupillary Reflexes

The tests and experiments performed so far have been concerned primarily with the role of the lens and retina in vision. The following experiments illustrate the roles of the pupil in adjustment of the eye to distance and light intensity.

Accommodation to Distance Three events take place when the eyes change their focus from a distant object to a close object:

1. Eye muscles react to achieve convergence of the eyes.
2. The lens becomes more convex.
3. A change occurs in the size of the pupil.

Working with your laboratory partner or a selected member of the class, perform the following experiment to observe what happens to the pupils when the eyes focus on a near object after looking at a distant object.

This experiment works best if the subject has pale blue eyes. If your laboratory partner does not qualify in this respect, request some other class member to be the subject and allow several other students to observe the results. Have the subject look at the wall on the opposite side of the room. The eyes are now relaxed and are focused at infinity.

While closely watching the pupils, place the printed page of your laboratory manual within six inches of the subject's face and ask the individual to focus on the print. Incidentally, the intensity of light on the printed page and distant wall should be equal. Do the pupils remain the same size when looking at the printed page? Repeat the experiment several times and record your results on the Laboratory Report.

Accommodation to Light Intensity Sudden exposure of the retina to a bright light causes immediate reflex contraction of the pupil in direct proportion to the degree of light intensity. The pupil contracts to approximately 1.5 mm. when the eye is exposed to intense light and it enlarges to almost 10 mm. in complete darkness. This approximates a total difference in pupillary area of about 40 times. In this reflex, impulses pass from the retina via the optic nerve through two centers in the brain and then return to the sphincter of the iris through the ciliary ganglion.

Perform this simple experiment to learn a little more about the pathway of this reflex. As in the previous experiment, select a subject with pale blue eyes so that the pupil size is more easily observed. Have the subject hold this laboratory manual, vertically, with the spiral binding close to the forehead and extending downward along the bridge of the nose. Now, position an unlighted laboratory lamp about six inches from the right eye. While watching the pupil of the left eye, turn on the lamp for one second and then turn it off. Make sure that no light spills over the from the right side of the book. Did the pupil that was not exposed to light become smaller? What does this reaction tell us with respect to the pathways of the nerve impulses?

Laboratory Report

Complete the last half of combined Laboratory Report 30, 31.

In this exercise, three aspects of the ear will be studied: (1) the components of the ear as related to hearing; (2) the physiology of hearing; and (3) hearing tests. The vestibular apparatus, which functions in equilibrium, will be studied in the next exercise. A brief statement pertaining to the characteristics of sound will precede our study of the ear. It is recommended that the sections on ear anatomy and physiology of hearing be completed prior to performing the hearing tests in the laboratory.

Characteristics of Sound

Sound waves moving through the air are propagated in wave forms which possess characteristics relative to the vibratory motion that generates them. The ear is able to distinguish tones that differ in pitch, loudness, and quality. **Pitch** is determined by the frequency of vibration; **loudness** pertains to the intensity of the vibration; and **quality** relates to the nature of the vibrations as revealed by the wave shape.

Figure 32.1 illustrates several curves depicting both the shapes of sound waves and the characteristics of the vibrations which produce them. The sine curves A and B in illustration I differ in frequency, with B producing the tone of higher pitch. Curves A and C in the middle group are of the same frequency; they differ only in amplitude, or loudness. Although the amplitude of A is twice that of C, the loudness of A will be four times that of C. This is because the intensity (I) of sound is directly proportional to the square of the amplitude (a):

$$I = 2\pi^2 \, Vf^2 \, a^2 \, d$$

V = velocity of wave propagation
f = frequency
d = density of medium

Waves A and D in illustration III differ in shape, D having some components of higher frequency that are not present in A. These curves represent sounds of different quality.

The ability of the ear to distinguish differences in pitch, amplitude, and quality depend on: (1) the conduction and pressure amplification of sound waves through the ossicles of the middle ear; (2) the stimulation of receptor cells in the cochlea; and (3) the conveyance of action potentials in the cochlear nerve to the auditory centers in the brain for interpretation. The interference of any component of this system results in hearing loss.

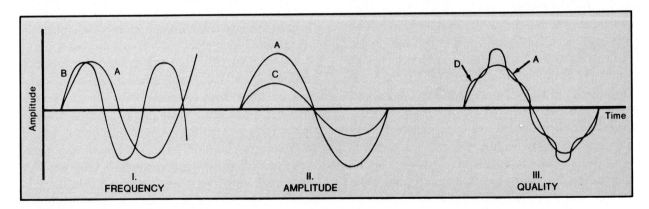

Figure 32.1 Differences in sound waves.

The types of hearing loss will be studied as hearing tests are performed. Let us now explore the anatomy of the ear as it functions in normal anatomy.

Components of the Ear

Anatomically, the human ear consists of three distinct divisions: the external ear, the middle ear and the internal ear. Figure 32.2 is a diagrammatic representation of its various components.

The External Ear

The outer or external ear consists of two parts: the auricle and external auditory meatus. The **auricle,** or *pinna,* is the outer shell of skin and elastic cartilage that is attached to the side of the head. The **external auditory meatus** is a canal about one inch in length that extends from the auricle into the head through the temporal bone. The skin lining the canal contains some small hairs and modified apocrine sweat glands that produce a waxy secretion called *cerumen.* The inner end of the external auditory meatus terminates at the **tympanic membrane,** or eardrum. The auricle serves to collect and direct sound waves into the tympanic membrane through the meatus.

The Middle Ear

This division of the ear consists of a small cavity in the temporal bone between the tympanic membrane and the inner ear. It contains three small bones *(ossicles)* that are united to form a lever system. The outermost ossicle, which is attached to the tympanic membrane, is the **malleus,** a hammer or club-shaped bone. The middle bone, an anvil-shaped structure, is the **incus.** The innermost bone, which fits into the **oval window** of the inner ear, is stirrup-shaped and is called the **stapes.** It is the role of these ossicles to transfer the forces from the eardrum to the cochlear fluids of the inner ear through the oval window. Although most sound waves reach the inner ear via the ossicles *(ossicular conduction),* some sounds, specifically very loud ones, reach the cochlea through the bones of the skull *(bone conduction).*

Leading downward from the middle ear to the nasopharynx is a duct, the **Eustachean tube,** which allows the pressure of the air in the middle ear to be equalized with the outside atmosphere. A valve at the nasopharynx end of the tube keeps the tube closed. Acts of yawning or swallowing cause it to open temporarily for pressure equalization.

The Internal Ear

This part of the ear consists of two labyrinths: the osseous and membranous labyrinths. The **osseous labyrinth** is shown in illustration B. It is the hollowed out portion of the temporal bone that contains an inner tubular structure of membranous tissue, the **membranous labyrinth.** The entire membranous labyrinth is shown in figure 33.1. Some portions of the membranous labyrinth are shown in cutaway portions of the osseous labyrinth in illustration B, figure 32.2. Within the membranous labyrinth is a fluid called the **endolymph.** Between the membranous and osseous labyrinths is a different fluid, the **perilymph.** These fluids act as conduction media for the forces involved in hearing and maintaining equilibrium.

The osseous labyrinth consists of three semicircular canals, the vestibule, and the cochlea. The **vestibule** is that portion that has the oval window on its side into which the stapes fits. The three **semicircular canals** branch off the vestibule to one side, and the **cochlea,** which is shaped like a snail's shell, emerges from the other side.

On the osseous labyrinth are two nerves, which are branches of the eighth cranial, or statoacoustic, nerve. The **vestibular nerve** is the upper nerve branch, which passes from the sensory areas of the semicircular ducts, saccule and utricle. The other branch is the **cochlear nerve,** which emerges from the cochlea.

The cochlea consists of a coiled, bony tube with three chambers extending along its full length. Illustration C in figure 32.2 reveals a cross-section of the cochlear tube. The upper chamber, or **scala vestibuli** (label 16), is so-named because it is continuous with the vestibule. The lower larger chamber is the **scala tympani.** The **round window** (label 1) is a membrane-covered opening that is on the osseous wall of the scala tympani. Between these two chambers is a triangular cross-section that represents the **cochlear duct.** This duct is bounded on its upper surface by the **vestibular**

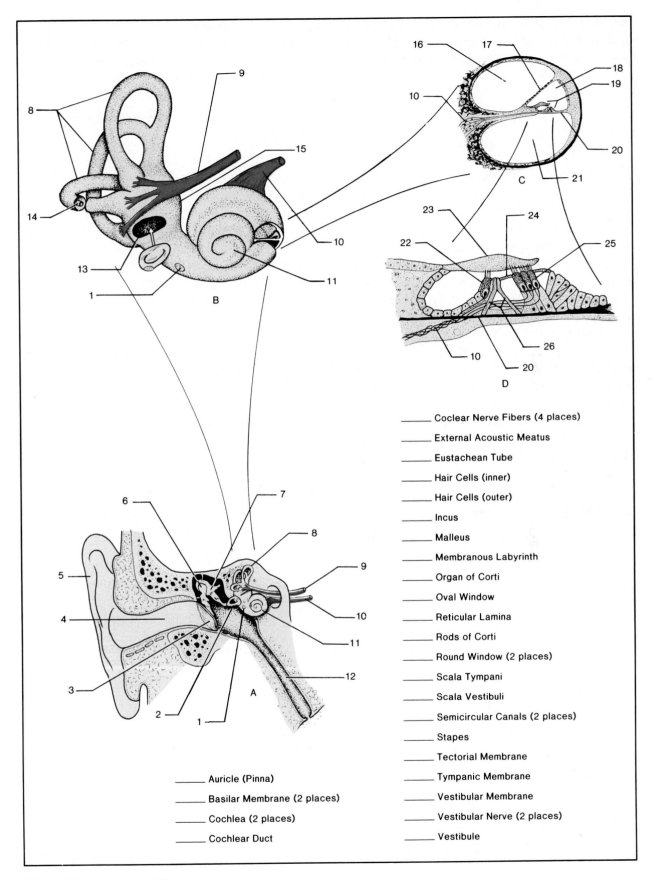

_____ Coclear Nerve Fibers (4 places)

_____ External Acoustic Meatus

_____ Eustachean Tube

_____ Hair Cells (inner)

_____ Hair Cells (outer)

_____ Incus

_____ Malleus

_____ Membranous Labyrinth

_____ Organ of Corti

_____ Oval Window

_____ Reticular Lamina

_____ Rods of Corti

_____ Round Window (2 places)

_____ Scala Tympani

_____ Scala Vestibuli

_____ Semicircular Canals (2 places)

_____ Stapes

_____ Tectorial Membrane

_____ Tympanic Membrane

_____ Vestibular Membrane

_____ Vestibular Nerve (2 places)

_____ Vestibule

_____ Auricle (Pinna)

_____ Basilar Membrane (2 places)

_____ Cochlea (2 places)

_____ Cochlear Duct

Figure 32.2   Anatomy of the ear.

membrane and on its lower surface by the **basilar membrane.** On the upper surface of the basilar membrane lies the **organ of Corti,** which contains the receptor cells of hearing. All of these chambers contain fluid: perilymph in the scala vestibuli and scala tympani, endolymph in the cochlear duct.

Illustration D reveals the detailed structure of the organ of Corti. Note how the hair tips of the **hair cells** are embedded in a gelatinous-like flap, the **tectorial membrane.** Leading from each hair cell is a nerve fiber that passes through the basilar membrane and becomes part of the cochlear nerve. The upper margins of the hair cells are held in place by the **recticular lamina.** Between the recticular lamina and the basilar membrane are reinforcing structures called the **rods of Corti.** The significance of all these structures must be taken into account in any plausible theory of hearing.

Assignment:

Label figure 32.2.

The Physiology of Hearing

As stated previously, hearing occurs when action potentials are received by the auditory centers of the brain. These action potentials, which pass along the cochlear nerve fibers of the eighth cranial nerve, are initiated by the hair cells in the organ of Corti. Activation of the hair cells depends on forces within the cochlear fluids, basilar membrane structure, secondary energy transfer, and endocochlear potential.

Role of Cochlear Fluids In figure 32.3 the cochlea has been uncoiled to reveal the relationships of the scala tympani, scala vestibuli, and cochlear duct. When the stapes moves in and out of the oval window, sound wave energy moves through the scala vestibuli and into the scala tympani. Since the perilymph is incompressible, the elasticitiy of the round window membrane accommodates the pressure changes by moving in and out in synchrony with the movements of the stapes.

The energy of sound waves moving through the cochlea reach the round window in two ways: (1) directly into the scala tympani through the helicotrema at the apex of the cochlea, and (2) through the flexible cochlear duct to the scala tympani. Arrows in figure 32.3 illustrate both routes of energy propagation.

Basilar Membrane Structure The first structure of the organ of Corti that reacts to vibrations of different sound frequencies is the basilar membrane. Within this membrane are approximately 20,000 fibers that project from the bony center of the cochlea toward the outer wall. The fibers closest to the stapes are short (.04 mm.) and stiff; they vibrate when stimulated by vibrations caused by high frequency sounds. The fibers at the end of the cochlea near the helicotrema are long (.5 mm.), more limber, and vibrate in harmony with low frequency sounds. These fibers are elastic reed-like structures that are not fixed at their distal ends. Because they are free at one end, they are able to vibrate like reeds of a harmonica to specific frequencies.

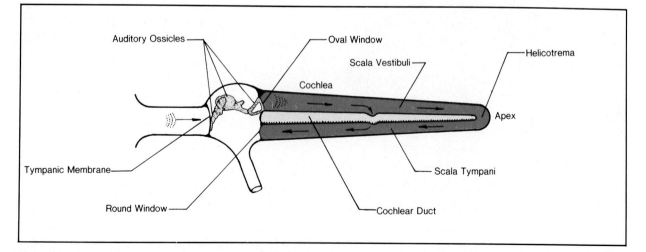

Figure 32.3 Pathway of sound wave transmission in ear.

Secondary Energy Transfer Once fibers at a particular point on the basilar membrane are set into motion by a specific sound frequency, the rods of Corti transfer the energy from the basilar membrane to the reticular lamina, which supports the upper portion of the hair cells (see illustration D, figure 32.2). This secondary energy transfer causes all these components to move as a unit, with the end result that the hairs of the hair cells are bent and stressed. The back and forth bending of the hairs, in turn, causes alternate changes in the electrical potential across the hair cell membrane. This alternating potential, known as the *receptor potential* of the hair cell, stimulates the nerve endings that are at the base of each hair cell and an action potential in the cochlear nerve fibers is initiated.

Endocochlear Potential The sensitivity of the hair cells to depolarization is greatly enhanced by the chemical differences between the perilymph and endolymph. Perilymph has a high sodium–to–potassium ratio; endolymph has a high potassium–to–sodium ratio. These ionic differences result in an *endocochlear potential* of 80 millivolts. It is significant that the tops of the hair cells project through the reticular lamina into the endolymph of the scala media, but the lower portions of these cells lie bathed in perilymph. It is believed that this potential difference of 80 mv. between the top and bottom of each hair cell greatly sensitizes the cell to slight movement of the hairs.

Frequency Range The variability in length and stiffness of the fibers within the basilar membrane enables the ear to differentiate sound waves as low as 30 cycles per second near the apex of the cochlea, and as high as 20,000 cycles per second near the base of the cochlea. In between these two extremes, at various spots on the basilar membrane, the membrane responds to the in-between frequencies. This method of pitch localization by the basilar membrane is called the *place principle*.

Loudness Determination As noted in figure 32.1, the loudness of a sound is reflected in the sound wave. Increased loudness of sounds causes an increase in the amplitude of vibration of the basilar membrane. With an increase in basilar membrane vibration, more and more hair cells become stimulated, causing **spatial summation** of impul-

ses; that is, more nerve fibers are carrying more impulses, and this is interpreted by the brain as increased loudness.

Assignment:

Answer the questions on the Laboratory Report that pertain to the physiology of hearing.

Hearing Tests

Although deafness may have many different origins, there are essentially two principal kinds of deafness: nerve and conduction deafness. If the cochlea or cochlear nerve is damaged, the condition is referred to as **nerve deafness.** Damage to the eardrum or ossicles, on the other hand, will result in **conduction deafness.** While conduction deafness can usually be remedied with surgery or hearing aids, nerve deafness cannot be corrected.

The three kinds of hearing tests outlined here are ones that are most frequently used. Each type of test serves a specific function. Perform those tests for which equipment is available.

Watch Tick Method

The use of a spring-wound pocket watch to screen patients for hearing loss has merit in simplicity. The inability of a patient to hear a ticking watch at prescribed distances from the ear can alert the examining physician to a hearing problem. Further tests using tuning forks or the audiometer can then be brought into play.

To perform this type of hearing test, team up with two other students to test each other's ears. One person will be the subject while one manipulates the watch and the other measures the distances and records information on the Laboratory Report.

Success in all hearing tests demands a quiet room. The best procedure is for groups of students to disperse as much as possible and to keep conversation at the whisper level.

Materials:

spring-wound pocket watch
cotton plugs
meter stick or measuring tape

1. Seat the subject comfortably in a chair and plug one ear with cotton.

2. Instruct the subject to use the index finger to signal when sound is first heard as the watch approaches or when sound disappears as the watch is moved away from the ear.

3. With the subject looking straight ahead, place the watch at a position that is out of hearing range of the ear, usually about three feet. Keep the face of the watch parallel to the side of the head.

4. Move the watch toward the ear at *a speed of about ½'' per second*. This is a very slow movement. The subject should signal when the **first tick** is heard.

5. Measure the distance with the meter stick or tape measure, record the information on the Laboratory Report, and repeat the test two or three times. On repeat tests change the approaches to the ear to prevent the factor of subject anticipation.

6. Reverse the procedure by starting close to the ear and receding from it. In this test, the subject signals the **last tick** that is heard.

7. Transfer the cotton plug to the other ear and follow the same procedure.

Tuning Fork Methods

The distinction between nerve and conduction deafness can readily be made with two tuning fork methods: the *Rinne* and *Weber* tests. In one test the base of the tuning fork is applied to the mastoid process behind the ear; in the other test the fork is placed on the forehead. In both tests the tuning fork is set in vibration by bouncing the fork off the heel of the hand as shown in figure 32.4. Avoid striking hard surfaces, such as the table top. These tests may be performed independently or with the aid of an examiner.

Materials:

    tuning fork
    cotton ear plugs

The Rinne Test This test enables you to identify the cause of hearing loss by placing the base of a tuning fork against the mastoid process. The method readily differentiates between nerve and conduction deafness.

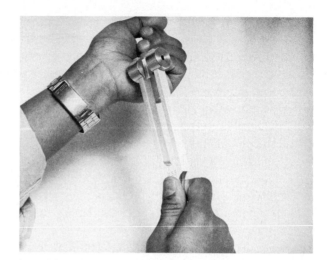

Figure 32.4 The tuning fork is activated by striking the heel of the hand.

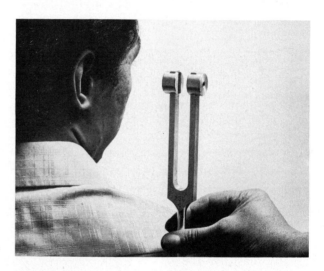

Figure 32.5 Rinne test. A vibrating tuning fork is held six inches away from the ear.

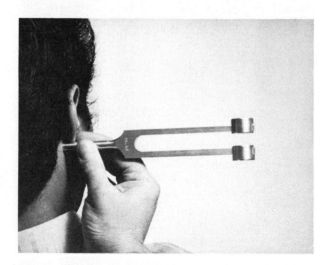

Figure 32.6 Rinne test. The stem of the tuning fork is placed on the mastoid process.

1. Plug one ear with cotton.
2. Set a tuning fork in motion by striking it on the heel of the hand. Place it in front of the unplugged ear with the tine facing the ear, as shown in figure 32.5; 3″ to 6″ from the ear is adequate distance.

   If there is a minimum of hearing loss, the vibrating sound will be heard immediately. As the sound intensity diminishes, a point will be reached when the sound finally disappears. At this instant, place the stem of the tuning fork against the mastoid process as in figure 32.6. If the sound vibrations reappear, *conduction deafness* exists.

   If considerable hearing loss is present, and the tuning fork vibrations cannot be heard for long or at all, place the stem of the vibrating fork against the mastoid process and note if the sound can be heard through bone conduction. If the sound is loud through the bone, *conduction deafness* exists. If no sound is heard, *nerve deafness* exists.
3. Repeat the test on the other ear.
4. Record results on the Laboratory Report.

The Weber Test When the stem of a vibrating tuning fork is placed on the forehead of a person with normal hearing, the sound of the fork is heard at equal intensity in both ears. If an individual with conduction deafness is tested in this manner, the sound will be heard louder in the deaf ear than in the normal ear. The reason is that the deaf ear, which is normally not activated by sound waves through the air, is more acutely atuned to sound waves being conducted to the cochlea through bone.

On the other hand, if this test is performed on a person who has one normal ear and one ear with nerve deafness, the sound will be more intense in the normal ear than in the deaf one.

Place a vibrating tuning fork on your forehead and compare the sounds heard in each ear. Report your conclusions on the Laboratory Report.

Audiometry

The audiometer is an instrument used for determining hearing losses in the audible range of normal speech. It measures the ability of the ear to hear sounds in the **frequency range** of 125–8,000 cycles per second. Although the hearing range of the human ear may be as broad as 30–20,000 c.p.s., it is the 125–8,000 c.p.s. range that is important in hearing the spoken word.

Figure 32.7 The audiometer.

This instrument also measures the **level of intensity** of sounds in the audible range that one can hear. The sensation of loudness experienced by the ear is not related in a simple way to the intensity of sound that strikes the tympanic membrane. While the range of sensitivity between a faint whisper and the loudest noise is one trillion times, the ear interprets this great difference as approximately a 10,000-fold change. What happens here is that the scale of intensity is compressed by the action of the tympanic membrane, ossicles, and cochlear duct, enabling the ear to have a much broader sensitivity range.

Because of this extreme range in sound intensities, it is necessary to express the intensity in terms of the logarithms of their actual intensities. The basic unit is called a **bel** (after Alexander Bell). Sounds that differ one bel in intensity differ by ten times. A **decibel** is one tenth of a bel. It is in decibel units that hearing is measured, mainly because this is the smallest unit that the human ear is able to differentiate.

The audiometer is, essentially, an electronic oscillator with earphones. As illustrated in figure 32.7, it has one control that regulates the frequency and another control that regulates the intensity, or loudness, of the sound. Two separate switches, or tone controls, near the bottom are used to release the tone into the earphones; the left switch (blue) is for the left earphone and the right (red) one is for the right earphone.

The Hearing Level Control is calibrated so that the zero intensity level of sound at each frequency is the loudness that can be barely heard by a normal person. The numbers on this control represent decibels. If it is necessary to adjust this control to 30 at 125 c.p.s. frequency, it means that the patient has a hearing loss of 30 decibels.

In this experiment, work with your laboratory partner to plot an audiogram on the Laboratory Report for each ear. Proceed as follows:

Materials:

> audiometer
> red and blue pencils

1. Place the headset securely on the subject's head so that the ear cushions make good contact over the subject's ears. Make sure that the ear cushions are not obstructed by clothing, earrings, etc.

Important: Make sure that the earphone with a red cord is placed over the right ear.

2. To enable the subject to become familiar with the tone, set the Frequency Control on 1,000 c.p.s. and the Hearing Level Control at 50 db; then, depress the right (red) Tone Control so that the subject can hear the tone in the **right ear.**

3. Rotate the Frequency Control throughout all frequencies at this 50 db. level to let the subject hear the frequencies prior to testing. Depress the Tone Control only when the dial is set on each frequency.

4. Return the Frequency Control to 1,000 c.p.s. and, while depressing the red tone control, rotate the Hearing Level Control counterclockwise slowly until you reach the point where the subject is just barely able to recognize the tone. If, for instance, the subject can hear the tone at 30 decibels, but cannot hear the tone at 25 decibels, the subject's threshold is established at 30 decibels for that frequency setting. Have the subject signal with a hand signal when the tone is heard.

5. Record this measurement on the audiogram chart of the Laboratory Report by marking a red "0" on the vertical line for 1,000 c.p.s., where it intersects the line for the decibel value. (In the above example, it would be where the 30 decibel line intersects the 1,000 c.p.s. line.)

6. Now rotate the Frequency Dial to 2,000 c.p.s. following the above procedures. After testing at this frequency, test the right ear at the remaining higher frequencies up to 8,000 c.p.s. Finally, finish the tests on the right ear by testing at frequencies of 250 and 500 c.p.s.

7. Test the **left ear,** following steps 2 through 6 above. Use the blue Tone Control switch for testing the left ear. Record each threshold level with an "X" on the appropriate frequency line with a blue pencil.

8. Connect the points on the chart with appropriately colored lines to produce a hearing graph for each ear.

Laboratory Report

Complete the Laboratory Report for this exercise.

The Ear: Its Role in Equilibrium

The part of the ear that functions in equilibrium is known as the **vestibular apparatus.** It incorporates the semicircular ducts, saccule, and utricle. These portions were more or less ignored in Exercise 32. The anatomical and physiological features of this unique part of the ear will be studied in this exercise.

Although the vestibular apparatus plays an important role in the maintenance of equilibrium, three other factors are also important: (1) visual evaluation of horizon position; (2) proprioceptor sensations in joint capsules; and (3) cutaneous sensations through certain exteroceptors. Sensations from these four sources are continuously subconsciously synthesized at the cortical level to provide the correct reflex responses necessary to maintain equilibrium.

In this exercise, we will evaluate the relative significance of visual, proprioceptive, and vestibular sensations in postural control. Some clinical tests for evaluating static and dynamic equilibrium mechanisms will also be explored.

Two types of equilibrium are identifiable: static and dynamic. *Static equilibrium* pertains to the effect of gravity on receptors. *Dynamic equilibrium* compensates for angular movements of the body in different directions. The physiology of each type follows.

Static Equilibrium

The sensory receptors of the vestibular apparatus that respond to gravitational forces are located in

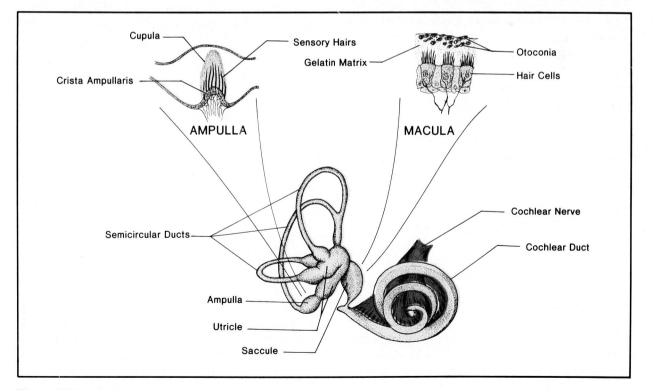

Figure 33.1   The membranous labyrinth.

the saccule and utricle. Note in figure 33.1 that the **saccule** connects directly to the cochlear duct and the **utricle** lies at the base of the semicircular canals. These two structures, being a part of the entire membranous labyrinth, contain **endolymph.**

On the inner walls of the saccule and utricle lie two sensory structures, the **maculae.** Each macula consists of **hair cells,** a **gelatinous matrix,** and calcium carbonate crystals called **otoconia.** The otoconia are embedded in the gelatinous matrix. Fibers of the vestibular nerve carry impulses from the hair cells to the brain.

When the head is tilted slightly in any direction, the otoconia are pulled by gravitational forces, causing the cilia of the hair cells to be bent by the action of the gelatinous matrix against them. Bending of the cilia in one direction causes impulse traffic in the vestibular nerve fibers to increase markedly; bending in another direction decreases the impulse traffic to a point of no reaction at all. This system is so sensitive to malequilibrium that as little as ½° shifting of the head in any direction from vertical is detectable.

Simply tilting the head, however, does not bring about a sense of malequilibrium. Fortunately, proprioceptors in the joints of the neck transmit inhibitory signals to the brain stem that neutralize the effects of the vestibular receptors. It is only when the whole body becomes disoriented that the hair-cell initiated impulses are not inhibited.

Balancing Test One of the simplest tests for determining the integrity of this static equilibrium mechanism is to have the subject stand perfectly still with eyes closed. Work with your partner to test each other's sense of balance. If there is any damage to this system, the subject will waver and tend to fall. Individuals with long term damage, however, are often able to stand fairly well due to well-developed proprioceptive mechanisms.

Dynamic Equilibrium

The receptor hair cells for dynamic equilibrium are contained in the **ampullae** of the **semicircular ducts.** These hair cells are arranged in a crest, the **crista ampullaris,** within each ampulla. The hair tufts, in turn, are embedded in a gelatinous mass to form a structure called the **cupula.**

The three semicircular ducts are arranged at right angles to each other in three different planes. When the head moves in any direction, the semicircular duct in the plane of directional movement moves relative to the endolymph within it; in other words, the fluid is stationary as the duct moves. This movement causes the cupula to act as a trap door, moving in either direction due to the force of the endolymph. The hair cells in this structure act much like the hair cells of the macula, in that bending in one direction produces increased action potentials and bending in the other direction is inhibitory. Nerve impulses along the vestibular nerve reflexly excite the appropriate muscles to maintain equilibrium.

Nystagmus

The principal function of the semicircular ducts is to maintain equilibrium in the initial stages of angular or rotational movement. As soon as the individual begins to change direction, the semicircular ducts trigger reflexes to the proper muscles in anticipation of malequilibrium. A reflex movement, called *nystagmus,* functions to produce fixed momentary images on the retina. It always occurs when the body is rotated.

Nystagmus is characterized by rapid and slow jerky movements of the eyes as the body is rotated. It is caused by reflexes transmitted through the vestibular nuclei, the cerebellum, and the medial longitudinal fasciculus to the ocular nuclei.

If the head is slowly rotated to the right, the eyes will, at first, move slowly to the left. This reflex eye movement, which is called the *slow component* of nystagmus, is caused by the backward rotation of endolymph in the left horizontal semicircular duct and forward movement of this fluid in the right horizontal semicircular duct. When the eyes have moved as far to the left as they possibly can, the eyes quickly shift to the right, producing what is called the *fast component* of nystagmus. This reflex involves nuclei in the brain stem associated with the abducens and oculomotor nerves. Thus, we see that the first slow component of nystagmus is initiated by the vestibular apparatus and the second fast component involves the brain stem.

Group Demonstration Proceed as follows to demonstrate nystagmus. Work in teams of five students, using one member as a subject. The subject, seated in a swivel chair, will be rotated by the other four members. By using several individuals to anchor and rotate the chair, the subject will be accorded maximum safety. Prior to beginning the test, the *chair's back should be securely tightened* to prevent backward tilting.

Materials:

swivel type armchair

1. Have the subject sit in the chair, crosslegged and gripping the arms of the chair.
2. The other four members of the team will form a circle around the chair, each member placing one foot against the base of the chair to prevent chair base movement.
3. Revolve the subject to the right *(clockwise)* and observe the eye movement as the subject gazes straight ahead during the rotation. Make ten revolutions at a rate of one revolution per second.
4. Allow the subject to remain seated for one or two minutes after the test to regain stability.
5. Repeat the procedure with another subject, but rotate this individual in a *counterclockwise* direction.
6. With the remaining two team members, follow the same procedure but have the subjects keep their eyes closed during rotation. Immediately upon stopping the chair, have the subjects open their eyes for observation.
7. Record all observations on the Laboratory Report.

Caution: The next test must not be performed here without special permission. It can induce vomiting.

Ice Water Test *(Caloric Stimulation)* A simple test that may be used to determine whether or not the vestibular apparatus of one ear is functioning is to perfuse the external ear canal with a small amount of ice water. If the vestibular apparatus is intact and functional, the subject will experience a discomforting sensation of rotation, and nystagmus will immediately be initiated. If the subject experiences no discomfort or nystagmus, it may be assumed that damage to the vestibular

apparatus or the vestibular nerve has occurred.

The physiological explanation of this reaction is that the cold water increases the density of the endolymph in a portion of the semicircular canals. With some of the endolymph heavier than other endolymph, convection currents are created in the semicircular canals that cause a sense of malequilibrium.

This test is often used to test for vestibular nerve damage resulting from overdosage of certain antibiotics such as streptomycin.

Proprioceptive Influences

To observe the relative roles of proprioceptors, vision and vestibular reflexes in equilibrium, perform the following experiments. Retain the same five-member teams, as previously.

Materials:

swivel type armchair
pencil
blindfold

At-Rest Reactions Direct the subject to sit in the swivel chair and perform the following simple acts to establish norms for comparison.

1. With eyes closed, have the subject place the heel of the right foot on the toes of the left foot.
2. With eyes closed, have the subject bring the index finger of the right hand to the tip of the nose from an extended arm position.
   Question: How do proprioceptors in the appendages function to accomplish these two feats? Record your conclusions on the Laboratory Report.
3. Next, have the subject, with eyes open, raise his or her hand from the right knee vertically and forward to point his or her finger at the eraser of a pencil held about two feet directly in front of the subject. Have the subject repeat this pointing six times, once per second on the command of the person holding the pencil. After each pointing, the hand is returned to rest on the right knee.
4. Have the subject repeat step #3, with the eyes closed. How close does the subject approach the eraser with eyes closed? Report results on the Laboratory Report.

Effects of Rotation The role of the eyes in kinesthetic efficiency will now be determined in tests similar to those above. Proceed as follows:

1. Direct the subject to sit crosslegged and keep the eyes open. Revolve the chair at one revolution per second for ten full turns. Halt the chair in exactly the same starting position. Ask the subject to point to the pencil eraser held in the same position as in the previous test. As before, the subject points to the eraser one time per second for a total of six times. After each pointing, the hand is returned to rest on the right knee. Record observations on the Laboratory Report.

2. Blindfold the subject, orient the chair to the same starting position, and, by feel, show the subject where the pencil eraser is located.

3. Now, rotate the subject for ten full turns, stopping the chair at the same starting point. As soon as the chair has come to rest, ask the subject to point to the spot where he or she thinks the eraser is, **but don't let subject touch the eraser.** Repeat the pointing a total of six times. Remember the pencil eraser must be held in the same position as previously, and must not be touched.

4. Repeat the entire procedure with another subject, but reverse the direction of rotation.

Laboratory Report

Complete the Laboratory Report for this exercise.

Part 8 Hematology

Hematological information accumulated from a series of blood tests can be of great value in many ways. Prior to surgery, this information can reveal certain types of problems that might occur in the operating room. In diagnostic studies, the presence of infection, and even the nature of an infection, may be revealed by certain types of tests. The presence of anemia and the causes of anemia can usually be determined with relatively simple tests.

Of the many kinds of blood tests that are performed in laboratories today, we have selected eight of the more commonly used tests for this unit. If all of these tests are performed, considerable information is at hand for diagnostic purposes.

A convenient way to utilize this section is for the student to perform each blood test on his or her own blood, recording all the information on tables I and II of combined Laboratory Report 34–38. Normal values for each test are given in the tables. To perform these tests, however, students will have to work in pairs. A significant feature about these tests is that no venipuncture is required; instead, only a drop of blood from a punctured finger is necessary for each test.

34 Blood Cell Counts

The formed elements of blood consist of *erythrocytes* (red blood cells), *leukocytes* (white blood cells), and *blood platelets*. Each component serves a specific function and, consequently, has a different clinical significance in medical diagnostics. In many pathological conditions, it is essential that the physician know precisely what the numerical count is for each constituent in the blood. In this exercise, we will employ three routinely used, standardized blood cell counting techniques to study the various kinds of cells.

Skill and reliability at performing cell counts comes only after performing many counts. It is not expected that your first counts will be as diagnostically reliable as counts made by an experienced technician who performs them daily. In some of these procedures, a 10% error is anticipated for even the experienced worker.

Our principal concerns here will be to (1) learn how to distinguish the various kinds of cells; (2) understand the significance of deviations from normal cell counts; and (3) become familiar with the basic procedures in performing cell counts. The following three blood cell counts will be made of your own blood. It will be necessary for you to work with a laboratory partner.

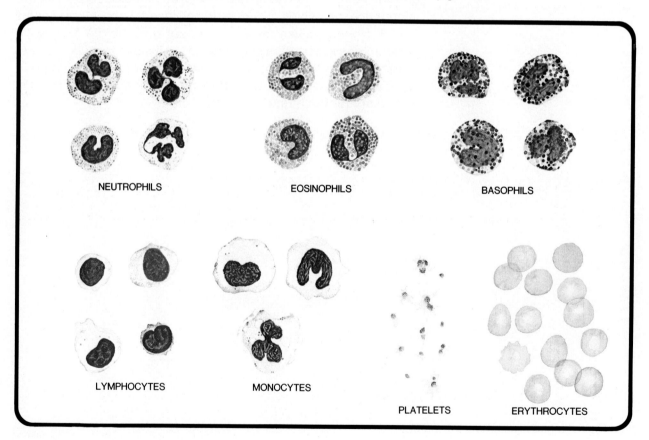

NEUTROPHILS   EOSINOPHILS   BASOPHILS

LYMPHOCYTES   MONOCYTES   PLATELETS   ERYTHROCYTES

Figure 34.1   Formed elements of blood.

Differential WBC Count

Human blood contains five basic types of leukocytes. Three of them, the neutrophils, eosinophils, and basophils, have conspicuous granules in their cytoplasm; these leukocytes are designated as *granulocytes*. The monocytes and lymphocytes do not exhibit pronounced granules; these cells are called *agranulocytes*. To make the nuclei and granules of these cells show up distinctly under the microscope, it is necessary to stain them with Wright's stain. Figure 34.1 reveals how these cells appear when stained with this special stain.

A *differential white blood cell count* is performed on a blood smear slide to determine the percentages of each type of leukocyte in a sample of blood. The clinical value of such a slide can be considerable. To make a blood count of this type, it is necessary to identify at least 100 leukocytes and record the numbers of each type that are seen on the slide. Since these cells are quite small, it is necessary to examine them with oil immersion optics.

The *normal* percentage ranges for the various leukocytes are as follows: **neutrophils** 50%–70%; **lymphocytes** 20%–30%; **monocytes** 2%–6%; **eosinophils;** 1%–5%; and **basophils** 0.5%–1%. Deviations from these percentages may indicate serious pathological conditions. High neutrophil counts, or *neutrophilia*, often signal localized infections such as appendicitis or abcesses in some other part of the body. *Neutropenia*, a condition in which there is a marked decrease in the numbers of neutrophils, occurs in typhoid fever, undulant fever, and influenza. *Eosinophilia* (high eosinophil count) may indicate allergic conditions or invasions by parasitic roundworms such as *Trichinella spiralis*, the *pork worm*. Counts of eosinophils may rise as much as 50% in cases of trichinosis. High lymphocyte counts, or *lymphocytosis*, are present in whooping cough and some viral infections.

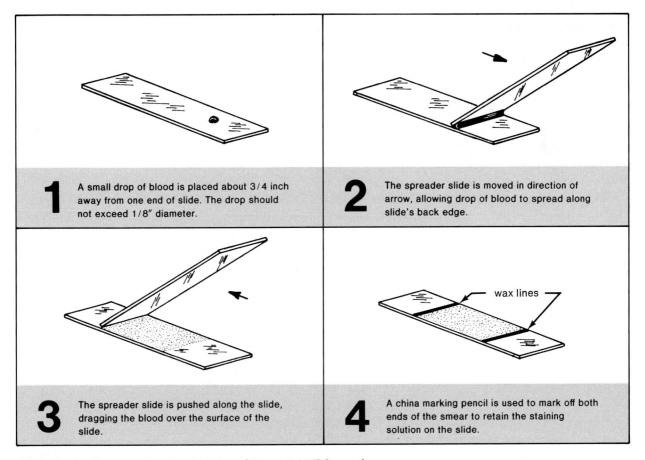

1 A small drop of blood is placed about 3/4 inch away from one end of slide. The drop should not exceed 1/8″ diameter.

2 The spreader slide is moved in direction of arrow, allowing drop of blood to spread along slide's back edge.

3 The spreader slide is pushed along the slide, dragging the blood over the surface of the slide.

4 A china marking pencil is used to mark off both ends of the smear to retain the staining solution on the slide.

wax lines

Figure 34.2 Smear preparation technique (differential WBC count).

Preparation of Slide

In the preparation of a suitable stained blood slide, it is essential that the smear be thick at one end and drawn out to a feather-thin edge. This type of preparation will provide a gradient of cellular density that will make it possible to choose an area which is ideal for counting. The angle at which the spreading slide is held in making the smear will determine the thickness of the smear. It may be necessary for you to make more than one slide to get an ideal one. Figure 34.2 illustrates the spreading technique.

Materials:

    clean microscope slides
    sterile disposable lancets
    sterile absorbent cotton
    Wright's stain
    distilled water in dropping bottle
    70% alcohol
    wax pencil
    bibulous paper

1. Clean three or four slides with soap and water. Handle them with care to avoid getting their flat surfaces dirty with the fingers. Although only two slides may be used, it is often necessary to repeat the spreading process, thus the extra slides.
2. Scrub the middle finger with 70% alcohol and stick it with a lancet. Put a drop of blood on the slide ¾″ from one end and spread with another slide in the manner illustrated in figure 34.2. *Note that the blood is dragged over the slide, not pushed.* Do not pull the slide over the smear a second time. If you don't get an even smear the first time, repeat the process on a fresh clean slide. To get a smear that will be the proper thickness, hold the spreading slide at an angle somewhat greater than 45°.
3. Draw a line on each side of the smear with a wax pencil to confine the stain which is to be added.
4. Cover the film with Wright's stain, *counting the drops* as you add them. Stain for **4 minutes** and then add the same number of drops of distilled water to the stain and let stand for another **10 minutes.** Blow gently on the mixture every few minutes to keep the solutions mixed.
5. Gently wash off the slide under running water for 30 seconds and shake off the excess. Blot dry with bibulous paper.

Performing the Cell Count

As soon as the slide is completely dry, scan it with the low power objective to find that area of the slide that has the best distribution of cells. Avoid the excessively dense areas. Place a drop of oil near the edge of the smear in the selected area and examine with the oil immersion objective.

Remove your Laboratory Report sheet from the back of the manual and record each type of leukocyte encountered, as you move the slide. Follow the path indicated in figure 34.3. For identification of each type of cell, refer to figure 34.1. Your Laboratory Report sheet indicates how the cells are to be tabulated.

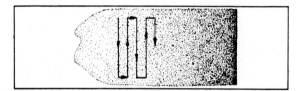

Figure 34.3   The path of the cell count.

Total White Blood Cell Count

While the differential count indicates relative numbers of types of leukocytes, it does not reveal how many white blood cells there are in a given volume of blood. This information is needed to facilitate a more accurate interpretation of the differential count. Although the number of white blood cells may vary with the time of day, exercise, and other factors, a range of 5,000 to 9,000 cells per cubic millimeter is considered normal. The count in children tends to be higher with a greater number of lymphocytes present.

When the total number of leukocytes in an adult exceeds 9,000 per cu. mm. by more than 10%, the individual is said to have *leukocytosis.* The value of the differential count in relationship to this type of count is that one can determine what cells are causing the high count.

The reverse of leukocytosis is *leukopenia.* In this case, white blood cell counts may be substantially less than 5,000 per cu. mm. If this condition is severe and persistant, the patient is in peril due to the lack of protection against bacteria. Leukopenia may result from some kinds of bacterial infections, poisons, antibiotic therapy, X-ray therapy, and many other causes.

To determine the number of leukocytes in a cubic millimeter of blood, a measured amount of blood is diluted with a weak acid solution to give a dilution of 1 part in 20. A counting chamber *(hemacytometer)* is then charged with this diluted blood and examined under the lower power objective of a microscope. All the cells that are seen in the four "W" areas of figure 34.8 are counted. This count is then fitted into a formula to determine the number of cells per cubic millimeter.

Dilution of Blood

Working with your laboratory partner, assist each other to prepare a diluted sample as follows:

Materials:

hemacytometer
cover glass
WBC diluting pipette
rubber tubing
WBC diluting fluid
cotton
alcohol
lancets
mechanical hand counter
clean cloth
pipette cleaning solutions

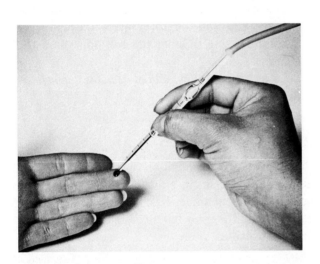

Figure 34.4 The blood is drawn up to the 0.5 mark by suction. If the blood goes past the mark, excess is removed by blotting the tip on paper.

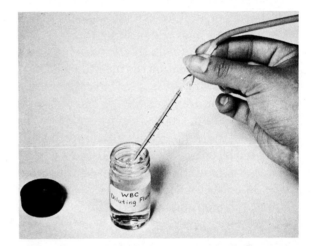

Figure 34.5 WBC diluting fluid is drawn up to the 11.0 mark. The pipette should be rotated between the fingers as fluid is being drawn up.

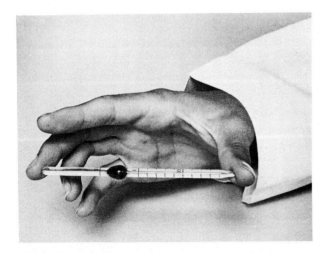

Figure 34.6 The charged WBC pipette is held parallel to the table top when shaken for mixing. Two to three minutes of shaking is necessary.

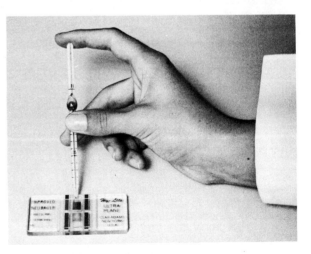

Figure 34.7 After discarding one-third of pipette contents, the hemacytometer chamber is charged by slowly releasing fluid onto glass slope.

1. Wash the hemacytometer and cover glass with soap and water, rinse well, and dry with a clean cloth or Kimwipes.
2. Produce a free flow of blood, wipe away the first drop and draw the blood up into the diluting pipette to the 0.5 mark. Refer to figure 34.4. If the blood goes a little past the mark, touch the end of the pipette to a piece of blotting paper to draw it back to the mark. If the blood goes substantially past the 0.5 mark, discharge the blood, wash the pipette in the four cleansing solutions, as in figure 34.9, and start over. The ideal way is to draw up the blood exactly to the mark on the first attempt. (Blood may be drawn up to the 1.0 mark if leukopenia is suspected. This would produce a dilution of 1:10 instead of 1:20.)
3. As shown in figure 34.5, draw the WBC diluting fluid up into the pipette until it reaches the 11.0 mark, rotating the pipette as the fluid is being drawn up.
4. Place the third finger over the end of the pipette, remove the rubber tubing, and place the thumb over the other end of the pipette.
5. Mix the blood and diluting fluid in the pipette for **2–3 minutes** by holding it as shown in figure 34.6. The pipette should be held parallel to the tabletop and moved through a 90° arc, with the wrist held rigidly.

Charging the Hemacytometer

Now that you have the blood diluted 1:20, position the cover glass on the hemacytometer and charge the chamber, observing these steps:

1. Discharge one-third of the bulb fluid from the pipette by allowing it to drop onto a piece of paper toweling.
2. While holding the pipette as shown in figure 34.7, deposit a tiny drop on the polished surface of the counting chamber next to the edge of the cover glass. *Do not let the tip of the pipette touch the polished surface for more than an instant.* If it is left there too long, the chamber will overfill. A properly filled chamber will have diluted blood filling only the space between the cover glass and counting chamber. No fluid should run down into the moat.
3. Charge the other side if the first side was overfilled.

Performing the Count

Place the hemacytometer on the microscope stage and bring the grid lines into focus under the **low power** (10X) objective. It will be necessary to reduce the light somewhat to make both the cells and lines visible.

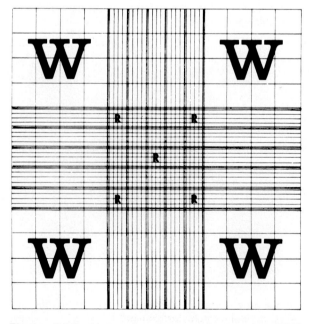

Figure 34.8 Hemacytometer counting areas: WBC counts are made in the four large "W" areas and RBC counts are made in the five small "R" areas.

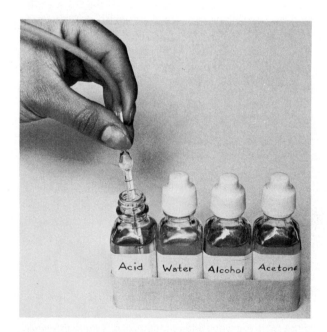

Figure 34.9 Pipettes must be cleaned with successive rinses of acid, water, alcohol, and acetone before reusing or storing.

Locate one of the "W" (white) sections (figure 34.8) to be counted and note if the cells are evenly distributed. Since the acid has destroyed the erythrocytes, only the leukocytes will be seen as small dots. If the distribution is poor, charge the other half of the counting chamber after further mixing. If the other chamber had been previously charged unsuccessfully by overflooding, wash off the hemacytometer and cover glass, shake the pipette for 2–3 minutes, and recharge it.

Count all the cells in the four "W" areas. To avoid over-counting of cells at the boundaries, **count the cells that touch the lines on the left and top sides,** but *not the ones that touch the boundary lines on the right and bottom sides.* This applies to the boundaries of the entire "W" areas.

Cleaning Pipette

Discharge the contents of the pipette and rinse it out by drawing the following fluids up into it: acid, water, alcohol, and acetone. Refer to figure 34.9.

Calculations

To determine the number of leukocytes per cubic millimeter, *multiply the total number of cells counted in the four "W" areas by 50.*

The factor of 50 is the product of the volume correction factor and dilution factor, or,

$$2.5 \times 20 = 50$$

The *volume correction factor of 2.5* is arrived at in this way: Each "W" area is exactly one square millimeter by 0.1 millimeter deep. Therefore, the volume of each "W" section is 0.1 cu. mm. Since four "W" sections are counted, the total amount of diluted blood that is examined is 0.4 cu. mm. Since we are concerned with the number of cells in one cu. mm. instead of 0.4 cu. mm., we must multiply our count by 2.5 derived from dividing 1.0 by 0.4.

Record your results on the Laboratory Report.

Red Blood Cell Count

Normal red blood cell counts for adult males average around 5,400,000 ($\pm$600,000) cells per cubic millimeter. The normal average for women is 4,600,000 ($\pm$500,000) per cubic millimeter. This difference in the sexes does not exist prior to puberty. At high altitudes, higher values will be normal for all individuals.

Anemia may be defined, simply, as a condition in which the oxygen-carrying capacity of the blood is reduced. A low RBC count will result in this condition. However, this does not mean that an individual with a normal RBC count cannot be anemic. Since the oxygen carrying capacity of the blood is primarily a function of the hemoglobin present, an individual with small red blood cells, but a normal count, will be anemic.

Polycythemia, a condition characterized by above normal RBC counts, may be due to living at high altitudes *(physiological polycythemia)* or red marrow malignancy *(polycythemia vera).* In physiological polycythemia, counts may run as high as 8,000,000 per cu. mm. In polycythemia vera, counts of 10–11 million are not uncommon.

The general procedures used for the WBC count are essentially the same as for counting erythrocytes. The only difference is that an RBC pipette must be used with RBC diluting fluid and different mathematics are involved. The RBC pipette has 101 instead of 11 scribed above the bulb. The RBC diluting fluid may be one of several isotonic solutions such as physiological saline (0.9% sodium chloride) or Hayem's solution. To perform a red blood cell count, follow this procedure:

Materials:

RBC diluting pipette
rubber tubing
RBC diluting fluid
other supplies used for WBC count

1. Draw the blood up to the 0.5 mark and the diluting fluid up to the 101 mark. Mixing is performed in the same manner as for the WBC count.
2. Charge the chamber using the same procedures as for the WBC count.
3. Count all the cells in the five "R" areas (see figure 34.8) using the **high power** objective. Observe the same rules as for the WBC count pertaining to *line counts.*
4. Multiply the total count of five areas by 10,000 (dilution factor is 200, volume correction factor is 50).
5. Rinse the pipette with acid, water, alcohol, and acetone.

Laboratory Report

Record your results on parts B and D of Laboratory Report 34–38.

35 Hemoglobin Percentage Measurement

Since hemoglobin is the essential oxygen-carrying ingredient of erythrocytes, its quantity in the blood determines whether or not a person is anemic. We learned in Exercise 34 that one may have a normal RBC count and still be anemic if the erythrocytes are smaller than normal. Such a condition, which is called *microcytic anemia,* results in one having an insufficient amount of hemoglobin. In other cases, a normal count in which the erythrocytes have a low hemoglobin percentage *(hypchromic cells)* may also result in anemia. Thus, it is apparent that the amount of hemoglobin present in a unit volume of blood is the critical factor in determining whether or not an individual is anemic.

Determination of the hemoglobin content of blood can be made by various methods. One of the oldest methods and, incidentally, a most inaccurate one, is the Tallqvist scale. In this technique a piece of blotting paper that has been saturated with a drop of blood is compared with a color chart to determine the percentage of hemoglobin. Very few, if any, physicians utilize this method today. A rapid, accurate method is to insert a cuvette of blood into a photocolorimeter that is calibrated for blood samples. Such a method, however, requires a rather expensive piece of electronic equipment. It also requires a considerable quantity of blood.

A relatively inexpensive instrument for hemoglobin determinations is the *hemoglobinometer.* Such a device compares a hemolyzed sample of blood with a color standard. Figures 35.1 through 35.4 illustrate the major steps to follow in using the American Optical hemoglobinometer *(Hb-Meter).* It is this piece of equipment that we will use in this test in determining hemoglobin content of a blood sample.

The hemoglobin content of blood is expressed in terms of grams per 100 ml. of blood. Three different standards are used, depending on the community where the test is performed. For our purposes we will use the 15.6 gms./100 ml. as standard. For adult males, the normal on this scale is 14.9 ± 1.5 gms./100 ml.; for adult females, 13.7 ± 1.5; for children at birth, 21.5 ± 3; for children at 1 year, 12 ± 1.5 and for children at 4 years, 13 ± 1.5 gms./100 ml. A conversion scale exists on the side of the A/O Hb-Meter which allows one to determine hemoglobin percentages from the grams Hb/100 ml. Proceed as follows to determine your own hemoglobin percentage.

Materials:

American Optical Hb-Meter
hemolysis applicators
lancets
cotton
alcohol

1. Disassemble the blood chamber by pulling the two pieces of glass from the metal clip. Note that one piece of glass has an H-shaped moat cut into it. This piece will receive the blood. The other piece of glass has two flat surfaces and serves as a cover plate.
2. Clean both pieces of glass with alcohol and Kimwipes. Handle by edges to keep clean.
3. Reassemble the glass plates in the clip so that the grooves on the moat plate face the cover plate. The moat plate should be inserted only halfway to provide an exposed surface to receive the drop of blood. See figures 35.1 and 35.2.

4. Disinfect and puncture the finger with a disposable lancet.
5. Place a drop of blood on the exposed surface of the moat plate, as shown in figure 35.1.
6. Hemolyze the blood on the plate by mixing the blood with the pointed end of a hemolysis applicator as shown in figure 35.2. It will take 30–45 seconds for all the red blood cells to rupture. Complete hemolysis has occurred when the blood loses its cloudy appearance and becomes a transparent red liquid.
7. Push the moat plate in flush with the cover plate and insert the sample into the side of the instrument, as in figure 35.3.
8. Place the eyepiece to your eye with the left hand in such a manner that the left thumb rests on the light switch button on the bottom of the hemoglobinometer.
9. While pressing the light button with the left thumb, move the slide button on the side of the instrument back and forth with the right index finger until the two halves of the split field match. The index mark on the slide knob indicates the grams of hemoglobin per 100 ml. of blood. Read the Percent Hemoglobin on the 15.6 scale.

Laboratory Report

Complete the portion of the Laboratory Report for this exercise.

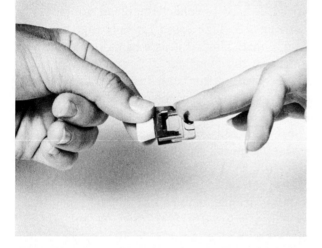

Figure 35.1 A fresh drop of blood is added to the moat plate of the blood chamber assembly. The blood must flow freely. Avoid squeezing the finger excessively.

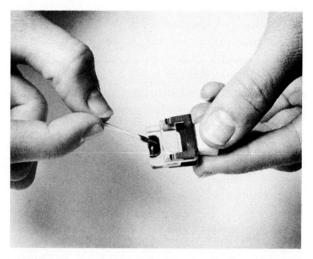

Figure 35.2 The blood sample is hemolyzed on the moat plate with a wooden hemolysis applicator. Thirty-five to forty-five seconds are required for complete hemolysis.

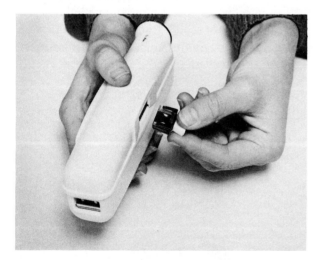

Figure 35.3 The charged blood chamber is inserted into the slot of hemoglobinometer. Before insertion, the unit should be tested to make sure the batteries are active.

Figure 35.4 Blood sample is analyzed by moving slide button with right index finger. When the two colors match in density, the grams / 100 ml. is read on the scale.

36 Packed Red Cell Volume

In Exercises 34 and 35, we employed two different tests for determining the presence or absence of anemia: (1) the RBC count and (2) the hemoglobinometer. We have seen that anemia may be due to a dilution of the number of erythrocytes or to a deficiency of hemoglobin. It was observed that the most significant factor is the amount of hemoglobin that is present in a given volume of blood. A third method that can be used in anemia diagnosis is to determine the volume percentage of blood that is occupied by the red blood cells that have been packed by centrifugation. This method is called the **VPRC**, or volume of packed red cells.

The VPRC is determined by centrifuging a blood sample in a special centrifuge tube called a *hematocrit,* or in a special type of capillary tubing. The centrifuge must be specifically designed to accommodate the hematocrit or capillary tubing. The speed of the centrifuge and the radius of its head must fall within rigid specification standards. The mathematical relationship of centrifuge head radius and speed is as follows:

$$RPM = \frac{202,146,700}{r \ (cm.)}$$

In this exercise, we will utilize a micro method for determining the VPRC. As illustrated in figure 36.1, only a drop of blood is needed to perform the test. Capillary tubing, which has been heparinized, is used to collect the blood sample. It is centrifuged at high speed for only four minutes. (Macro methods that utilize hematocrits, usually centrifuge the blood for 30 minutes at a much slower speed.) After centrifugation, the percentage of total volume occupied by the packed red blood cells is read on a special tube reader (figure 36.6), or directly on the head of the centrifuge, depending on the type of centrifuge.

An interesting relationship exists between grams of hemoglobin per 100 ml. and the VPRC: *the VPRC is usually three times the gm. Hb/100 ml.* In men, the normal VPRC range is between 40% and 54%, with 47% as average. In women, the normal range is between 37% and 47%, with 42% as average. The great advantages of this method over the two previous methods are (1) simplicity, (2) speed, and (3) high degree of accuracy. Although this micro method is not as accurate as macro methods, it still functions in an

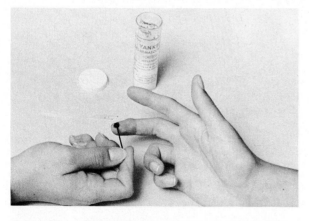

Figure 36.1 Blood is drawn up into heparinized capillary tube for hematocrit determination.

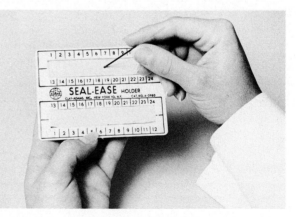

Figure 36.2 The end of the capillary tube is sealed with clay.

accuracy range of ±2% for most blood samples. Proceed as follows to determine the VPRC of your own blood.

Materials:

    lancets
    cotton
    alcohol
    micro-hematocrit centrifuge
    tube reader
    heparinized capillary tubes
    Clay-Adams Seal-Ease

1. Produce a free flow of blood on the finger. Wipe away the first drop of blood.
2. Place the marked end (red) of the capillary tube into the drop of blood and allow the blood to be drawn up about two-thirds of the way into the tube. Holding the open end of the tube

downward from the blood source will cause the tube to fill more rapidly.
3. Seal the blood end of the tube with Seal-Ease. See figure 36.2.
4. Place the tube into the centrifuge with the sealed end against the ring of rubber at the circumference. Load the centrifuge with even number of tubes (2, 4, 6, etc.) to properly balance the load.
5. Secure the inside cover with a wrench (figure 36.4) and fasten down the outside cover.
6. Turn on the centrifuge, setting the timer for **four minutes.**
7. Determine the VPRC by placing the tube in the mechanical tube reader. Instructions for reading are on the instrument.

Laboratory Report

Record your results on table II, Part D, of Laboratory Report 34–38.

Figure 36.3 The capillary tubes are placed in the centrifuge with the sealed end toward the perimeter.

Figure 36.4 The safety lid is tightened with a lock wrench.

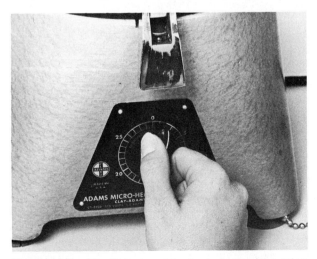

Figure 36.5 The timer is set for four minutes by turning the dial clockwise.

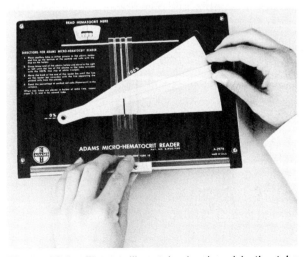

Figure 36.6 The capillary tube is placed in the tube reader to determine the hematocrit.

37 Coagulation Time

The coagulation of blood is a complex phenomenon involving over thirty substances. The majority of these substances inhibit coagulation and are called *anticoagulants;* the remainder, which promote coagulation, are designated as *procoagulants.* Whether or not the blood will coagulate depends on which group predominates in a given situation. Normally, the anticoagulants predominate, but when a vessel is ruptured, the procoagulants in the affected area assume control, causing a clot to form in a relatively short period of time.

Although considerable research has been devoted to the series of reactions that occurs in blood clotting, some gaps still exist in our understanding of the entire process. Almost all hematologists agree, however, on the following sequence of events:

First, a **prothrombin activator** is released by a damaged blood vessel or fractured blood platelets. Secondly, **prothrombin,** a plasma alpha globulin, is converted to the enzyme **thrombin** by the prothrombin activator and calcium ions. Finally, the **thrombin** converts **fibrinogen** to **fibrin.**

It is obvious from the above series of events that three substances must be present in the plasma: prothrombin, fibrinogen, and calcium ions. The prothrombin and fibrinogen are secreted by the liver into the blood; calcium is a normal constituent of the plasma. Prothrombin is an unstable protein which readily breaks down into thrombin and other compounds in the presence of calcium and the activator. The action of thrombin on fibrinogen is to remove two peptides from each molecule to form a single molecule of **fibrin monomer.** As soon as fibrin monomers contact each other, they polymerize to form the characteristic long fibrin threads that make up the blood clot.

Of the various methods that have been devised to evaluate the rate of blood coagulation, the

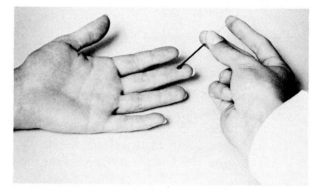

Figure 37.1 Blood is drawn up into a nonheparinized capillary tube.

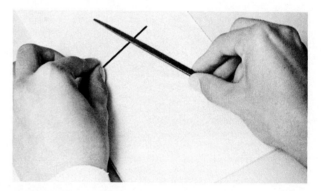

Figure 37.2 At one minute intervals, sections of the tube are filed and broken off for the test.

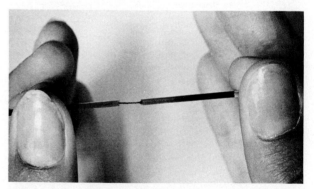

Figure 37.3 When the blood has coagulated, strands of fibrin will extend between broken ends of tube.

one outlined here is the simplest and most reliable. As illustrated in figures 37.1 through 37.3, blood is allowed to coagulate in a capillary tube. By breaking the tube at intervals of one minute, the approximate time of fibrin formation can be determined. Proceed as follows to determine the clotting time of your own blood.

Materials:

    lancets
    cotton
    alcohol
    capillary tubes (0.5 mm. diam.)
    3-cornered file

1. Puncture the finger to expose a free flow of blood. **Record the time.** Place one end of the capillary tube into the drop of blood. Hold the tube so that the other end is lower than the drop of blood so that the force of gravity will aid the capillary action.

2. At one-minute intervals, break off small portions of the tubing by scratching the glass with a file first. *Important: Separate the broken ends slowly and gently* while looking for coagulation. Coagulation has occurred when threads of fibrin span the gap between the broken ends. **Record the time** as that time from which the blood first appeared on the finger to the formation of fibrin.

Laboratory Report

Record your results on table II, Part E, of Laboratory Report 34–38.

38 Blood Typing

Red blood cells may contain various types of proteins which have been designated as A, B, O, C, D, E, c, d, e, M, N, etc. The presence or absence of these various proteins determines the type of blood possessed by an individual. Since the presence of specific proteins is genetic, an individual's blood type is the same in old age as at birth. It never changes. The only factors that we are concerned with here in this exercise are the A, B, O, and D(Rh) factors since they are most commonly involved in transfusion reactions.

To determine an individual's blood type, drops of blood typing sera are added to suspensions of red blood cells to detect the presence of **agglutination** (clumping) of the cells. ABO typing may be performed at room temperature with saline suspensions of red blood cells as shown in figure 38.1. Rh typing (D factor), on the other hand, requires higher temperatures (around 50°C.) and whole blood instead of diluted blood. For ABO typing, the diluted blood procedure is preferable. For convenience, however, the warming box method may be used for combined ABO and Rh typing.

ABO Blood Typing

Materials:

> small vial (10 mm. diam. × 50 mm. long)
> disposable lancets (*B-D Microlance, Serasharp, etc.*)
> 70% alcohol and cotton
> wax pencil and microscope slides
> typing sera (anti-A and anti-B)
> applicators or toothpicks
> saline solution (0.85%)
> 1 ml. pipettes

1. Mark a slide down the middle with a marking pencil, dividing the slide into two halves (see figure 38.1). Write *anti-A* on the left side and *anti-B* on the right side.
2. Pour approximately 1 ml. of saline solution into a small vial or test tube.
3. Scrub the middle finger with a piece of cotton saturated with 70% alcohol and pierce it with a sterile disposable lancet. Allow two or three drops of blood to mix with the saline by holding the finger over the end of the vial and washing it with the saline by inverting the tube several times.
4. Place a drop of this red cell suspension on each side of the slide.
5. Add a drop of anti-A serum to the left side of the slide and a drop of anti-B serum to the right side. *Do not contaminate the tips of the serum pipettes with the material on the slide.*
6. After mixing each side of the slide with *separate* applicators or toothpicks, look for agglutination. The slide should be held about 6″ above an illuminated white background and rocked gently for two or three minutes. Record your results on the Laboratory Report as of three minutes.

Combined ABO and Rh Typing

As stated, Rh typing must be performed with heat on blood that has not been diluted with saline. A warming box such as the one in figure 38.2 is essential in this procedure. In performing this test, two factors are of considerable importance: first, only a small amount of blood must be used, (a drop of about 3 mm. diam. on the slide) and second, proper agitation must be executed. The agglutination that occurs in this antibody-antigen reaction results in finer clumps; therefore, closer examination is essential. If the agitation is not properly performed, agglutination may not be as

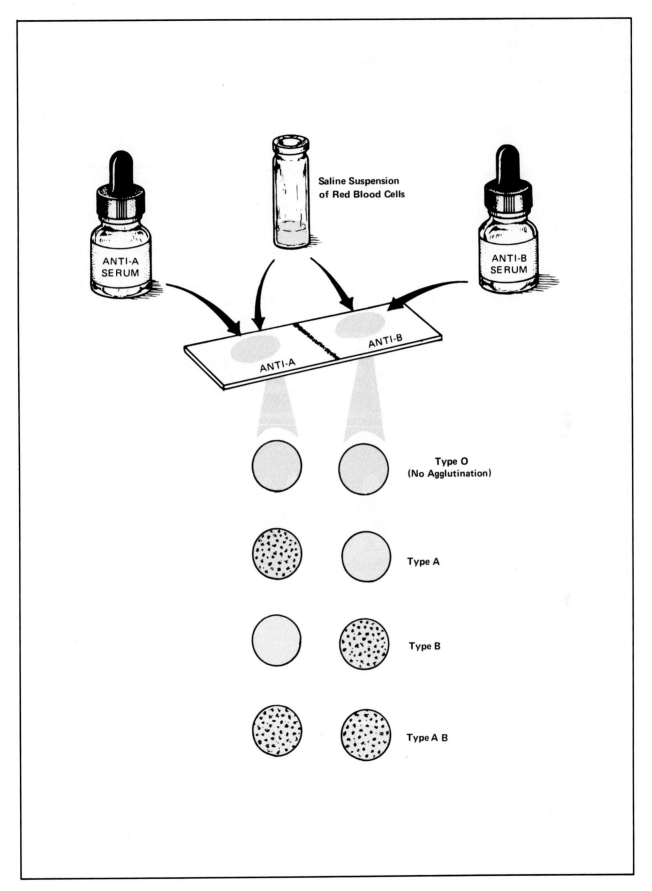

Figure 38.1 Blood-typing ABO groups.

apparent as it should be. In this combined method, we will use the whole blood for the ABO typing also. Although this method works out satisfactorily as a classroom demonstration for the ABO groups, it is *not as reliable* as the previous method in which saline and room temperature are used, *and should not be used clinically.*

Materials:

slide warming box with a special marked
    slide
anti-A, anti-B, and anti-D typing sera
applicators or toothpicks
70% alcohol
cotton
disposable sterile lancets *(B-D Microlances, Sera-sharp, etc.)*

1. Scrub the middle finger with a piece of cotton saturated with 70% alcohol and pierce it with a sterile disposable lancet. Place a small drop in each of three squares on the marked slide on the warming box. To get the proper proportion of serum to blood, do not use a drop larger than 3 mm. diameter on the slide.

2. Add a drop of anti-D serum to the blood in the anti-D square, mix with a toothpick, and note the time. **Only two minutes should be allowed for agglutination.**

3. Add a drop of anti-B serum to the anti-B square and a drop of anti-A serum to the anti-A square. Mix the sera and blood in both squares with *separate fresh* toothpicks.

4. Agitate the mixtures on the slide by slowly rocking the box back and forth on its pivot. At the end of two minutes, examine the anti-D square carefully for agglutination. If no agglutination is apparent, consider the blood to be Rh negative. By this time, the ABO type can also be determined.

Laboratory Report

Complete Laboratory Report 34–38.

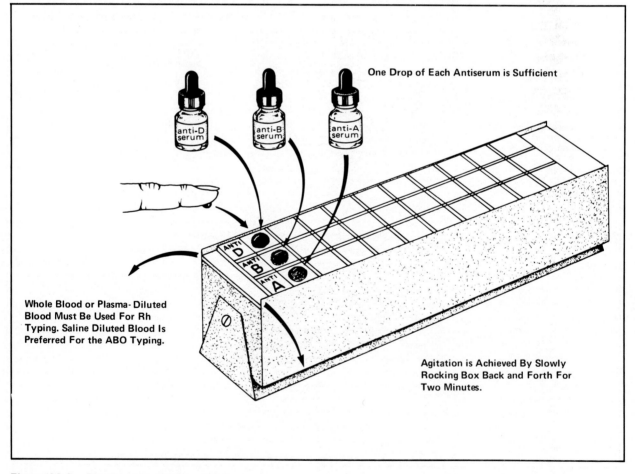

One Drop of Each Antiserum is Sufficient

Whole Blood or Plasma-Diluted Blood Must Be Used For Rh Typing. Saline Diluted Blood Is Preferred For the ABO Typing.

Agitation is Achieved By Slowly Rocking Box Back and Forth For Two Minutes.

Figure 38.2 Blood-typing with warming box.

Part **9** The Circulatory System

This unit consists of nine exercises. Four of these exercises are anatomical in nature; five are physiological. The anatomical exercises (39, 44, 45, and 46) pertain to the heart, arteries, veins, lymphatic circulation, and fetal circulation. The physiological exercises (40, 41, 42, 43, and 47) pertain to cardiovascular sounds, ECG monitoring, pulse monitoring, peripheral circulation control, and blood pressure monitoring.

Since so much of circulation physiology pertains to the heart, considerable emphasis is placed here on the anatomy of this organ. It is anticipated that with the completion of Exercise 39 (Heart Anatomy) the student should not only be able to proceed more intelligently with the circulation experiments in this section, but also acquire a clearer understanding of the goals of much of today's open heart surgery. Many of the experiments in this unit are of practical clinical value.

39 Anatomy of the Heart

In this study of the anatomy of the heart we will use a sheep heart for dissection, since its size and anatomy are so similar to the human heart. Before beginning the dissection, however, the anatomy of the heart will be studied in figures 39.1 and 39.2.

Internal Anatomy

Figure 39.1 reveals the internal anatomy of the human heart. The blue and red colors of the vessels signify whether those vessels carry oxygenated or deoxygenated blood; red signifies oxygenated blood and blue represents deoxygenated blood.

Chambers of the Heart

Note that the heart has four chambers: two small upper atria and two larger ventricles. The atria receive all the blood that enters the heart and the ventricles pump all the blood out. Note that the **right atrium** in the frontal section is on the left side of the illustration and the **left atrium** is on the right side. Observe, also, that the **right ventricle** is somewhat larger than the **left ventricle,** but that the left ventricle has a thicker wall. Separating the two ventricles is a partition, the **ventricular septum.**

Wall of the Heart

The wall of the heart consists of three layers: the myocardium, the endocardium, and the epicardium. The **myocardium** is the muscular portion of the wall that is composed of cardiac muscle tissue. Note in the enlarged section of the wall that the myocardium makes up the bulk of the wall thickness.

Lining the inner surface of the heart is the **endocardium.** It is a thin serous membrane that is continuous with the endothelial lining of the arteries and veins. An infection of this membrane is called *endocarditis.*

Attached to the outer surface of the heart is another serous membrane, the **epicardium,** or **visceral pericardium.** The production of serous fluid by this membrane on the outer surface of the heart enables the heart to move freely within the pericardial sac. Note that the **pericardial sac,** or **parietal pericardium,** consists of two layers: an inner serous layer and an outer fibrous layer. Between the heart and the parietal pericardium is the **pericardial cavity** (label 19). (Its dimension is exaggerated here.)

Vessels of the Heart

The major vessels of the heart are the two vena cavae, the pulmonary artery, the four pulmonary veins, and the aorta. The **superior vena cava** is the upper blue vessel on the right atrium; it conveys deoxygenated blood to the right atrium from the head and arms. The **inferior vena cava** is the lower blue vessel that empties deoxygenated blood from the trunk and legs into the right atrium. From the right atrium blood passes to the right ventricle, where the blood leaves the heart through the **pulmonary artery.** This artery branches into the **right** and **left pulmonary arteries,** which carry blood to the lungs.

Blood is drained from the lungs by means of four **pulmonary veins** (4 small red vessels), which carry it to the left atrium. This blood passes into the left ventricle and finally out through the **aorta.** The initial emergence of the aorta is obscured, but it can be seen as the large red vessel at the top of the heart. The aorta carries oxygenated blood to the systemic circulation.

Note what appears to be a short vessel between the pulmonary artery and the aorta. It is called the **ligamentum arteriosum.** During prenatal life this structure is a functional blood vessel, the *ductus arteriosum,* that allows blood to pass from the pulmonary artery to the aorta. After birth, when breathing begins, this vessel becomes nonfunctional and is converted to connective tissue.

_____ Anterior Interventricular Artery

_____ Aorta

_____ Aortic Semilunar Valve

_____ Apex

_____ Bicuspid Valve

_____ Chordae Tendineae

_____ Circumflex Artery

_____ Coronary Sinus

_____ Endocardium

_____ Inferior Vena Cava

_____ Left Atrium

_____ Left Coronary Artery

_____ Left Pulmonary Artery

_____ Left Ventricle

_____ Ligamentum Arteriosum

_____ Myocardium

_____ Opening to Coronary Arteries

_____ Opening to Coronary Sinus

_____ Papillary Muscle

_____ Parietal Pericardium

_____ Pericardial Cavity

_____ Pulmonary Artery

_____ Pulmonic Semilunar Valve

_____ Pulmonary Veins

_____ Right Atrium

_____ Right Coronary Artery

_____ Right Pulmonary Artery

_____ Right Ventricle

_____ Superior Vena Cava

_____ Tricuspid Valve

_____ Ventricular Septum

_____ Visceral Pericardium

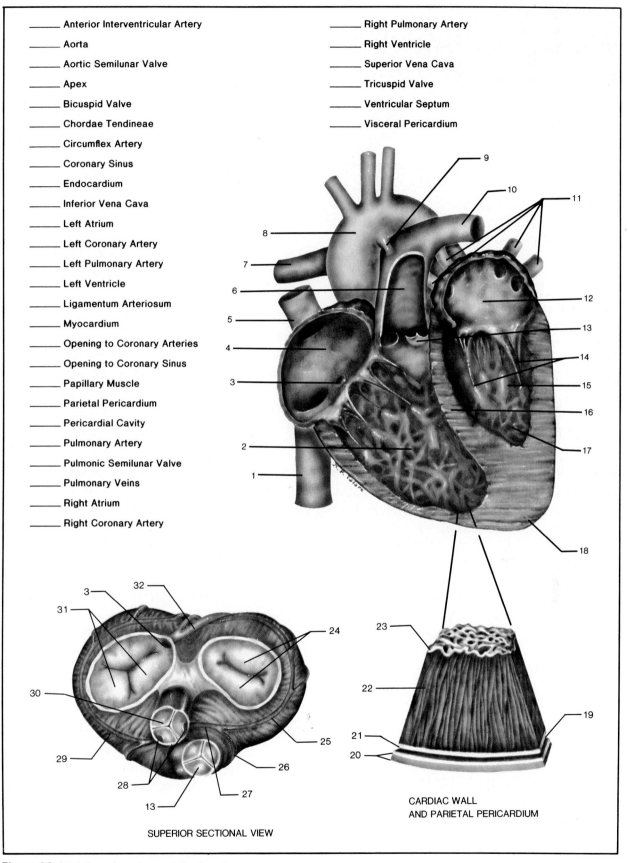

SUPERIOR SECTIONAL VIEW

CARDIAC WALL
AND PARIETAL PERICARDIUM

Figure 39.1   Internal anatomy of the heart.

Valves of the Heart

The heart has two atrioventricular and two semi-lunar valves. Between the right atrium and right ventricle is the **tricuspid valve.** Between the left atrium and the left ventricle is the **bicuspid valve,** or *mitral valve.* The differences between these two atrioventricular valves are seen in the superior sectional view of figure 39.1. Note that the tricuspid valve has three flaps or cusps, and the bicuspid valve has only two cusps. Observe, also, that the edges of the cusps have fine cords, the **chordae tendineae,** which are anchored to **papillary muscles** on the wall of the heart (see frontal section). These cords and muscles prevent the cusps from being forced up into the atria during systole (ventricular contraction).

The other two valves are the **pulmonic semilunar** and **aortic semilunar valves.** They are located at the bases of the pulmonary artery and the aorta. They prevent blood in those vessels from flowing back into the heart during diastole (relaxation phase). Note in the superior sectional view that each of these valves has three small cusps, and that they are shown closed for purposes of illustration.

The Coronary Circulation

The vessels that supply the heart muscle with blood comprise the *coronary circulatory system.* Oxygenated blood in the aorta passes into right and left coronary arteries through two small openings at a point just superior to the aortic semilunar valve. These **openings to the coronary arteries** are seen in the wall of the aorta in the superior sectional view in figure 39.1. The **left coronary artery** has two principal branches: a **circumflex artery** that passes around the heart in the left atrioventricular sulcus and the **anterior interventricular artery,** which lies in the interventricular sulcus on the anterior surface of the heart. The **right coronary artery** lies in the right atrioventricular sulcus and has branches that supply the posterior and anterior surfaces of the ventricular muscle.

Once the coronary blood has been relieved of its oxygen and nutrients it is picked up by various veins that parallel the arteries and empty into the **coronary sinus** (label 2, posterior aspect, figure 39.2). Blood in the coronary sinus is emptied into the right atrium. Its point of entry can be seen in the superior sectional and frontal sectional views of figure 39.1.

Assignment:

Label figure 39.1.

External Anatomy

The study of the external anatomy of the heart is relatively simple once the internal anatomy is understood. Figure 39.2 reveals two external views of the heart. The upper illustration shows the anterior aspect of the heart as it lies in the mediastinum between the two lungs. The lower illustration is of the posterior surface.

Anterior Aspect

Note that the heart lies within the **mediastinum** at a slight tilt to the left so that the **apex,** or tip of the heart, is somewhat on the left side. The obscured appearance of the coronary vessels in this view is due to the fact that the **parietal pericardium** lies intact over the organ. The clarity of the structures on the posterior view is due to the fact that the parietal pericardium has been removed.

The atria in both of these views are colored to designate the state of oxygen content in those chambers: thus, a blue right atrium and a red left atrium. When the sheep heart is dissected, these color distinctions will be lacking.

Differentiating the right ventricle from the left ventricle is best determined by locating the *intraventricular sulcus* first. This sulcus is a slight depression over the ventricular septum that contains the **anterior interventricular artery** and the **great cardiac vein.** The yellow material seen in this sulcus and other areas of the heart consists of fatty deposits. The left ventricle is on the left side of the interventricular sulcus, and the right ventricle is to the right of it.

Observe that the aorta forms an **aortic arch** over the heart and descends behind the heart. That part that is near the heart is called the **ascending aorta.** The part that passes down behind the heart is called the **descending aorta.** Only a short portion of the descending aorta is seen at the bottom of the illustration.

Locate the right and left pulmonary arteries, which emerge from under the aortic arch. Also,

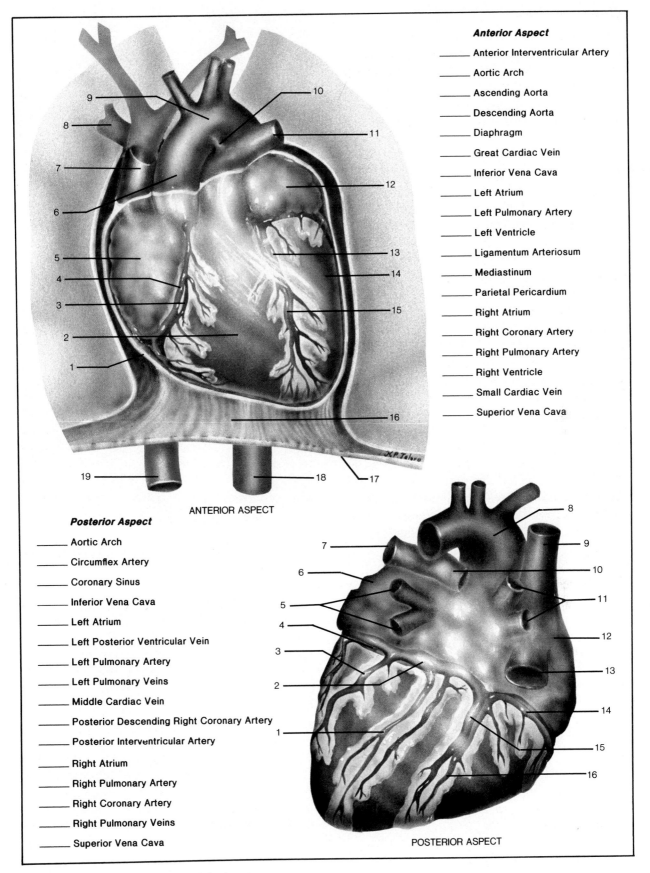

Figure 39.2 External anatomy of the heart.

identify the ligamentum arteriosum between the pulmonary artery and the aorta. The superior vena cava shows up well near the top of the right atrium. Only a short portion of the inferior vena cava is seen at the bottom of this view. The **right coronary artery** and the **small cardiac vein** are visible in the atrioventricular sulcus between the right atrium and right ventricle.

Posterior Aspect

This view of the heart in figure 39.2 reveals more clearly the position of the four pulmonary veins. The two right pulmonary veins are located near the vena cavae. Note how close together the vena cavae are located. The appearance of a single blue vessel lying below the aorta is actually the site of division of the pulmonary artery into right and left branches.

The following coronary arteries are seen on this side of the heart: (1) the **right coronary artery,** which lies in the right atrioventricular sulcus; (2) the **posterior descending right coronary artery,** which is a downward extension of the right coronary artery; (3) the **posterior interventricular artery** (label 1); and (4) the **circumflex artery.** The major coronary veins on this side are: (1) the **middle cardiac vein,** which parallels the posterior descending right coronary artery; (2) the **left posterior ventricular vein** (label 3); and (3) the **posterior interventricular vein.** All these coronary veins empty into the **coronary sinus** (label 2).

Assignment:

Label figure 39.2.

Sheep Heart Dissection

As stated earlier, the sheep heart makes an excellent dissection substitute for the human heart since it is very similar in structure. Our purpose in this dissection is to identify as many of the structures as possible that were studied in figures 39.1 and 39.2.

Materials:

sheep heart, fresh or preserved
dissecting instruments and tray

1. Rinse the heart with cold water to remove excess preservative or blood. Allow water to flow through the large vessels to irrigate any blood clots out of its chambers.

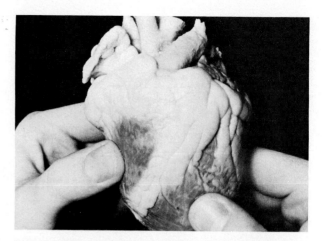

Figure 39.3   The anterior surface of the sheep heart. The left ventricular wall feels firmer than the right when squeezed.

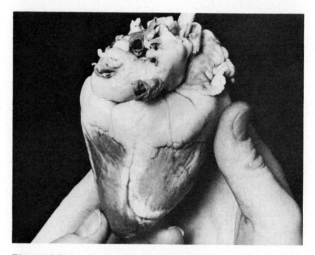

Figure 39.4   The posterior aspect of the heart. Note the four thin-walled pulmonary veins protruding from the fatty area.

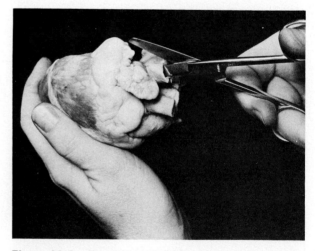

Figure 39.5   The first cut is started in the superior vena cava and extended down through the right atrium and right ventricle.

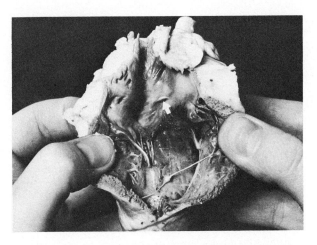

Figure 39.6 The interior of the heart as revealed by the first cut. White arrow points to coronary sinus opening.

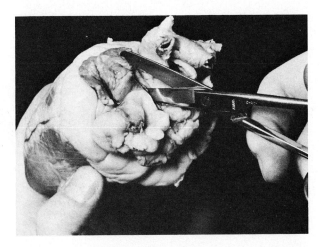

Figure 39.7 The second cut is started in the left atrium and extended down into the left ventricle.

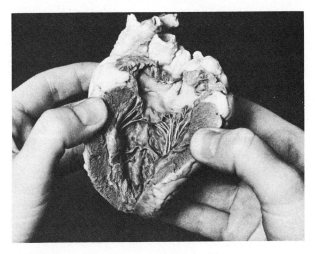

Figure 39.8 The interior of the heart as seen after the second cut. Note how much thicker the myocardium is in this section than that seen in figure 39.6.

2. Look for evidence of the **parietal pericardium,** which surrounds the heart. This fibroserous membrane is usually absent from laboratory specimens, but there may be remnants of it attached to the large blood vessels of the heart.

3. Attempt to isolate the **visceral pericardium** *(epicardium)* from the outer surface of the heart. Since it consists of only one layer of squamous cells it is very thin. By using a sharp scalpel you may be able to peel a small portion of it away from the myocardium.

4. Identify the **right** and **left ventricles** by squeezing the walls of the heart, as shown in figure 39.3. The right ventricle will have a thinner wall since it pumps blood only through the lungs, where resistance is low. The thicker wall of the left ventricle is needed to force blood to the head, arms, trunk, and legs.

5. Identify the **anterior interventricular sulcus** (between the two ventricles), which is usually covered with fatty tissue. Carefully trim away the fat from this sulcus to expose the **anterior interventricular artery** and the **great cardiac vein.**

6. Locate the thin-walled **right** and **left atria** of the heart. The right atrium is seen on the left side of figure 39.3, straight up from the left thumb.

7. Identify the **aorta,** which is the large vessel just to the right of the right atrium as seen in figure 39.3. It may be necessary to *very carefully* peel away the excess fat in this area to expose the artery, taking care not to damage the **ligamentum arteriosum.**

8. Locate the **pulmonary artery,** which is the large vessel between the aorta and the left atrium as seen when looking at the anterior side of the heart. If the vessel is of sufficient length, trace it to where it divides into the **right** and **left pulmonary arteries.**

9. Examine the posterior surface of the heart. It should appear as in figure 39.4. Note that only the right atrium and two ventricles can be seen on this side. The left atrium is obscured by fat from this aspect. Look for the four thin-walled **pulmonary veins** that are embedded in the fat. Probe into these vessels and you will see that they lead into the left atrium.

10. Locate the **superior vena cava,** which is attached to the upper part of the right atrium.

Insert one blade of your dissecting scissors into this vessel, as shown in figure 39.5, and cut through it into the atrium to expose the **tricuspid valve** between the right atrium and right ventricle. Don't cut into the ventricle at this time.

11. Fill the right ventricle with water, pouring it in through the tricuspid valve. Gently squeeze the walls of the ventricle to note the closing action of the cusps of this valve.

12. Drain the water from the heart and continue the cut with scissors from the right atrium through the tricuspid valve down to the apex of the heart.

13. Open the heart and flush it again with cold water. Examine the interior. The open heart should appear as in figure 39.6.

14. Examine the interior wall of the right atrium. This inner surface has ridges, giving it a comblike appearance; thus, it is called **pectinate muscle** (*pecten:* comb). Between the inferior vena cava and the tricuspid valve you should see the **opening to the coronary sinus.** It is through this opening that blood of the coronary circulation is returned to the venous circulation. This opening can be seen in figure 39.6.

15. Insert a probe under the cusps of the mitral valve. Are you able to see three separate flaps?

16. Locate the **papillary muscles** and **chordae tendineae.** How many papillary muscles do you see in the right ventricle? Identify the **moderator band,** which is a reinforcement cord

between the ventricular septum and the ventricular wall. Its presence prevents excessive stress from occurring in the myocardium of the right ventricle.

17. With scissors, cut the right ventricular wall up along its lower margin parallel to the anterior interventricular sulcus to the pulmonary artery. Continue the cut through the exit of the right ventricle into the pulmonary artery. Spread the cut surfaces of this new incision to expose the **pulmonic semilunar valve.** Wash the area with cold water to dispel blood clots. Study this semilunar valve with a probe, identifying its three cusps.

18. Insert one blade of your scissors into the left atrium, as shown in figure 39.7. Cut through the atrium into the left ventricle. Also, cut from the left ventricle into the aorta, slitting this vessel longitudinally.

19. Examine the **bicuspid valve,** which lies between the left atrium and left ventricle. With a probe, identify the two cusps.

20. Examine the pouches of the **aortic semilunar valve.** Compare the structure and number of pouches of this valve with the pulmonic valve.

21. Look for the two **openings to the coronary arteries,** which are in the walls of the aorta just above the aortic semilunar valve. Force a blunt probe into each hole. Note how they lead into the two coronary arteries.

Laboratory Report

Complete the Laboratory Report for this exercise.

Cardiovascular Sounds

The deceptively simple nature of heart sounds belie their actual diagnostic value. To the cardiologist, a wealth of information can be derived by listening for (1) variations in the rate and rhythm of the beat; (2) the intensity and nature of the sounds; and (3) the presence of murmurs and other extraneous sounds. Although it is beyond the scope of this course to delve into a complete analysis and interpretation of cardiovascular sounds, a generalized study will be made.

Three Sounds

Ordinarily, two heart sounds are produced with each cardiac cycle. In children and young adults, a third sound also occurs, but this sound disappears with aging. Cardiovascular sounds are created by the contraction of the myocardium and the movement of the valves. They can be monitored with a stethoscope, audio monitor, or phonocardiograph.

The **first sound** (lub) coincides with ventricular systole and accompanies the cardiac impulse, or apex beat. This sound is low, dull, and lasts longer than the second sound. It is caused by the action of the ventricular myocardium and the simultaneous closure of the *mitral* and *tricuspid valves.* The intensity of the sound will vary with the heart rate due to the change in the abruptness of valve closure. The first sound usually lasts about 150 milliseconds.

The **second sound** (dup) follows the first sound after a brief pause. This sound has a snapping quality of higher pitch and shorter duration. The sound is created by the sudden, virtually simultaneous, closure of the *aortic* and *pulmonic valves* at the end of ventricular systole.

During inspiration, a *splitting of the second sound* is often heard. This is perfectly normal. This phenomenon is caused by the increased venous return of blood to the right side of the heart. Since the right ventricle gets more blood during inspiration, it needs more time to pump its contents to the lungs; thus, the pulmonic valve closure is delayed. The duration of the second sound is approximately 120 milliseconds.

The **third sound** is lower pitched and is loudest near the apex of the heart. It is detected more easily when the subject is recumbent. It appears to be caused by the vibration of the ventricular walls and the A-V valves during diastole. The vibration is probably caused by the rapid inrush of blood into the ventricles from the atriae.

Abnormalities

Abnormal heart sounds are, collectively, referred to as **murmurs.** They are often due to damaged valves. The disease that causes most heart damage is rheumatic fever, and the valve that is most often affected by this disease is the mitral valve. The aortic valve is second in susceptibility. A damaged valve may impede the flow of blood, causing a condition called *stenosis;* or it may not close properly, permitting backward flow, or *regurgitation,* of blood. Stenosed valves produce a nozzle effect that causes blood to squirt jet-like at high velocity through the valve during systole. The sound from a stenosed aortic valve is harsh and can sometimes be heard several feet from the patient. It can also be felt by simply placing the hand on the chest.

Auscultatory Areas

Listening to sounds of the body, usually with the aid of a stethoscope, is called *auscultation.* Figure 40.1 illustrates the areas of the chest wall from which the heart sounds from each valve are usually best auscultated. Although the heart sounds

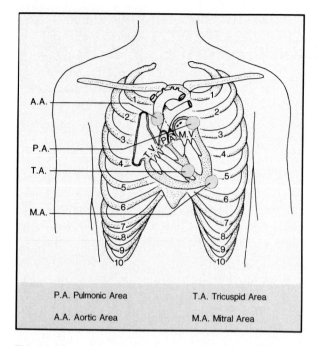

P.A. Pulmonic Area   T.A. Tricuspid Area

A.A. Aortic Area   M.A. Mitral Area

Figure 40.1 Auscultatory areas for heart sounds.

are transmitted to the entire precordium, the valvular sounds are not loudest directly over their anatomic locations; instead, the sounds are projected to specific areas as shown in figure 40.1. Note that the **mitral area** coincides with the region of the cardiac apex, which is at the intersection of the fifth intercostal space and the midclavicular line. The **tricuspid area** is about 2″ to 3″ medial and slightly superior to the mitral area at a point where the ensiform cartilage of the sixth rib joins the body of the sternum. The **aortic area** is on the right border of the sternum in the space between the second and third ribs. The **pulmonic area** is at approximately the same vertical level as the aortic area, only it is on the left side, about the same distance from the midline to the left.

Examination Methods

Three types of equipment will be used here to study the cardiovascular sounds: the stethoscope, an audio monitor, and the Unigraph. Room noise must be kept at a minimum when using the stethoscope. It will also be necessary to place the stethoscope next to the skin, since the sounds are usually too muffled through clothing.

The *audio monitor,* such as the one illustrated in figure 40.2, will be used first as a class demonstration to listen to these sounds. This device utilizes a small microphone transducer to pick up

the sounds. The unit contains a speaker, amplifier, and controls to filter out extraneous sounds.

A *phonocardiogram* will be produced by recording the different sounds on a Unigraph. Figure 40.3 illustrates the setup. Note that a special adapter is used to couple a microphone and pressure transducer to the Unigraph. Although the pressure transducer is not used in this experiment, it must be attached to the adapter to balance out the microphone electronically.

Proceed as follows to study the various cardiovascular sounds.

Materials:

> stethoscopes
> microphones
> audio monitor (Grass AM7)
> Unigraph
> adapter for Unigraph (Trans/Med 6605)
> Statham P23AA pressure transducer

Audio Monitor Demonstration The instructor will demonstrate the heart sounds of one member of the class to the entire group. The sensitivity of this setup is such that baring the chest is usually not necessary. After the demonstration, students may work with this unit in a separate room. Sounds from this unit are distracting to students using stethoscopes in the laboratory.

When attempting to demonstrate second sound splitting, hand signals should be used instead of verbalization. Instruct the subject to inspire when you raise your hand and to expire when you lower your hand.

Before the audio monitor is plugged in, set the controls as follows: Power OFF, Volume 0, Lo Filter 10 Hz., Hi Filter 15 Hz., Input OFF, Clipper OFF. Plug the microphone jack into socket #1.

During monitoring set the controls as follows: Power ON, Volume 0–4 as needed, Lo Filter up to 100 as needed, Hi Filter down to 100 as needed, and Clipper OFF. The volume and filter controls are adjusted accordingly to get the clearest sounds of sufficient amplitude.

When monitoring is completed, shut down the equipment by setting the controls as follows: Power OFF, Volume 0, Lo Filter 10 Hz., Input OFF, Clipper OFF. Unplug the unit.

Stethoscope Auscultation Work with a laboratory partner to auscultate each other's heart sounds. When fitting the ear pieces of the stethoscope into the ears, direct them inward and upward. Keep in mind that a cardiologist distinguishes the sounds from the different valves by a process of elimination; that is, the cardiologist moves the stethoscope from one area to another noting the loudness and characteristics of the sounds in different spots. Gradually, the cardiologist picks out the sound components from each valve.

Compare the sounds in sitting and supine positions. Can you locate the apex of the heart? Remember, the apex of the heart is closest to the body wall and raps against the precordium during the cardiac cycle.

Attempt to detect the splitting of the second sound during deep inspiration. Use hand signals instead of conversation: raised hand for inspiration, lowered hand for expiration.

Finally, compare the sounds before and after exercise, recording your observations on the Laboratory Report. For exercise, leave the room to run up and down the stairs, or jog for a short distance. Avoid making noise in the laboratory. If you encounter sounds that appear to be murmurs, consult with your laboratory instructor.

Unigraph Setup To produce a phonocardiogram of your heart sounds proceed as follows:

1. Check the Unigraph to see if a Trans/Med 6605 adapter is attached to the input end of the unit. If none is present, get one from the demonstration table and attach it to the unit.

2. Attach the Statham transducer and microphone jack to the 6605 adapter.

3. Plug in the power cord of the Unigraph and adjust the controls as follows: Power switch ON, chart control lever at STBY, heat control knob at 2:00 position, speed control lever at slow position, Gain at 1 MV/CM, and Mode on TRANS.

4. Place the microphone against one of the auscultatory areas and start the chart moving by placing the c.c. lever at Chart On. Adjust the temperature control knob to get a good recording. Center the tracing with centering control knob and adjust sensitivity to produce at least 5 mm. stylus displacement.

5. Move the speed control lever 180° to the high speed position and check the intensity of the line on the paper again, making any necessary temperature knob adjustments.

6. Make recordings over each auscultatory area. Be sure to record the specific valve names on the chart. Attach the phonocardiograms to the Laboratory Report and answer all questions.

7. When finished with this equipment, deactivate the Unigraph by setting the following controls: Power Switch OFF, heat control knob completely counterclockwise, Gain at 2 MV/CM, speed control lever at slow position.

Laboratory Report

Complete the Laboratory Report for this exercise.

Figure 40.2 The audio monitor with microphone is used for auscultating heart sounds.

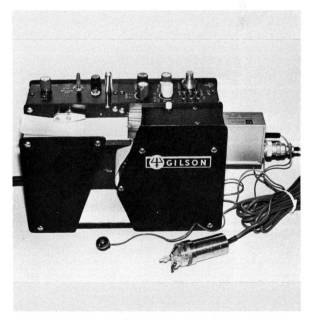

Figure 40.3 The Unigraph setup is used for producing a phonocardiogram of heart sounds.

41 The Electrocardiogram

It was observed in the last exercise that the SA node initiates depolarization of the atria, which spreads to the ventricles via the AV node, bundle of His, and Purkinje fibers. The electrical potential changes resulting from this depolarization and repolarization sequence can be monitored to produce a record called an **electrocardiogram.** The abbreviation for the electrocardiogram was first designated as EKG, a carryover from its German derivation. For a time in the United States, this abbreviation was widely used. It still is used to some extent; however, since instrument manufacturers in this country are using the abbreviation ECG, the anglicized abbreviation is more in vogue.

To produce an ECG, it is necessary to attach two or three electrodes to the body. If one electrode is placed slightly above the heart and to the right, and another electrode is placed slightly below the heart to the left, one will record the wave form at its maximum potential. It is along this axis that the electrical potential of the heart is generated.

In clinical practice, the cardiologist records several ECGs with the electrodes placed at different positions on the body. This is done for di-

agnostic reasons. For our purposes here, however, we will use only one hookup arrangement, as shown in figure 41.4: one on the right wrist, another on the left wrist, and the third one on the left ankle.

The wave form of a normal ECG is illustrated in figure 41.2. The initial depolarization of the SA node, which causes atrial contraction, manifests itself as the **P wave.** This depolarization is immediately followed by repolarization of the atria. Atrial repolarization is not usually seen with surface electrodes because it is masked by the overwhelming influence of ventricular depolarization. When implanted electrodes are used, a **TA wave** is evidence of atrial repolarization.

Within 120–220 msec. the electrical stimulation reaches the AV node, causing it to depolarize. The passage of the depolarization wave from the AV node down the bundle of His spreads into the myocardial tissue. The depolarization of the myocardial tissue is seen as the **QSR complex** of the ECG. Depolarization of the ventricles at this time results in ventricular systole. As soon as depolarization is completed here, repolarization takes place. Repolarization of the ventricles manifests

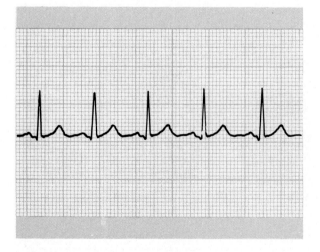

Figure 41.1 An electrocardiogram.

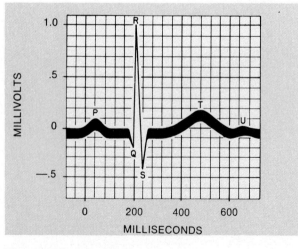

Figure 41.2 The ECG wave form.

itself as the **T wave.** Some ECG wave forms show an additional wave occurring after the T wave. It is designated as the **U wave.** This small wave is attributed to the gradual repolarization of the papillary muscles.

The value of the ECG to the cardiologist is immeasurable. Almost all serious abnormalities of the heart muscle can be detected by analyzing the contours of the different waves of the ECG wave form. Interpretation of ECG wave forms to determine the nature of heart damage is a highly-developed technical skill involving vector analysis and goes considerably beyond the scope of this course; however, figure 41.3 illustrates three examples of abnormal ECGs in which heart damage has occurred.

The Unigraph will be used in the experiment to produce an ECG. Students will work in teams of three or four, with one individual being the subject, and the others making the electrode hookups and operating the controls. Failure in this experiment usually results from poor skin contact, damaged cables, or incorrect hookup.

Materials:

 Unigraph and patient cable
 ECG electrodes (3)
 Scotchbrite pad
 70% alcohol
 electrode paste
 cot for subject (optional)

Unigraph Calibration

While one member of your team attaches the electrodes to the subject prepare the Unigraph as follows:

1. While the main power switch is at the OFF position, set the controls as follows: chart control switch at STBY, speed control lever at the slow position, gain selector knob at 1 MV/CM, sensitivity knob completely counter-clockwise, mode selector control at CC-Cal, and the Hi Filter and Mean switches on NORM.
2. Turn on the power switch and rotate the stylus heat control to the 2:00 position.
3. Place the chart control switch on CHART ON and observe the trace and center the stylus, using the centering control.
4. Increase the sensitivity by rotating the sensitivity control clockwise one-half turn.
5. Depress the 1 MV button, hold it down for about 2 seconds, and then release it. The upward deflection should measure 1 cm. The downward deflection should equal the upward deflection (figure 41.6). If the travel is not 1 cm., adjust the sensitivity control and depress the 1 MV button until exactly 1 cm. is achieved.

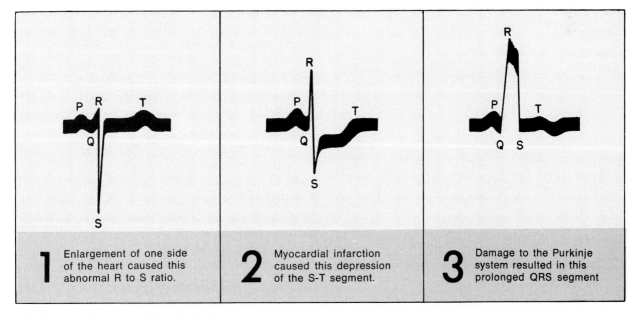

1. Enlargement of one side of the heart caused this abnormal R to S ratio.
2. Myocardial infarction caused this depression of the S-T segment.
3. Damage to the Purkinje system resulted in this prolonged QRS segment

Figure 41.3 Abnormal electrocardiograms.

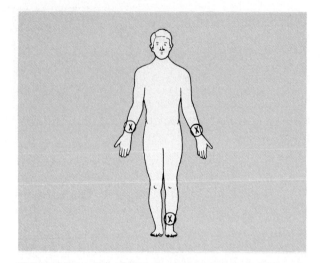

Figure 41.4 Electrodes are fastened to both wrists and the left ankle.

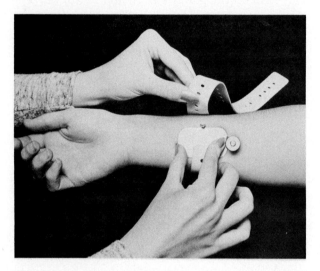

Figure 41.5 After applying electrode paste to the electrode, it is strapped in place.

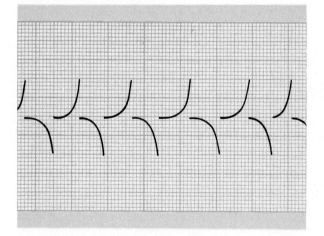

Figure 41.6 When the Unigraph is properly calibrated, the tracing should look like this.

6. Place the chart control switch on STBY and label the chart with the date and subject's name.
7. Place the mode selector control on ECG and move the speed selector lever to the high speed position (25 mm. per second). The Unigraph is now ready to accept the patient cable.

Preparation of Subject

While the equipment is being calibrated, the electrodes should be attached to the subject by other members of the team. Although it is not absolutely essential, it is desirable to have the individual recumbent on a comfortable cot.

To achieve good electrode contact, rub the contact areas of the two wrists and left ankle with a Scotchbrite pad, disinfect the skin with 70% alcohol, and add a little electrode paste to the flat surface of each electrode prior to attaching them to the limbs. Figure 41.5 illustrates how the rubber strap of each electrode is secured. When all three electrodes are in place, attach the leads of the cable to the electrodes. The ground lead (unnumbered, black) is attached to the ankle. Attach one numbered lead to the electrode on the right wrist and the other numbered lead to the left wrist. The subject is now ready for monitoring.

Monitoring

Turn on the Unigraph and observe the trace. Adjust the stylus heat control so that optimum recording is achieved. Compare the ECG with figures 41.1 and 41.2. Is the QRS spike up or down (positive or negative)? If it is negative (downward), reverse the leads to the two wrists.

After recording for about 30 seconds, stop the paper and study the ECG. Does the wave form appear normal? Determine the pulse rate by counting the spikes between two margin lines and multiplying by two.

Disconnect the leads to the electrodes and have the subject exercise for 2–5 minutes by leaving the room to run up and down stairs or jog outside the building. After reconnecting the electrodes, record for another 30 seconds and compare this ECG with the previous one. Save the record for attachment to the Laboratory Report.

Laboratory Report

Complete the Laboratory Report for this exercise.

When the ventricles of the heart undergo systole, a surge of blood flows into the arterial tree, which manifests itself as the **pulse** in the extremities. Since the rate and strength of the pulse indicate cardiovascular function, physicians always monitor the pulse in routine medical examinations. This physiological parameter is usually determined simply by placing the fingers over one of the subject's arteries in the wrist or neck region.

There are several types of transducers that have been developed for monitoring the pulse. In this exercise, we will utilize a photoelectric plethysmograph on the index finger, as illustrated in figure 42.1. The recording produced with this transducer is called a **plethysmogram.** The details of construction and manner of operation of the

plethysmograph are described in Exercise 16. It operates on the principle of a light beam that is projected into the tissues of the finger. When the beam is altered by a surge of blood passing through the finger, a photoresistor is activated by the scattered light to produce a signal that can be recorded. A review of its construction and operation in Exercise 16 should be made prior to doing this experiment.

In addition to determining the pulse rate, this setup has merit in detecting *indirect and relative pulse pressure* differences. Although the plethysmogram is, essentially, a measure of blood volumes flowing through the finger, it does tell us something about blood pressure. You have to keep in mind that blood pressure is affected by the

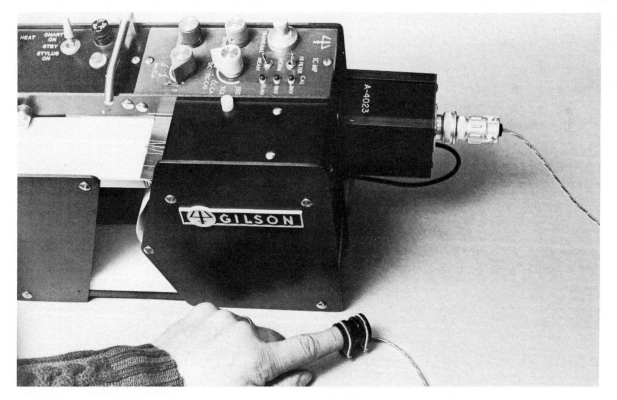

Figure 42.1   Monitoring the pulse.

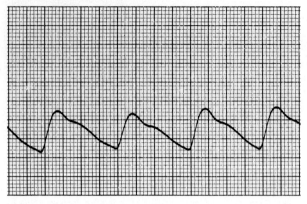

Figure 42.2 A plethysmogram.

amount of blood flowing through a point at a given time. In spite of this close relationshp between blood volume and blood pressure, it is quite difficult to accurately calibrate the Unigraph in such a way that we can take relative blood pressure readings from the plethysmogram; approximations of blood pressure levels can be determined, however, by observing baseline shifts such as those displacements that correspond to breathing.

We hope to accomplish several things with this experiment: (1) determine the exact pulse rate; (2) explore what factors affect the pulse rate; (3) identify the dicrotic notch and its significance; and (4) determine the range of pulse rates among class members. Proceed as follows:

Materials:

Unigraph
A4023 adapter (Gilson)
Photoresistor pulse pickup (Gilson T4020)

1. Attach the A4023 adapter to the receiving end of the Unigraph. Make sure that the lock nut is tightened securely and that the phone jack is also plugged in.
2. Secure the jack of the pulse pickup to the adapter. Plug in the power cord.
3. Place the pulse pickup on the index finger of the left hand with the light source facing the pad of the fingertip.
4. Set the Unigraph controls as follows: speed control lever at SLOW position, stylus heat control at 2:00 position, gain control at 2 MV/CM, sensitivity control completely counterclockwise, and Mode selector on DC.
5. Turn on the power switch and set the chart control switch at STYLUS ON. Wait about 15 seconds for the stylus to warm up and then place the chart control switch at CHART ON. Observe the trace.

6. Adjust the centering control to get the trace near the center of the paper. Now, increase the sensitivity by turning the sensitivity control clockwise until you get 1.5 or 2 cm. deflection on the chart. If this amount of deflection cannot be achieved, turn the sensitivity control down again and increase the gain to 1 MV/CM. Now increase the sensitivity with the sensitivity control to get the desired deflection.
7. Make a tracing for about 2 minutes, stop the paper (c.c. on STBY), and determine the pulse rate. (Since the space between two margin marks represents 30 seconds, simply count the spikes and fractions thereof between two marks and multiply by two.)
8. Change the speed to 25 mm. per second and record the pulse for about 1 minute while holding the hand perfectly still. Place the c.c. switch on STBY and examine the plethysmogram. Is the strength of the pulse consistently the same? Are the intervals between each pulse identical? Do you see a **dicrotic notch** on the descending slope of the curve? This interruption of the curve is due to sudden closure of the aortic valve of the heart during diastole.
9. Return the speed control lever to the slow position and turn on the chart again. Record for about 10 seconds with the hand at the level of the heart, and then raise the hand to a position above your head for about 10 seconds. Now lower the hand to its former position for another 10 seconds. Stop the chart. What happened to the strength of the pulse in the elevated position as compared to the other position?
10. If you are a smoker, direct someone to light a cigarette for you. With the chart set at the slow speed, record the pulse for about 3 or 4 minutes while smoking and inhaling as you would normally. Did smoking appear to affect the pulse rate or the strength of the pulse?
11. Disconnect the pulse pick-up, jog in place, use a treadmill, or work out sufficiently for a few minutes to establish a rapid heart rate. Reattach the pulse pick-up and record about 30 seconds at fast speed initially and 3 minutes at slow speed subsequently to record recovery.

Laboratory Report

Complete the Laboratory Report for this exercise.

Peripheral Circulation Control (Frog) 43

The arterial and venous divisions of the circulatory system are united by an intricate network of *capillaries* in the tissues. Capillaries are short thin-walled vessels of approximately 9 micrometers in diameter. They are so numerous that all living body cells are no more than two to three cells away from one of these vessels.

The vast number of capillaries makes their combined diameter so great that blood flows slowest through them. This slow flow of blood is of great importance because it allows sufficient time for the exchange of materials between the blood and the tissue cells. The pathway of this exchange may be expressed as follows:

$$\frac{capillary}{blood} \rightarrow \frac{tissue}{fluid} \rightarrow cells$$

Whether or not blood enters a capillary is determined by the size of its preceding arteriole and the action of its *precapillary sphincter muscle*. The arteriole has the capacity of vasodilation and vasoconstriction to affect the amount of blood flow through it. The precapillary sphincter, which guards the entrance to the capillary, can also restrict or permit blood flow. Both the arterioles and sphincters are under neural control of the *vasomotor center* of the medulla oblongata through the autonomic nervous system. The vasomotor center is influenced by many factors, but carbon dioxide concentration of the blood is of major importance.

Locally, carbon dioxide, histamine, and epinephrine may influence the volume of blood flowing through capillaries. High CO_2 concentration

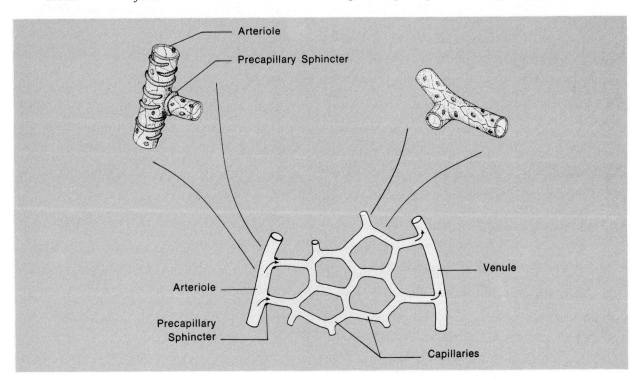

Figure 43.1 Capillary circulation.

in certain tissues will cause an increased flow of blood in these tissues with a decrease of blood in capillaries of other areas not being subjected to high CO_2 concentrations. Increased epinephrine levels in the blood cause vasodilation in certain areas such as muscles, heart, and lungs, and vasoconstriction in other areas. *Histamine,* a hormone that is present in large quantities in allergic reactions, causes extensive vasodilation in the peripheral circulation. Such reactions may cause a precipitous drop in blood pressure, and death.

In this exercise, we will study some of the factors that influence capillary circulation. A frog will be strapped to a board so that the capillaries in the webbing between its toes can be studied under a microscope.

Materials:

    frog, small size
    frog board
    cloth straps (4″ × 12″)
    string
    pins
    rubber bands
    epinephrine (1:1000)
    histamine (1:10,000)

1. Obtain a frog from the stock table and strap it to a frog board as illustrated in figure 43.2.

Wrapping a piece of cloth around the body and head and tying it securely with string should hold the animal in place.

2. Pin the webbing of the foot over the hole in the board. Keep the webbing moistened with water.

3. If necessary, secure the board to the microscope stage with a large rubber band.

4. Position the foot webbing over the light source and focus on it with the 10X objective. Observe the blood flowing through the vessels. Locate an **arteriole,** with its characteristic pulsating, rapid flow, and a **venule,** with its slightly slower, steady flow. Also, identify a **capillary,** with its smaller diameter and blood cells moving through it slowly and in single file.

5. Study a capillary more closely to see if you can locate the juncture of the capillary and arteriole, which is the site of the **precapillary sphincter.** You probably won't be able to see the sphincter, but you can observe the irregular flow of blood cells into the capillary, which results partly from the sphincter muscle's action.

6. Now, remove most of the water from the frog's foot by blotting it with paper towelling and add several drops of 1:10,000 **histamine** so-

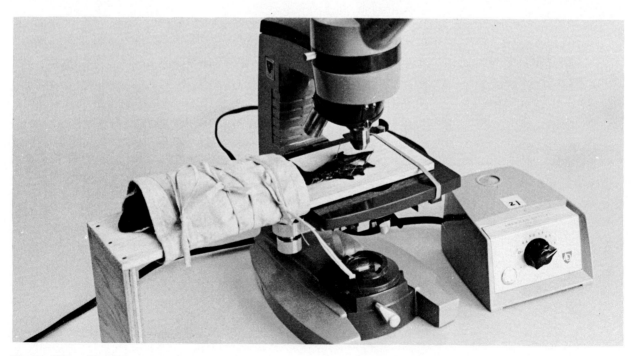

Figure 43.2 Blood flow setup.

lution. Observe the change in blood flow and record your observations on the Laboratory Report.

7. Wash the foot with water, blot it, and then add several drops of 1:1000 **epinephrine** solution. Observe any change in blood flow and record your results on the Laboratory Report.

8. Wash the foot again, remove the frog from the board, and return it to the stock table. Clean the microscope stage, if necessary.

Laboratory Report

Complete the Laboratory Report for this exercise.

44 The Arteries and Veins

In this exercise the circulation of the blood will be studied. The first portion of the exercise pertains to the general plan and involves figure 44.1. The remainder of the exercise pertains to specific arteries and veins. Although approximately 54 arteries and veins are listed in this exercise, only the major ones have been included. There are many smaller blood vessels that are beyond the scope of this cursory study.

The Circulatory Plan

Figure 44.1 is an incomplete flow diagram of the heart and major regions of the body. The problem of this assignment is to connect the blood vessels of the heart with the various organs and regions so that all parts are properly supplied and drained of blood. After all the blood vessels have been drawn in, those that contain oxygenated blood should be **colored with red** and those with deoxygenated blood should be **colored blue.** The following description explains the circulatory plan.

The circulatory system consists of three separate circuits, or "systems": the pulmonary, systemic, and coronary. The *pulmonary system* carries blood from the heart through the lungs and back to the heart. The *systemic system* supplies blood to all parts of the body except the lungs and heart muscle. The *coronary system* is the shortest circuit which supplies only the myocardium of the heart. Figure 44.1 shows only the pulmonary and systemic circuits.

Blood enters the right atrium (left side of illustration) from the **superior** and **inferior vena cavae,** which have collected blood from all parts of the body. This blood is dark colored because it is low in oxygen and high in carbon dioxide content. From the right atrium the blood passes to the right ventricle. When the heart contracts, blood leaves the right ventricle through the **pulmonary artery** to the lungs, where it picks up oxygen and gives off carbon dioxide. The blood leaves the lungs by way of the **pulmonary veins** and passes back to

the left atrium of the heart. Blood in the pulmonary veins is brightly colored due to its high oxygen content. The pulmonary arteries, veins, and capillaries constitute the pulmonary system. The systemic circulatory system includes the remainder of the circulatory system discussed in the next two paragraphs.

From the left atrium the blood passes to the left ventricle. When the heart contracts, blood leaves the left ventricle through the **aortic arch.** This blood, which is rich in oxygen, passes to all parts of the body. The aortic arch has branches that go to the head and arms. (For simplicity, only one blood vessel is shown passing to these regions.) Passing downward on the right side of the illustration, the aortic arch becomes the **aorta,** which has branches going to the liver, digestive organs, kidneys, pelvis, and legs. The branch that enters the liver is the **hepatic artery.** The intestines are supplied by the **superior mesenteric artery.** The kidneys receive blood through the **renal arteries.** The pelvis and legs are supplied by several arteries, but only one is shown for simplicity.

The blood leaving most of the organs of the trunk empties directly into a large collecting vein, the inferior vena cava. The **renal veins** from the kidneys and **hepatic vein** from the liver empty into the inferior vena cava. Several veins drain the legs and pelvic region. Blood from the intestines does not go directly to the inferior vena cava; instead, all blood from this region passes to the liver by way of the **portal vein.** This route of the blood from the intestines to the liver is called the **portal circulation.**

Assignment:

After completing and coloring all the blood vessels in figure 44.1, add arrows to the diagram to indicate the direction of blood flow. Label all blood vessels, also.

Answer questions on the Laboratory Report pertaining to the circulatory plan.

The Arterial Division

The principal arteries of the body are shown in the left-hand illustration of figure 44.2. Blood leaving the left ventricle of the heart is carried in a large curved artery, the **aortic arch.** From the upper surface of this vessel emerge three arteries: the left subclavian, the left common carotid, and the innominate.

The **left subclavian** artery is the artery that passes from the aortic arch into the left shoulder behind the clavicle. In the armpit *(axilla)* the subclavian becomes the **axillary** artery. The axillary, in turn, becomes the **brachial** artery in the upper arm. This latter artery divides into the radial and ulnar arteries in the forearm. The **radial** artery follows the radial bone and the **ulnar** artery follows the ulnar bone.

The **innominate** artery is the short vessel coming off the aortic arch on the right side of the body. It gives rise to two arteries. The branch which extends upward to the head is the **right common ca-**

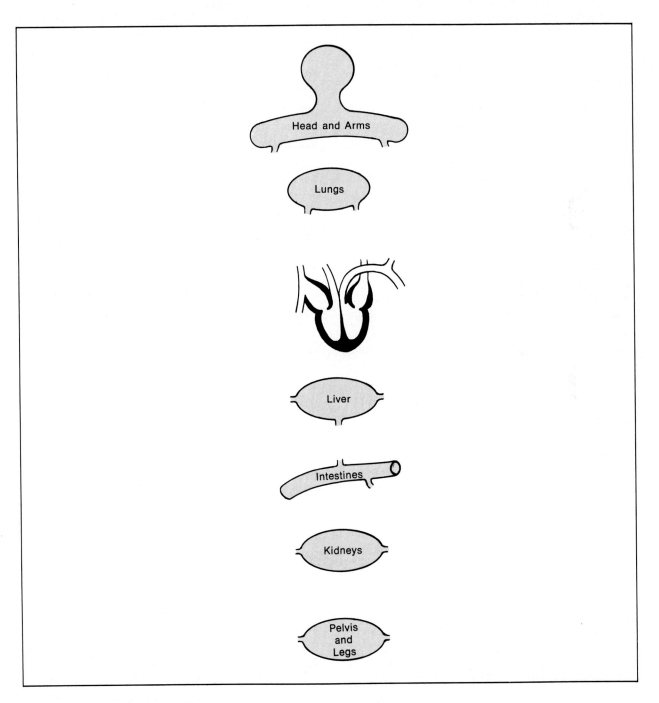

Figure 44.1   The circulatory plan.

rotid artery. It furnishes the right side of the head with blood. The **right subclavian** artery is an outward extension of the innominate artery from the base of the right common carotid. It passes behind the right clavicle.

The blood vessel which extends upward from the aortic arch between the left subclavian and innominate arteries is the **left common carotid.** It supplies the left side of the head with oxygenated blood.

The aortic arch becomes the **aorta** as it passes down through the thorax into the abdominal cavity. It is the main trunk of the arterial system, having many arteries that pass from it to various internal organs. Only a few of these arteries are shown in figure 44.2. The first branch, which is seen just below the heart, is the **celiac** artery. It supplies the stomach, liver, and spleen. Just below the celiac is the **superior mesenteric** artery, which supplies most of the small intestine and part of the large intestine. Below this latter artery are the right and left **renal** arteries that pass into the right and left kidneys. The single branch below the renal arteries is the **inferior mesenteric** artery, which supplies part of the large intestine and rectum with blood.

In the lumbar region the aorta divides into the right and left **common iliac** arteries. Each common iliac passes downward a short distance and then divides into a smaller inner branch, the **internal iliac** *(hypogastric)* artery, and a larger branch, the **external iliac** artery, which continues on down into the leg. The external iliac becomes the **femoral** artery in the upper three-fourths of the thigh. In the general region of the origin of the femoral artery, the **deep femoral** artery branches off of it. This artery courses backward and downward along the medial surface of the femur. In the knee region the femoral becomes the **popliteal** artery. Just below the knee the popliteal divides into the **posterior tibial** artery and the **anterior tibial** artery.

Assignment:

Label the arterial portion of figure 44.2.

The Venous Division

The major veins of the body are seen in the right-hand illustration of figure 44.2. Emptying into the upper portion of the right atrium of the heart is the **superior vena cava,** which receives blood from the head and arms. The **inferior vena cava** is the large vein that empties into the right atrium from below the heart, carrying blood from the remainder of the body to the heart.

In the neck region are seen four veins: two internal jugulars and two external jugulars. The **internal jugular** veins are larger than the **external jugular** veins and are closer to the median line of the body. The larger internal jugular veins empty into the innominate veins, and the external jugulars empty into the subclavian veins. The **innominate** veins are short veins that empty into the superior vena cava.

Three major veins collect blood in the upper arm, carrying it toward the heart. They are the cephalic, basilic, and brachial veins. The **cephalic** vein courses along the lateral aspect of the arm and empties into the **subclavian** vein, which lies behind the clavicle. The **basilic** vein lies on the medial side of the arm, and the **brachial** vein lies along the posterior surface of the humerus. Between the basilic and cephalic veins in the elbow region is seen the **median cubital** vein. The **accessory cephalic** vein lies on the lateral portion of the forearm and empties into the cephalic in the elbow region. The brachial and basilic veins empty into a short **axillary** vein in the armpit region. The axillary, in turn, empties into the subclavian at the same place where the cephalic enters.

Blood in the legs is returned to the heart by superficial and deep sets of veins. The superficial veins are just beneath the skin. The deep veins accompany the arteries. Both sets are provided with valves, which are more numerous in the deep than in the superficial ones. The **posterior tibial** vein (label 24) is one of the deep veins that lies behind the tibia. This vein collects blood from the calf region and foot. The posterior tibial becomes the **popliteal** vein in the knee region. Above the knee this vessel becomes the **femoral** vein, which, in turn, empties into the **external iliac** vein in the upper leg region. The external and **internal iliac** *(hypogastric)* veins (label 11) empty into the **common iliac** vein. The inferior vena cava, thus, receives blood from the two common iliacs, renal veins, and many others not shown in this diagram. One other vein of the leg, the **great saphenous,** a superficial vein, originates from the **dorsal venous arch** on the superior surface of the foot. The great saphenous passes up the medial surface of the leg and enters the femoral vein at the top of the thigh.

Although the inferior vena cava has many other veins emptying into it, only a few are shown

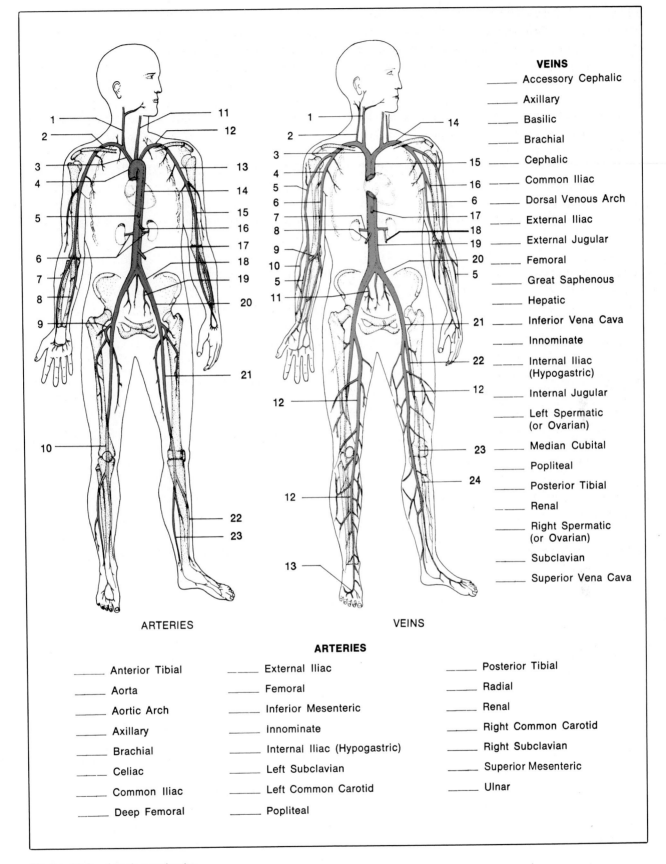

Figure 44.2 Arteries and veins.

VEINS

_____ Accessory Cephalic
_____ Axillary
_____ Basilic
_____ Brachial
_____ Cephalic
_____ Common Iliac
_____ Dorsal Venous Arch
_____ External Iliac
_____ External Jugular
_____ Femoral
_____ Great Saphenous
_____ Hepatic
_____ Inferior Vena Cava
_____ Innominate
_____ Internal Iliac
        (Hypogastric)
_____ Internal Jugular
_____ Left Spermatic
        (or Ovarian)
_____ Median Cubital
_____ Popliteal
_____ Posterior Tibial
_____ Renal
_____ Right Spermatic
        (or Ovarian)
_____ Subclavian
_____ Superior Vena Cava

ARTERIES

VEINS

ARTERIES

_____ Anterior Tibial
_____ Aorta
_____ Aortic Arch
_____ Axillary
_____ Brachial
_____ Celiac
_____ Common Iliac
_____ Deep Femoral

_____ External Iliac
_____ Femoral
_____ Inferior Mesenteric
_____ Innominate
_____ Internal Iliac (Hypogastric)
_____ Left Subclavian
_____ Left Common Carotid
_____ Popliteal

_____ Posterior Tibial
_____ Radial
_____ Renal
_____ Right Common Carotid
_____ Right Subclavian
_____ Superior Mesenteric
_____ Ulnar

in figure 44.2. Just under the heart is seen the **hepatic** vein, which receives blood from the liver. The **renal** veins return blood from the kidneys to the inferior vena cava. Note that the left renal vein has a branch, the **left spermatic** (or ovarian) that carries blood away from the left testis or ovary. The **right spermatic** (or ovarian) empties into the inferior vena cava at a point inferior to where the renal vein joins the inferior vena cava.

The presence of valves in veins can be vividly demonstrated on the back of your hand. If you apply pressure to a prominent vein on the back of your left hand near the knuckles with the middle finger of your right hand and then strip the blood in the vein away from the point of pressure with your index finger, you will note that blood does not flow back toward the point of pressure. This indicates that valves prevent backward flow.

Assignment:

Label the veins in figure 44.2.

Portal Circulation

It was noted in the study of the circulatory plan (figure 44.1) that blood from the intestines passes to the liver instead of going directly to the heart for recycling. The vessel that collects blood from the intestines (as well as the stomach, spleen, and pancreas) is the **portal vein.** Figure 44.3 illustrates the ramifications of this drainage system.

Blood leaving the digestive tract will vary considerably, from time to time, in composition. Following meals, the blood will be rich in nutrients; after long fasting it will be low in glucose and other nutrients. It is the function of the liver to establish

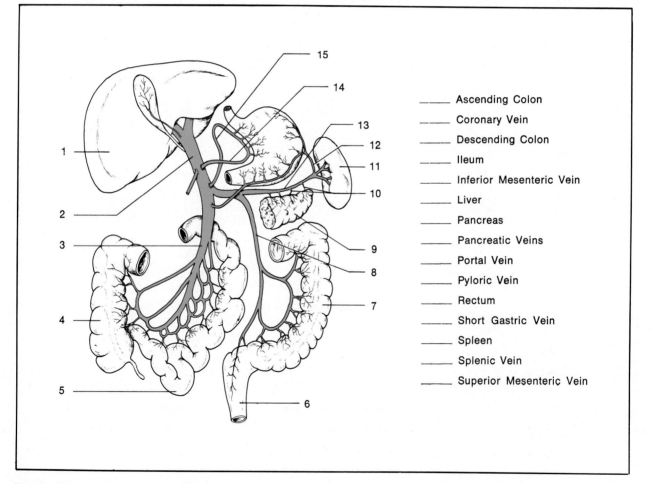

_____ Ascending Colon
_____ Coronary Vein
_____ Descending Colon
_____ Ileum
_____ Inferior Mesenteric Vein
_____ Liver
_____ Pancreas
_____ Pancreatic Veins
_____ Portal Vein
_____ Pyloric Vein
_____ Rectum
_____ Short Gastric Vein
_____ Spleen
_____ Splenic Vein
_____ Superior Mesenteric Vein

Figure 44.3 Portal circulation, human.

homeostasis by altering the blood's composition to desired levels.

Figure 44.3 reveals sections of the colon, small intestines, pancreas, spleen, and stomach. The large vessel that collects blood from the ascending colon and ileum is the **superior mesenteric** vein. It empties directly into the **portal** vein. Blood from the remainder of the colon and rectum is collected by the **inferior mesenteric,** which empties into the **splenic,** or *lienal,* vein. This latter blood vessel parallels the length of the pancreas, receiving blood from the pancreas via several **pancreatic** veins. Near the spleen, the splenic vein also receives blood from the stomach through the **short gastric** vein.

From the stomach a venous loop empties blood from its medial surface into the portal vein. The upper portion of the loop is the **coronary** vein; the lower portion, the **pyloric** vein.

Assignment:

Label figure 44.3.

Complete the Laboratory Report for this exercise.

45 Fetal Circulation

During human embryological development the circulatory system is necessarily somewhat different from that after birth. The fact that the lungs are nonfunctional before birth dictates that less blood should go to the lungs. Also, because the food and oxygen supply must come from the mother by way of the placenta, blood vessels to and from the placenta through the umbilical cord must be present and functional.

Figure 45.1 illustrates, in a diagrammatic way, the circulatory pathway through the fetal heart. It is the purpose of this short exercise to trace the blood through the veins and arteries of figure 45.1 so that you understand the pathway of the blood prior to birth and comprehend the changes that must occur after birth.

In the lower left-hand corner of the illustration is seen a large vessel with two smaller vessels wrapped around it. The large vessel is the **umbilical vein,** which carries blood from the placenta of the mother. It contains all the nutrients and oxygen required by the developing fetus. The two small vessels wrapped around the umbilical vein are the **umbilical arteries.** These arteries return blood laden with carbon dioxide and wastes to the placenta. Near the liver of the fetus the umbilical vein narrows to become a short narrow vessel, the **ductus venosus.** At the entrance to the ductus venosus is a **sphincter muscle** that closes off that vessel when the umbilical cord is severed at birth. This helps to prevent excessive loss of blood from the infant. The ductus venosus empties into the **inferior vena cava,** where rich oxygenated blood from the placenta is mixed with unoxygenated blood. From here the blood passes into the right atrium, where it is mixed with blood from the superior vena cava. Blood from the right atrium moves in two directions. Part of it goes into the right ventricle and part of it passes into the left atrium through an opening, the **foramen ovale.** Observe that this opening has a flap-like valve over it.

When the heart contracts, blood leaves the right ventricle through the pulmonary artery, which, in turn, divides to form the **right** and **left pulmonary arteries.** At the same time, blood leaves the left ventricle through the ascending aorta. Between the pulmonary arteries and the aortic arch is a short vessel, the **ductus arteriosus,** that carries blood intended for the pulmonary circulatory system over into the systemic system, thus bypassing the lungs. Of course, some blood goes to the lungs to supply the needs of growing lung tissue, but nowhere the amount that is needed after birth.

The remainder of the circulatory system resembles the adult plan in that the aorta passes down through the body and divides into two common iliac arteries. The umbilical arteries, which branch off the common iliacs, become the *internal iliacs,* or *hypogastric* arteries, in the child after birth.

Once the fetus is separated from the placenta at birth, changes occur in the heart, veins, and arteries to provide a new route for the blood. As stated, one of the first things that happens is the closure of the umbilical vein by the sphincter muscle to help in the prevention of blood loss. Coagulation of blood in this vessel, as well as ligature by the physician, also assists in this respect. As soon as the infant begins to breathe, more blood is drawn to the lungs via the pulmonary arteries. The abandonment of the path through the ductus arteriosus causes this vessel to collapse and begin its gradual transformation to connective tissue of the *ligamentum arteriosum.* With the increase in blood volume and pressure in the left atrium, due to a greater blood flow from the lungs, the flap on the foramen ovale closes this opening. Eventually, connective tissue completely seals off this valve. The umbilical vein becomes the *round ligament* of the liver. The ductus venosus becomes a fibrous band on the inferior surface of the liver.

Laboratory Report

Label figure 45.1 and answer the questions on Laboratory Report 45–47 that pertain to this exercise.

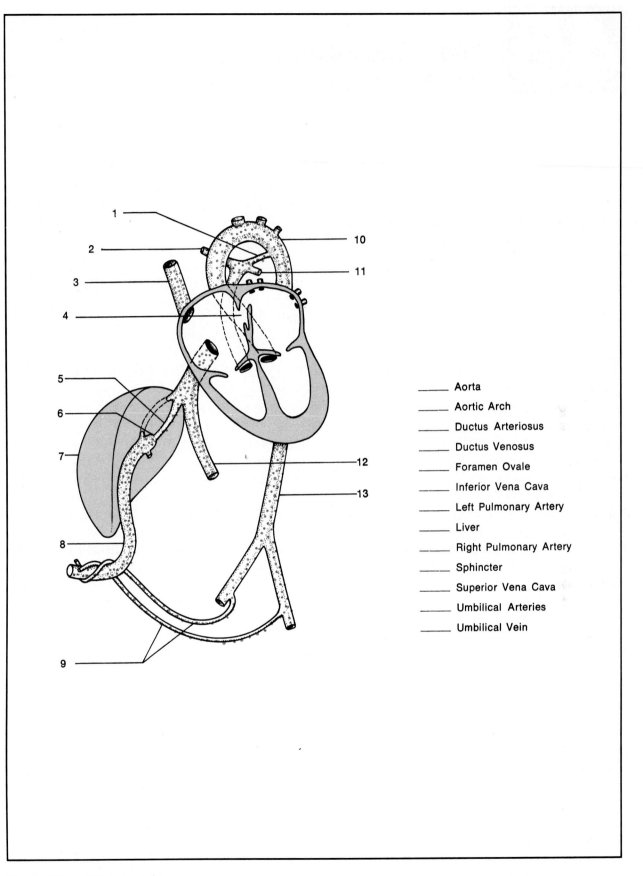

_____ Aorta

_____ Aortic Arch

_____ Ductus Arteriosus

_____ Ductus Venosus

_____ Foramen Ovale

_____ Inferior Vena Cava

_____ Left Pulmonary Artery

_____ Liver

_____ Right Pulmonary Artery

_____ Sphincter

_____ Superior Vena Cava

_____ Umbilical Arteries

_____ Umbilical Vein

Figure 45.1   Fetal circulation.

The Lymphatic System and the Immune Response

The lymphatic system consists of a drainage system of lymphatic vessels that returns tissue fluid from the interstitial cell spaces to the blood. On its journey back to the blood, the lymph in these vessels is cleansed of bacteria and other foreign material by macrophages (reticuloendothelial system) in the lymph nodes. In addition to macrophages, the lymph nodes are packed with lymphocytes that play a leading role in cellular and humoral immunity. In this exercise we will explore the genesis and physiology of lymphocytes as well as the overall anatomy of the lymphatic system.

Lymph Pathway

The smallest vessels of the lymphatic system are the **lymph capillaries.** These tiny vessels have closed ends, are microscopic, and are situated among the cells of the various tissues of the body. The lymph capillaries unite to form the **lymphatic vessels** *(lymphatics)* which are the visible vessels shown in the arms and legs in figure 46.1. The irregular appearance of the walls of these vessels is due to the fact that they contain many valves to restrict the backward flow of lymph.

As lymph moves toward the center of the body through the lymphatics, it eventually passes through one or more of the small oval **lymph nodes.** Note that these nodes are clustered in the neck, groin, axillae, and abdominal cavity; they are also found in smaller clusters in other parts of the body. They vary in size from that of a pinhead to an almond.

Two collecting vessels, the thoracic and right lymphatic ducts, collect the lymph from different regions of the body. The largest one is the **thoracic duct,** which collects all the lymph from the legs, abdomen, left half of the thorax, left side of the head, and left arm. Its lower extremity consists of a sac-like enlargement, the **cisterna chyli.** Lymph

from the intestines passes through lymphatics in the mesentery to the cisterna chyli. The lymph from this region contains a great deal of fat and is usually referred to as **chyle.** The thoracic duct empties into the left subclavian vein near the left internal jugular vein. The **right lymphatic duct** is a short vessel that drains lymph from the right arm and right side of the head into the right subclavian vein near the right internal jugular.

Lymph Node Structure

Each lymph node has a slight depression on one side, the **hilum,** through which blood vessels and the **efferent lymphatic vessel** emerge. **Afferent lymphatic vessels** may enter the lymph node at various points. Although each lymph node has only one efferent vessel, it may have several afferent lymphatic vessels. Surrounding the entire node is a **capsule** of fibrous connective tissue. Just inside the capsule lies the **cortex,** which consists of (1) sinuses reinforced with reticular tissue and (2) **germinal centers** *(nodules).* The germinal centers represent structural units of the node. They arise from small nests of lymphocytes or lymphoblasts and reach about one millimeter in diameter. The central portion of a lymph node is the **medulla.**

Assignment:

Label figure 46.1.

Genesis of Lymphocytes

The term *lymphocyte* usually refers to a family of cells characterized by the absence of specific granules and the presence of a centrally located nucleus. Lymphocytes in the blood are small to medium in size (4–10 μm); those in lymph are usually larger (up to 15 μm). Although all lymphocytes are morphologically similar, there are considerable physiological differences between individual cells. In addition to being present in blood, lymph, and lymph nodes, they are also seen in the bone marrow, thymus, spleen, and lymphoid masses associated with the digestive, respiratory, and urinary passages.

Two different kinds of lymphocytes account for what are known as humoral and cellular immunity. *Humoral immunity* is that type of immunity which is achieved through circulating **antibody.** *Cellular immunity* is immunity in which *sensitized lymphocytes* (**lymphoblasts**) act against

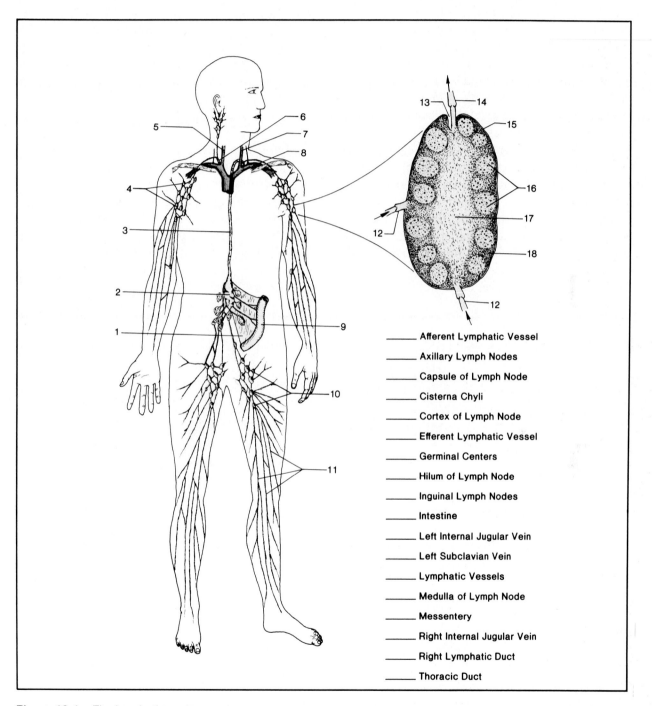

_____ Afferent Lymphatic Vessel

_____ Axillary Lymph Nodes

_____ Capsule of Lymph Node

_____ Cisterna Chyli

_____ Cortex of Lymph Node

_____ Efferent Lymphatic Vessel

_____ Germinal Centers

_____ Hilum of Lymph Node

_____ Inguinal Lymph Nodes

_____ Intestine

_____ Left Internal Jugular Vein

_____ Left Subclavian Vein

_____ Lymphatic Vessels

_____ Medulla of Lymph Node

_____ Messentery

_____ Right Internal Jugular Vein

_____ Right Lymphatic Duct

_____ Thoracic Duct

Figure 46.1   The lymphatic system.

foreign cells; these sensitized cells are responsible for tissue rejection in organ transplants and grafts. Both types of immunity are *acquired,* since they are not present at birth and since they develop in the presence of foreign biological agents.

The lymphocytes that are responsible for antibody formation are called **B-lymphocytes,** or B-cells (*B* for bursa). The lymphocytes that are responsible for cellular immunity are **T-lymphocytes,** or T-cells (*T* for thymus).

Both types of lymphocytes originate from **lymphocytic stem cells** in the bone marrow before birth and shortly after birth. The stem cells are released into the blood and follow one of two pathways for "processing", as illustrated in figure 46.2. The cells that settle in the thymus for modification become T-cells; the remaining cells are processed in some other unknown area to become B-cells. Birds possess a *bursa of Fabricius* on the hindgut, where B–cells are formed; in man an equivalent of this organ is being sought. At present, it is speculated that lymphoid tissue in the fetal liver or even in the bone marrow may be responsible for B-cell processing.

Once the T-lymphocytes have matured in the thymus, they leave the thymus via the blood and colonize in specific loci of the lymph nodes and spleen. T-cells have a long life. In the presence of antigens they are transformed to **lymphoblasts,** which are responsible for cellular immunity.

The stem cells that mature in the B-cell processing tissues leave these tissues through the blood as B-lymphocytes and colonize in specific regions of the lymph nodes and spleen also. When these B-cells come in contact with the appropriate antigens they are transformed into **plasma cells** which produce the antibodies of humoral immunity.

Laboratory Assignment

Materials:
    prepared slides of lymph node
    microscope

1. Examine a prepared slide of a lymph node and identify the structures illustrated in figure 46.1. If drawings are to be required make them on a separate sheet of paper.
2. Answer all questions on the combined Laboratory Report that pertain to this exercise.

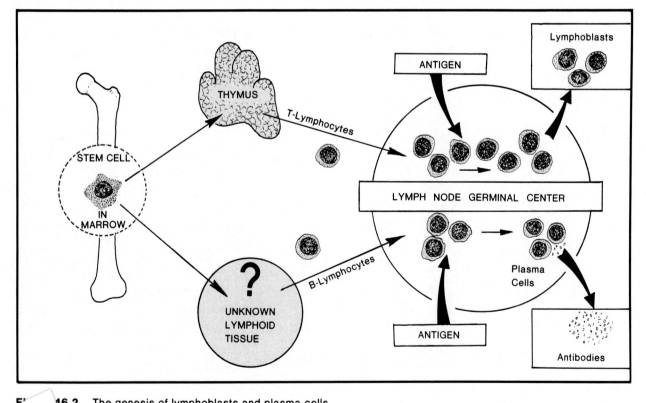

Fig. 46.2 The genesis of lymphoblasts and plasma cells.

Blood Pressure Monitoring  **47**

The contractions of the ventricles of the heart exert a propelling force on the blood which manifests itself as blood pressure in vessels throughout the body. Since the heart acts as a pump that forces out spurts of blood at approximately seventy times a minute while at rest, the pressure during a short interval of time will fluctuate up and down.

The force on the walls of the blood vessels is greatest during systole and is called the **systolic pressure.** When the heart relaxes (diastole) and fills up with blood in preparation for another contraction, the pressure falls to its lowest value. The pressure during this phase is called the **diastolic pressure.**

There are several different methods that one might use for determining blood pressure. Probably the most sensitive and precise way is to insert a hollow needle into a vessel, which allows the force of the blood pressure to act on a pressure transducer. The electronic signal from such a device can be fed into an amplifier and recording device for monitoring. A simpler electronic method using an arm cuff is described in Exercise 49.

In this exercise, we will use a conventional sphygmomanometer and stethoscope for determining blood pressure. In this method, the stethoscope is used to listen for the **Korotkoff** (or Korotkow) **sounds** produced by arterial blood rushing through a partially occluded brachial artery in the upper arm. Prior to using the stethoscope, your instructor will demonstrate these sounds using an audio monitor.

Figures 47.1 and 47.2 illustrate the procedures for applying the cuff of the sphygmomanometer to the upper arm. When the cuff is properly positioned, it is inflated to shut off the flow of blood in the brachial artery. The pressure within the cuff is then released slowly by unscrewing the valve on the bulb. As the air is slowly released from the cuff, the stethoscope is used to listen for the emergence of the Korotkoff sounds. As soon as the sounds become audible, the pres-

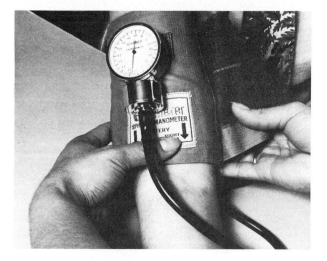

Figure 47.1 The sphygomomanometer cuff is wrapped around the upper arm, keeping the lower margin of cuff above the line through the epicondyles of the humerus.

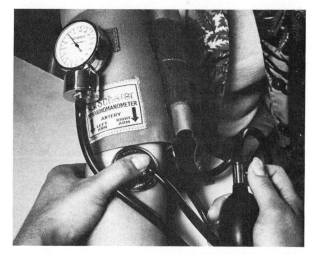

Figure 47.2 The inflated cuff shuts off the flow of blood in the brachial artery. Korotkoff's sounds are listened for as pressure is gradually lowered.

sure gauge is noted. The reading at this point is the **systolic pressure.** As air continues to slowly escape the cuff, the gauge is watched until the Korotkoff sounds disappear. The pressure at which the sounds just disappear is the **diastolic pressure.**

The systolic pressure of a normal adult is approximately 120 mm. Hg. while the diastolic pressure is usually around 80 mm. Hg. Such a blood pressure is designated as 120/80. Infants will have a pressure around 90/55. In old age, the pressure usually averages around 150/90.

The performance of this measurement presents no particular problems except that the room must be kept as quiet as possible to facilitate hearing the sounds with the stethoscope; therefore, it is essential that conversations be kept to a whisper. After the instructor has demonstrated the Korotkoff sounds with the audio monitor, work with your laboratory partner to take each other's blood pressure with a stethoscope and sphygmomanometer.

Materials:

    microphone
    audio monitor
    stethoscope
    sphygmomanometer

Demonstration

After activating the audio monitor, a sphygmomanometer will be applied to the upper arm of the subject. Before air is pumped into the sphygmomanometer, the microphone is slipped under the cuff by the subject and placed over the brachial artery. Figure 47.3 illustrates the setup. Air is then pumped into the cuff to achieve a pressure of at least 180 mm. Hg. and locked in by closing the valve on the bulb. The valve is released slowly and the students will observe the manometer or pressure gauge to note at what pressure the first Korotkoff sounds become audible. Continued release of the air pressure takes place until the Korotkoff sounds disappear.

Student Participation

N̶ ̶ ̶ ̶hat you understand what sounds will be
̶ ̶ ̶ ̶d how they relate to the pressure gauge,
̶ ̶ ̶ ̶s follows:

̶ ̶ ̶ ̶he cuff around the right upper arm.
̶ ̶ ̶ ̶hod of attachment will depend on the

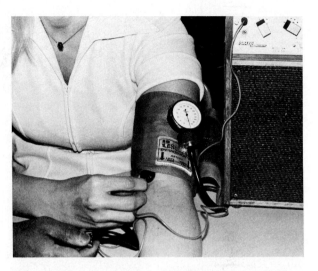

Figure 47.3 Auscultation of Korotkoff's sounds with an audio monitor is performed by slipping a microphone under the cuff over brachial artery.

type of equipment. Your instructor will show you how to use the specific type available in your laboratory. See that the subject's forearm rests comfortably on the table.

2. Close the metering valve on the neck of the rubber bulb. *Don't twist it so tight that you won't be able to open it!*

3. Pump air into the sleeve by squeezing the bulb in your right hand. Watch the pressure gauge. Allow the pressure to rise to about 180 and continue to hold the metering valve closed.

4. Position the bell of the stethoscope just below the cuff, at a point that is *midway between the epicondyles of the humerus.* This is the lowest extremity of the brachial artery. About 2.5 cm. distal to this point, the artery bifurcates to become the radial and ulnar arteries.

5. Slowly release the valve so that the pressure goes down slowly. Listen carefully as you watch the pressure fall. When you just begin to hear the pulsations, note the pressure on the gauge. This is the systolic pressure.

6. Continue listening as the pressure falls. Just when you are unable to hear the Korotkoff sounds anymore, note the pressure on the gauge. This is the diastolic pressure.

7. Repeat two or three times to see if you get consistent results.

8. Now, have the individual do some exercise, such as running up and down the stairs a few times. Measure the blood pressure again.

Laboratory Report

Record your results on Laboratory Report 45–47.

Part 10 The Respiratory System

This unit is comprised of four exercises which cover the anatomy of the respiratory system, respiratory regulation, lung capacities, and a diagnostic procedure often used to detect respiratory impairment. Spirometry experiments are included in two of the exercises. The roles of carbon dioxide, hydrogen ion concentration, and oxygen in regulating breathing are covered in Exercise 49. The effects of the Valsalva maneuver, deglutition, and submersion on respiratory reflexes are also studied in this exercise. The timed, forced expiratory volume (FEV_T) test in Exercise 51 is a clinically approved test used for screening patients who have impaired pulmonary function.

48 The Respiratory Organs

In this exercise we will study both gross and microscopic anatomy of the respiratory system. After labeling diagrams of the human respiratory system, sheep and frog materials will be used for dissection.

Materials:

> model of median section of head
> larynx model
> frog, pithed
> sheep pluck
> dissecting instruments and trays
> prepared slides of rat trachea and lung tissue
> Ringer's solution

Upper Respiratory Tract

In addition to breathing, the upper respiratory tract also functions in eating and speech; thus, a study of the respiratory organs in this region necessarily includes organs concerned with these other two activities. Figure 48.1 is of this area.

The two principal cavities of the head are the nasal and oral cavities. The upper **nasal cavity,** which serves as the passageway for air, is separated from the lower **oral cavity** by the *palate.* The anterior portion of the palate is reinforced with bone and is called the **hard palate.** The posterior part, which terminates in a finger-like projection, the **uvula,** is the **soft palate.**

The oral cavity consists of two parts, the oral cavity proper and the oral vestibule. The **oral cavity proper** is the larger cavity that contains the tongue; the **oral vestibule** is the smaller one between the lips and teeth.

During breathing, air enters the nasal cavity through the nostrils, or **nares.** The lateral walls of this cavity have three pairs of fleshy lobes, the superior, middle, and inferior **nasal conchae.** These lobes serve to warm the air as it enters the body.

Above and behind the soft palate is the *pharynx.* That part of the pharynx posterior to the tongue where swallowing is initiated is the **oropharynx.** Above it is the **nasopharynx.** Two openings from the **Eustachean tubes** are seen on the walls of the nasopharynx.

After air passes through the nasal cavity, nasopharynx and oropharynx, it passes through the larynx and trachea to the lungs. The **larynx,** or voice box, has two cartilages (labels 12 and 13) that form the reinforcement of its outer walls. The **thyroid cartilage** is the upper one which produces an external protuberance, the *Adam's apple.* The **cricoid cartilage,** a smaller one, lies inferior to the thyroid cartilage. A third cartilage of the larynx, the **epiglottis,** is situated diagonally over the opening *(glottis)* of the larynx. This cartilage prevents food from entering the larynx during swallowing. In the larynx is a pair of **vocal folds** which vibrate to produce sound waves when actuated by air movement. Posterior to the larynx and trachea lies the **esophagus,** which carries food from the pharynx to the stomach.

In several areas of the oral cavity and nasopharynx are seen islands of lymphoidal tissue called *tonsils.* The **pharyngeal tonsils,** or *adenoids,* are situated on the roof of the nasopharynx; the **lingual tonsils** are on the posterior inferior portion of the tongue; and the **palatine tonsils** are on each side of the tongue on the lateral walls of the oropharynx. These masses of tissue are as important as the lymph nodes of the body in protection against infection. The palatine tonsils are the ones most frequently removed by surgical methods.

Assignment:

Label figure 48.1.

Study models of the head and larynx. Be able to identify all structures.

Lower Respiratory Passages

Figure 48.2 illustrates the lower respiratory passages. The parts of the larynx in the throat region consist of the upper **epiglottis,** a large **thyroid cartilage** and the smaller **cricoid cartilage.** Below the larynx extends the **trachea** to a point in the center of the thorax where it divides to form two short

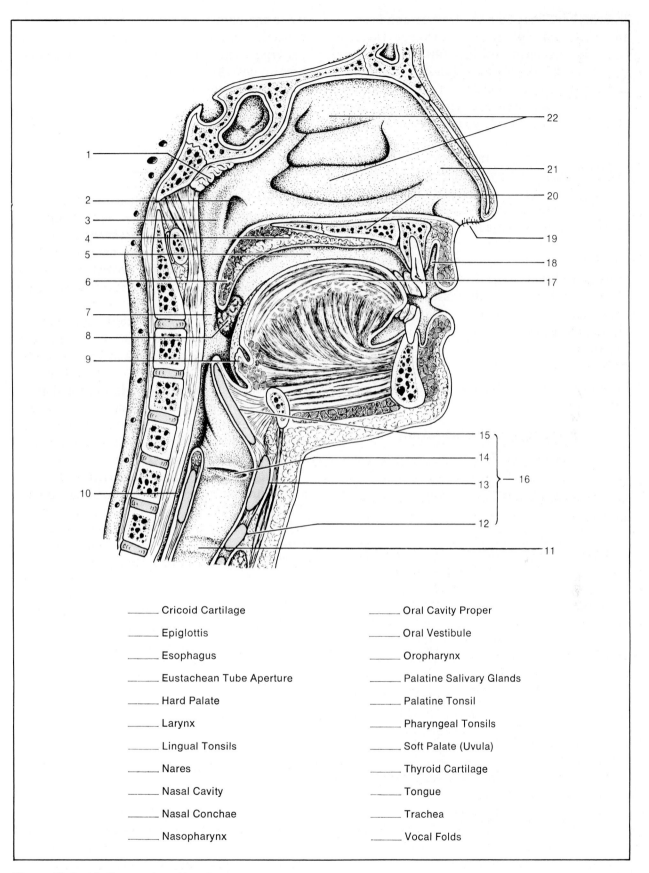

Figure 48.1   Median section of head.

_____ Cricoid Cartilage   _____ Oral Cavity Proper

_____ Epiglottis   _____ Oral Vestibule

_____ Esophagus   _____ Oropharynx

_____ Eustachean Tube Aperture _____ Palatine Salivary Glands

_____ Hard Palate   _____ Palatine Tonsil

_____ Larynx   _____ Pharyngeal Tonsils

_____ Lingual Tonsils   _____ Soft Palate (Uvula)

_____ Nares   _____ Thyroid Cartilage

_____ Nasal Cavity   _____ Tongue

_____ Nasal Conchae   _____ Trachea

_____ Nasopharynx   _____ Vocal Folds

bronchi. Note that both the trachea and bronchi are reinforced with rings of cartilage (hyaline type). Each bronchus divides further into many smaller tubes called **bronchioles.** At the terminus of each bronchiole is a cluster of tiny sacs, the **alveoli,** where gas exchange with the blood takes place. Each lung is made up of thousands of these sacs.

Free movement of the lungs in the thoracic cavity is facilitated by the pleural membranes. Covering each lung is a **pulmonary pleura** and attached to the thoracic wall is a **parietal pleura.** Between these two pleurae is a potential cavity, the **pleural** *(intrapleural)* **cavity.** Normally, the lungs are firmly pressed against the body wall with little or no space between the two pleurae. The bottom of the lungs rests against the muscular **diaphragm,** a principal muscle of respiration. The parietal pleura is attached to the diaphragm also.

Assignment:

Label figure 48.2. Color the pleural membranes **red.**

Sheep Pluck Dissection

A "sheep pluck" consists of the trachea, lungs, larynx, and heart of a sheep as removed during routine slaughter. Because of its size and similarity to the human respiratory tract it is ideal for dissection. Although it is as safe to handle as any material from a meat market, practice sanitary precautions. At the end of the period be sure to wash your hands thoroughly.

Materials:

sheep plucks, less the liver
dissecting instruments
straws or serological pipettes (5 ml. size)

1. Lay out a fresh sheep pluck on a tray. Identify the major organs, such as the **lungs, larynx, trachea, diaphragm** remnants, and **heart.**
2. Examine the larynx more closely. Can you identify the **epiglottis, thyroid** and **cricoid cartilages?** Look into the larynx. Can you see the **vocal folds?**
3. Cut a cross section through the upper portion of the trachea and examine a sectioned **cartilaginous ring.** Is the ring continuous all the way around?
4. Force your index finger down into the trachea, noting the smooth and slimy nature of the inner lining. Why is this inner membrane so smooth and slimy?
5. Note that each lung is divided into lobes. How many lobes are there in the right lung? On the left lung? Compare with the human.

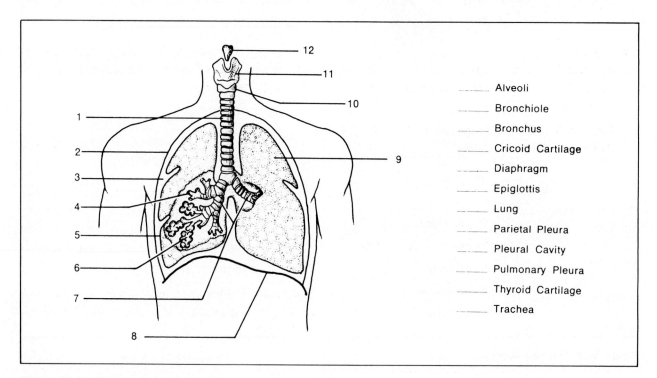

_____ Alveoli
_____ Bronchiole
_____ Bronchus
_____ Cricoid Cartilage
_____ Diaphragm
_____ Epiglottis
_____ Lung
_____ Parietal Pleura
_____ Pleural Cavity
_____ Pulmonary Pleura
_____ Thyroid Cartilage
_____ Trachea

Figure 48.2 The respiratory tract.

6. Rub your fingers over the surface of the lung. What membrane on the surface of the lung makes it so smooth?

7. Free the connective tissue around the *pulmonary artery* and expose its branches leading into the lungs. Also, locate the *pulmonary veins* that empty into the left atrium. Can you find the *ligamentum arteriosum* which is between the pulmonary artery and aorta?

8. With a sharp scalpel free the trachea from the lung tissue and trace it down to where it divides into the **bronchi.**

9. With a pair of scissors cut off the trachea at the level of the top of the heart.

10. Now, cut the trachea down its posterior surface on the median line with a pair of scissors. The posterior surface of the trachea is opposite the heart. Extend this cut down to where it branches into the two primary bronchi. (Observe that the upper right lobe has a separate bronchus leading into it. This bronchus branches off some distance above where the primary bronchi divide from the trachea.)

11. Continue opening up the respiratory tree until you get down deep into the center of the lung. Note the extensive branching.

12. Insert a plastic straw or serological pipette into one of the bronchioles and blow into the lung. If a pipette is used, *put the mouthpiece of the pipette into the bronchiole and blow on the small end.* Note the great expansive capability of the lung.

13. Cut off a lobe of the lung and examine the cut surface. Note the sponginess of the tissue.

14. Record all your observations on the Laboratory Report.

Important: Wash your hands with soap and water at the end of the period.

Frog Lung Observation

To study the nature of living lung tissue in an animal, we will do a microscopic study of the lung of a frog shortly after pithing and dissection. A frog lung is less complex than the human lung, but it is made of essentially the same kind of tissue. Proceed as follows:

Materials:

frog, recently pithed
dissecting instruments and dissecting trays
Ringer's solution

1. Pin the frog, ventral side up, in a wax-bottomed dissecting pan.

2. With your scissors make an incision through the skin along the midline of the abdomen.

3. Make transverse cuts in both directions at each end of the incision and lay back the flaps of skin. Now, carefully make an incision through the right or left side of the abdomen over the lung and parallel to the midline. Take care not to cut too deeply. The inflated lung should now be visible.

4. With a probe gently lift the lung out through the incision. Do not perforate the lung! From this point on keep the lung moist with Ringer's solution.

5. If the lungs of your frog are not inflated, insert a medicine dropper into the slit-like glottis on the floor of the oral cavity and blow air into them by squeezing the bulb. Deflated lungs may be the result of excessive squeezing during pithing.

6. Observe the shape and general appearance of the frog lung. Note that it is basically sac-like.

7. Place the frog under a dissecting microscope and examine the lungs. Locate the network of ridges on the inner walls. The thin-walled regions between the ridges represent the **alveoli.** Look carefully to see if you can detect blood cells moving slowly across the alveolar surface. Careful examination of the lungs may reveal the presence of parasitic worms. They are quite common in frogs.

8. When you have completed this study, dispose of the frog as directed and clean the pan and instruments.

Histological Study

Procure a slide of rat or mouse trachea and lung tissue and examine as follows:

Trachea Examine first under low power and then under high power. Identify the *ciliated epithelium, hyaline cartilage,* and *muscle layers.* Draw a small section of the various layers on a separate sheet of paper.

Lung Tissue Examine some lung tissue under high power and draw a few *alveoli,* a *bronchiole* and *arteriole.*

Laboratory Report

Complete the Laboratory Report for this exercise.

49 Regulation of Respiration

Breathing is a vital activity in which the muscles of inspiration and expiration automatically increase and decrease the volume of the thoracic cavity to ventilate the alveoli sufficiently to satisfy the exact oxygen requirements of all cells in the body. The center of control for this activity is the **respiratory center** of the medulla. The amount of CO_2, H^+, and O_2 in the blood and cerebrospinal fluid (CSF) are the chemical stimuli that act directly, or indirectly, on the respiratory center to regulate the muscles of respiration.

In this exercise, we will study the relative roles of CO_2, H^+, and O_2, and some of the other controlling factors in breathing. The effects of **hyperventilation, deglutition apnea,** the **Valsalva maneuver,** and the **diving reflex** will also be studied. Prior to performing any of these experiments, a review of respiratory center regulation will be covered.

The Significance of pH

The rate of alveolar ventilation is affected primarily by the amount of CO_2 and H^+ in the blood and secondarily by the amount of oxygen in the blood. Figure 49.1 illustrates those parts of the circulatory and nervous systems that are affected by these gases. Note that CO_2 and H^+ in the blood and cerebrospinal fluid (CSF) act directly on neurons of the respiratory center. The amount of oxygen in these fluids has no effect on these neurons.

Although increased levels of CO_2 in the blood cause a greater increase in alveolar ventilation than H^+, it has been established that it is actually the hydrogen ions that directly affect the respiratory center. The more pronounced effect of CO_2 is due to: (1) its greater ease of moving from the blood into the CSF and (2) its ability to increase the hydrogen ion concentration of the CSF by forming carbonic acid. Poor diffusion of H^+ from the blood into the CSF is due to the strong electrical charge of these ions.

Normally, air contains .04% carbon dioxide. If the CO_2 content of inspired air is increased to 1% or 2%, alveolar ventilation is generally unaffected. Levels of CO_2 above 5%, however, cause elevated CO_2 concentration in tissue fluids with increased labored breathing. Maximum ventilation is reached at 9% CO_2. Beyond this level, a further increase in CO_2 begins to depress the respiratory center, causing progressive reduction in the respiratory rate. At 15%–20%, an individual begins to go into a coma; at 50%, death occurs in only a few minutes.

The Role of Oxygen

As indicated in figure 49.1, it is possible for extremely low blood oxygen levels to trigger nerve impulses in the vagus and glossopharyngeal nerves. *Chemoreceptors* that are located in the carotid bodies, aortic arch, and right subclavian arteries are very susceptible to low oxygen levels; however, under normal conditions, oxygen levels never get low enough to trigger nerve impulses to induce increased alveolar ventilation. It is only in diseases such as *pneumonia* and *emphysema* where the oxygen level can become low enough to have a stimulatory effect on the respiratory center. Since the CO_2 and H^+ levels are at depressingly high levels in patients with these lung diseases, it is the stimulatory effect of low oxygen levels that keeps the respiratory muscles functioning. It should be added here, parenthetically, that high CO_2 and H^+ levels also stimulate the chemoreceptors of the carotid bodies and aorta, but this effect is masked by the greater effectiveness of these gases directly on the respiratory center.

Experimental Procedures

A study of hyperventilation, deglutition apnea, the Valsalva maneuver, and the diving reflex will now be performed. Note that the essential supplies and equipment are listed separately for each experiment

Deep Breathing Effects *(Hyperventilation)*

The pH of blood and tissue fluid during normal breathing is around 7.4. If forced deep breathing takes place, the pH of these fluids may be elevated to 7.5 or 7.6 as carbon dioxide is blown off. The reduced hydrogen ion concentration depresses the respiratory center, lessening the desire for increased alveolar ventilation. To observe the effects of this activity proceed as follows:

Materials:

    paper bag

1. Breathe very deeply at a rate of fifteen inspirations per minute for one or two minutes. Do not hurry the rate; concentrate on breathing deeply. Observe that the following symptoms occur:
   - It will become increasingly difficult to breathe deeply.
   - A feeling of dizziness will develop. Two events cause this dizziness. First, the blood pressure drops due to dilation of the splanchnic vessels, lessening the amount of blood to the brain. Second, the reduced CO_2 content of the blood causes vasoconstriction of cerebral blood vessels, further reducing the blood supply.

2. Now, place a paper bag over your nose and mouth, holding it tightly to your face and breathing deeply into it for about 3 minutes. Note how much easier it is to breathe into the bag than into free air. Why is this?

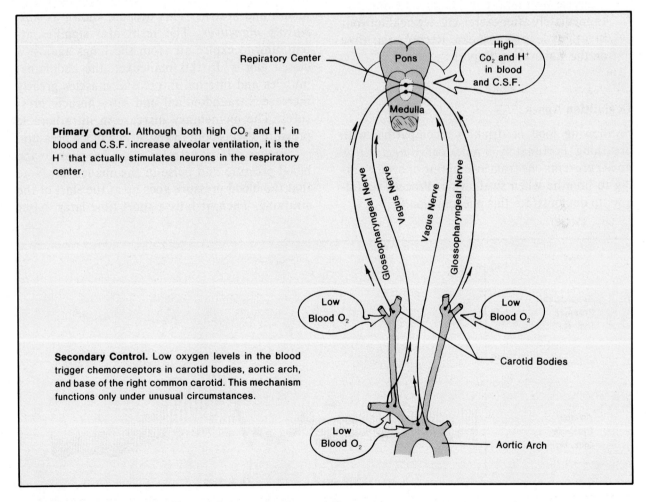

Primary Control. Although both high CO_2 and H^+ in blood and C.S.F. increase alveolar ventilation, it is the H^+ that actually stimulates neurons in the respiratory center.

Secondary Control. Low oxygen levels in the blood trigger chemoreceptors in carotid bodies, aortic arch, and base of the right common carotid. This mechanism functions only under unusual circumstances.

Figure 49.1 Roles of CO_2, $H+$, and O_2 in respiratory regulation.

Hyperventilation in individuals, caused by unconscious deep breathing or sighing, can cause extreme discomfort and even unconsciousness. All the symptoms are due to the washing out of CO_2 from the blood, causing *alkalosis*. This paper bag routine is a necessary maneuver used to reestablish the desired content of CO_2 in the blood.

3. Allow your breathing rate to return to normal and breathe, normally, for 3 or 4 minutes. At the end of a normal inspiration (without deep inhaling) pinch your nose shut with the fingers of one hand and hold your breath as long as you can. Do this three times, timing each duration carefully Record these times and calculate the average on the Laboratory Report.

4. Now, breathe deeply as in step #1 above for 2 minutes and then hold your breath as long as you can. Record your results on the Laboratory Report.

5. Exercise by running in place for 2 minutes. Immediately after exercising, try holding your breath as long as you can. Record your time on the Laboratory Report.

Deglutition Apnea

Swallowing food or drink is incompatible with breathing. Fortunately, a reflex called *deglutition apnea* prevents one from attempting or even wanting to breathe when swallowing. Proceed as follows to demonstrate this phenomenon.

Materials:

clean beaker of drinking water
straw or glass tubing

Hold your breath with your nose closed until the need for oxygen becomes almost unbearable; then, begin to sip water from the beaker through the straw. Does the desire to inhale disappear as you sip through the straw? Can you trace the pattern of nerve impulses in this reflex? Record your answer on the Laboratory Report.

The Valsalva Maneuver

We have seen that conditions in the blood profoundly affect the respiratory rate; conversely, respiratory activities can affect the circulatory system. Valsalva's maneuver will illustrate one of the effects of breathing on the circulatory system.

Antonio Valsalva, the Italian anatomist, discovered in 1723 that the middle ear could be filled with air if one expires forcibly against closed mouth and nostrils. This became known as *Valsalva's maneuver*. The term also signifies attempting to expire air from the lungs against a closed glottis. In this maneuver, the abdominal muscles and internal intercostal muscles greatly increase intraabdominal and intrathoracic pressures. The momentary increase in intrathoracic pressure affects both the pulse and blood pressure.

Figure 49.2 illustrates what happens to the blood pressure and pulse in this maneuver. Note that the blood pressure goes up at the start of the straining. Then, it falls a short time later. After

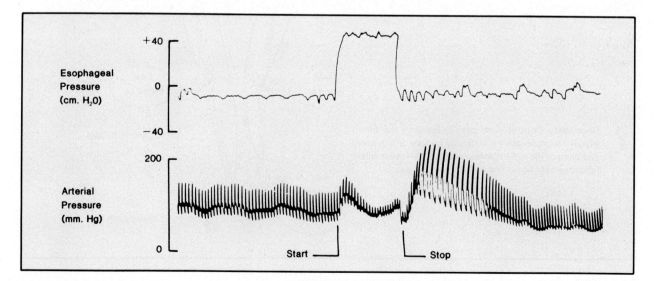

Figure 49.2 Blood pressure and pulse response in Valsalva's maneuver.

the maneuver, the blood pressure goes up considerably more and the heart rate slows somewhat. The initial rise in blood pressure is caused by the addition of intrathoracic pressure to the pressure within the aorta. The subsequent drop in blood pressure is caused by decreased cardiac output, resulting from reduced venous return. The reduced venous return is caused by the restricted flow of blood into the right side of the heart through the compressed vena cavae. The final surge in increased cardiac output is a compensation for the reduced prior output.

The Valsalva maneuver is more than just a stunt to observe in the laboratory. It has real merit in the diagnosis of two pathological conditions. First, patients suffering from *autonomic insufficiency* (a disease of unknown etiology) exhibit no pulse changes during this maneuver. Second, patients with *hyperaldosteronism* fail to show pulse rate changes and blood pressure rise after the maneuver. Patients that have had this latter disease exhibit normal responses in the Valsalva maneuver after aldosterone-secreting tumors are removed.

Work in teams of four or five to monitor these two parameters on one of the team members.

Either a Duograph or Polygraph can be used. To monitor the blood pressure, we will use a Statham pressure transducer that has internal calibration for 100 mm. Hg. No adapter needs to be used between this transducer and the Gilson IC-MP module. The pulse will be determined with a finger pulse pickup as in Exercise 42. Figure 49.3 illustrates the setup.

Caution: Although the Valsalva maneuver is perfectly safe for normal, healthy individuals, any team member who has a history of heart disease should not be used as a subject.

Materials:

    Duograph or Gilson Polygraph
    adult arm cuff
    Statham T-P231D pressure transducer
    stand and clamp for T-P231D transducer
    Gilson T-4020 finger pulse pickup
    Gilson A-4023 adapter for pulse pickup

Equipment Setup To the input end of the #1 module, attach an A-4023 adapter. Make sure that the locknut is tightened securely and that the free jack of the adapter is also plugged into the module. Se-

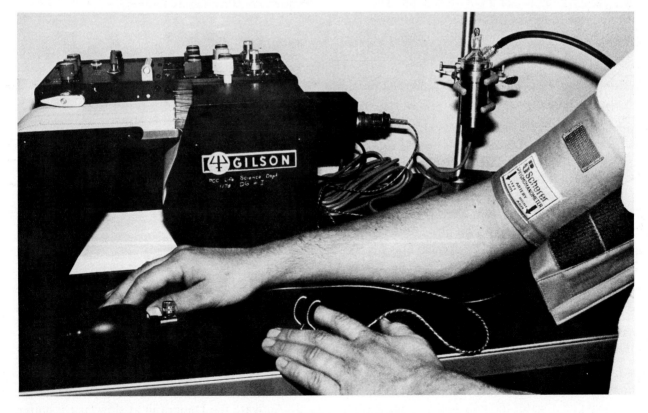

Figure 49.3 Monitoring pulse and blood pressure in Valsalva's maneuver.

cure the jack of the pulse pickup to the adapter. Into the input end of the #2 module, secure the jack of the Statham pressure transducer. If the stand and clamp is available, support the Statham transducer, as shown in figure 49.3. If no support is available, place the transducer in a secure spot where it will not be pushed off the table.

Place the cuff on the upper right arm, securing it tightly in place. Attach the open end of the pressure transducer to the tube that leads into the cuff. Test the cuff by pumping some air into it and noting if the pressure holds when the metering valve is closed on the bulb. Release the pressure for the comfort of the subject while other adjustments are made. Now, insert the index finger of the left hand into the finger pulse pickup.

Plug in the Duograph and turn on the power switch. Set the speed control lever in the SLOW position and the stylus heat control knobs of both channels in the 2:00 position. Set the Gain for channel #1 at 2MV/CM, and for channel #2 at 1 MV/CM, and turn the sensitivity control knobs of both channels completely counterclockwise (lowest value). Set the mode for channel #1 on DC and for channel #2 on TRANS. Calibrate the Statham transducer on channel #2 by depressing the TRANS button and simultaneously adjusting the sensitivity knob to get 2 cm. deflection. The calibrated line represents 100 mm. Hg. pressure. Adjust the sensitivity on channel #1 so that you get 1.5 to 2 cm. deflection on the pulse.

Monitoring. Proceed as follows to monitor the Valsalva maneuver:

1. Pump air into the cuff and establish a baseline for one minute.
2. Have the subject take a deep breath and perform a Valsalva maneuver, exerting as much internal pressure as possible. Just as the maneuver is begun, press down on the event marker button and hold it down until the subject ends the maneuver. Repeat this maneuver several times to provide enough chart material for each team member.
3. Repeat the experiment using other team members as subjects.

The Diving Reflex

Many aquatic, air breathing animals have evolutionary adaptions that enable them to remain submerged under water for quite long periods of time.

The *diving reflex* is one of these adaptations. In these animals there are prompt cardiovascular and respiratory adjustments that are triggered by partial entry of water into the air passages. These adjustments include a change in the heart rate, the shunting of blood from less essential tissues, and a cessation of breathing. In the human, these changes are less pronounced, but some adjustments are made.

To observe the existence of a less-developed diving reflex in the human, we will monitor the changes in respiration and pulse while the face is immersed in water. A subject for this experiment should be one who has no aversion to submersion (of only the face).

The equipment setup for this experiment is almost identical to the setup for the Valsalva maneuver, except that a chest bellows takes the place of the sphygmomanometer cuff. The immersion maneuver may take place at your laboratory station if large pans are available, or at a sink. If the sink is at a perimeter counter of the laboratory, it will be necessary to move the Duograph and other equipment to a spot adjacent to it.

Several variables will be studied in this experiment, such as the effect of temperature changes and the effect of hyperventilation prior to immersion. Proceed as follows:

Materials:

Duograph
finger pulse pickup
A-4023 adapter
Gilson T-4030 chest bellows
Statham T-P231D pressure transducer
sink filled with water, or large pan of water
ice
towel
thermometer

1. Attach the finger pulse pickup and chest bellows to the subject. Keep the chain of the bellows as high into the armpits as possible. Attach the tubing from the bellows to the Statham transducer. Arrange the equipment and connections in such a way that immersion movement does not tug on connections.
2. While the subject is being prepared, fill the pan or sink with tap water, adjusting the temperature to 70° F.
3. Activate the Duograph at slow speed, noting

the trace. Adjust the excursion for tidal breathing at about 1 cm. Establish a reference by recording for at least **one minute** before immersion. During this time, the subject will breathe regularly (tidal inspirations) without speaking or hyperventilating. **When suitable sensitivity and gain are obtained, do not readjust them during the experiment.**

4. To further establish reference values, direct the subject to hold the breath for **one minute** after a tidal inspiration, without prior deep breathing or hyperventilating. Allow the subject to recover for 2 to 3 minutes by normal breathing.

5. After recovery, have the subject completely submerge his or her face into the water. Depress the event button at the exact moment of immersion, holding it down during the full time of immersion and releasing it as soon as the head is raised from the water. **Immersion should last 15 to 30 seconds.** Use a large towel

to protect the subject's clothing from getting wet. Repeat the procedure.

6. Chill the water to 60°F. with ice cubes and repeat step #5. Do not leave ice floating on the water. Repeat the procedure.

7. Have the subject hyperventilate vigorously for 30 seconds and hold the breath for as long as possible. This is done without immersion. When regular breathing has resumed, hyperventilation is repeated and then the face is submerged and held under water **as long as possible.** Be sure to depress the event button during immersion.

8. Disconnect the subject, put away the apparatus, clean the work area, and return all materials to their proper places. Analyze the tracing as a team and label it with additional information as needed.

Laboratory Report

Complete the Laboratory Report for this exercise.

50 Spirometry: Lung Capacities

The volume of air that moves in and out of the lungs during breathing is measured with an apparatus called a **spirometer.** Two types are available: the hand-held (Propper) type (figures 50.2 and 50.3) and the tank-type recording spirometer, (figure 51.1). When a recording spirometer is used a **spirogram** similar to figure 50.1 can be made. The Propper spirometer is designed, primarily, for screening measurements of vital capacity.

In this exercise, we will use the Propper spirometer to determine individual respiratory volumes. Although this convenient device cannot measure volumes of inhalation, it is possible to determine most of the essential lung capacities. While most members of the class are working with the Propper spirometers, some students will be working with the recording spirometer to do the experiment in Exercise 51.

Materials:

Propper spirometer and disposable mouthpieces (Ward's Nat'l Science Est., Monterey, Ca., Cat. No. 14W5070)
70% alcohol
cotton

Tidal Volume (TV)

The amount of air that moves in and out of the lungs during a normal respiratory cycle is called the *tidal volume*. Although sex, age, and weight determine this and other capacities, the average normal tidal volume is around 500 ml. Proceed as follows to determine your tidal volume.

1. Swab the stem of the spirometer with 70% alcohol and place a disposable mouthpiece over the stem (figure 50.2).

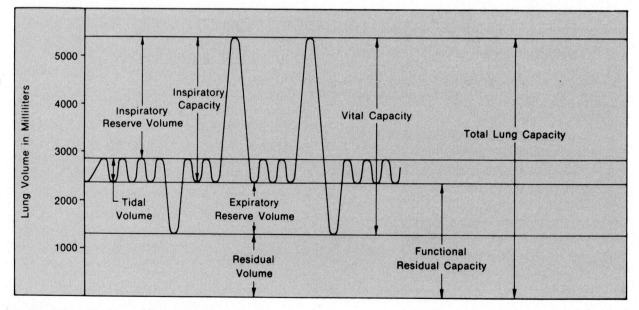

Figure 50.1 Spirogram of lung capacities.

2. Rotate the dial of the spirometer to zero (figure 50.3).

3. After three normal breaths, expire three times into the spirometer while inhaling through the nose. Do not exhale forcibly. *Always hold the spirometer with the dial upward.*

4. Divide the total volume of the three breaths by 3. This is your tidal volume. Record this value on the Laboratory Report.

Minute Respiratory Volume (MRV)

Your *minute repiratory volume* is the amount of tidal air that passes in and out of the lungs in one minute. To determine this value, count your respirations for one minute and multiply this by your tidal volume. Record this volume on the Laboratory Report.

Expiratory Reserve Volume (ERV)

The amount of air that one can expire beyond the tidal volume is called the *expiratory reserve volume.* It is usually around 1,100 ml.

To determine this volume, set the spirometer dial on 1,000 first. After making three normal expirations, expel all the air you can from your lungs through the spirometer. Subtract 1,000 from the reading on the dial to determine the exact volume. Record the results on the Laboratory Report.

Vital Capacity (VC)

If we add the tidal, expiratory reserve, and inspiratory reserve volumes, we arrive at the total functional or *vital capacity* of the lungs. This value is determined by directing an individual take as deep a breath as possible and exhaling all the air possible. Although the average vital capacity for men and women is around 4,500 ml., age, height, and sex do affect this volume appreciably. Even the established norms can vary as much as 20% and still be considered normal. Tables of normal values for men and women are given in Appendix A.

Set the spirometer dial on zero. After taking two or three deep breaths and exhaling completely after each inspiration, take one final deep breath and exhale all the air through the spirometer. A slow, even forced exhalation is optimum.

Repeat two or more times to see if you get approximately the same readings. Your VC should be within 100 ml. each time. Record the average on the Laboratory Report. Consult tables IV and V in Appendix A for predicted (normal) values.

Inspiratory Capacity (IC)

If you take a deep breath to your maximum capacity after emptying your lungs of tidal air, you will have reached your maximum *inspiratory capacity.* This volume is usually around 3,000 ml.

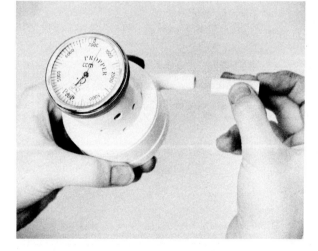

Figure 50.2 The sanitary disposable mouthpiece is applied to the stem of the spirometer.

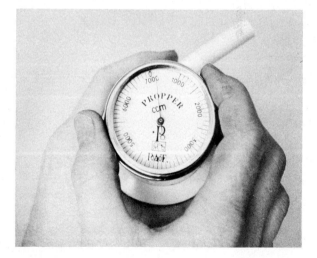

Figure 50.3 The dial face is rotated to zero prior to measuring exhalations.

Note from the spirogram that this volume is the sum of the tidal and inspiratory reserve volumes.

Since this type of spirometer cannot record inhalations, it will be necessary to calculate this volume, using the following formula:

$$IC = VC - ERV$$

Inspiratory Reserve Volume (IRV)

This is the amount of air that can be drawn into the lungs in a maximal inspiration after filling the lungs with tidal air. Since the IRV is the inspiratory capacity less the tidal volume, make this subtraction and record your results on the Laboratory Report.

Residual Volume (RV)

The volume of air in the lungs that cannot be forcibly expelled is the *residual volume*. No matter how hard one attempts to empty one's lungs, a certain amount, usually around 1,200 ml., will remain trapped in the tissues. The magnitude of the residual volume is often significant in the diagnosis of pulmonary impairment disorders. Although it cannot be determined by simple ordinary spirometric methods, it can be done by washing all the nitrogen from the lungs with pure oxygen and measuring the volume of nitrogen expelled.

Laboratory Report

After recording your lung capacities on Laboratory Report 50, 51, evaluate your results.

Spirometry: The FEV~T~ Test

51

Impairment of pulmonary function in the form of asthma, emphysema, and cardiac insufficiency (left-sided heart failure) can be detected with a spirometer. The symptom common to all these conditions is shortness of breath, or *dyspnea*. The lung capacity that is pertinent in these conditions is the vital capacity.

In several types of severe pulmonary impairment, the vital capacities may exhibit nearly normal vital capacities; however, if the *rate* of expiration is recorded on a kymograph and timed, the extent of pulmonary damage becomes quite apparent. This test is called the *Timed Vital Capacity or Forced Expiratory Volume (FEV$_T$)*. Figure 51.1 illustrates a spirometer (Collins Recording Vitalometer) which will be used in this experiment. The record produced on the kymograph in such a test is revealed in figure 51.2 and is called an *expirogram*.

To perform a timed vital capacity test, one makes a maximum inspiration and expels all the air from the lungs into the spirometer as fast as possible. The moving kymograph drum has chart paper on it that is calibrated vertically in liters and

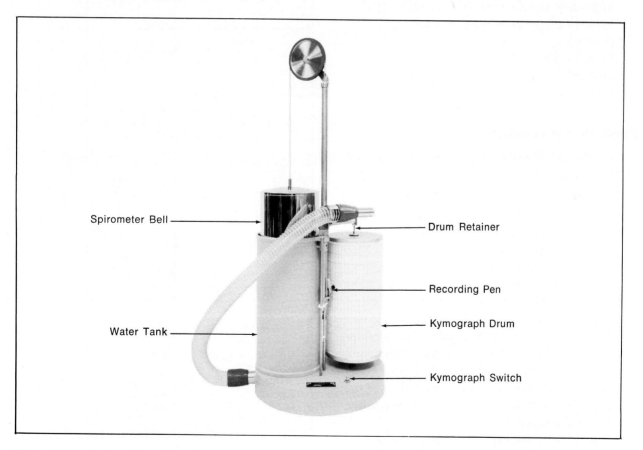

Figure 51.1   A recording spirometer.

horizontally in seconds. A pen produces a record on the chart paper which reveals how much air is expelled per unit of time.

Approximately 95% of a normal person's vital capacity can be expelled within the first three seconds. From a diagnostic standpoint, however, the *percentage of vital capacity that is expelled during the first second* is of paramount importance. An individual with no pulmonary impairment should be able to expel 75% of the lung's total capacity during the first second. Individuals with emphysema and asthma, however, will expel a much lower percentage due to "air entrapment." Given sufficient time to exhale, they might be able to do much better.

In this experiment, each member of the class will have an opportunity to determine his or her FEV$_T$. Disposable mouthpieces will be available for each person. Follow this procedure:

Materials:

> spirometer (Collins Recording Vitalometer)
> disposable mouthpieces (Collins #P612)
> kymograph paper (Collins #P629)
> noseclip
> Scotch tape
> dividers
> ruler

Preparation of Equipment

Prior to performing this test, the equipment should be readied as follows:

1. Remove the spirometer bell and fill the water tank to within 1½″ of the top. Before filling *be sure to close the drain petcock.* Replace the bell, making certain that the bead chain rests in the pulley groove.
2. Remove the kymograph drum by lifting the drum retainer first. Wrap a piece of chart paper around the drum and fasten with Scotch tape. Make sure that the right edge overlaps the left edge. Replace the kymograph drum, checking to see that the bottom spindle is in the hole in the bottom of the drum. Lower the kymograph drum retainer into the hole in the top of the drum.
3. Check the recording pen. Remove the protective cap which prevents it from drying out. If

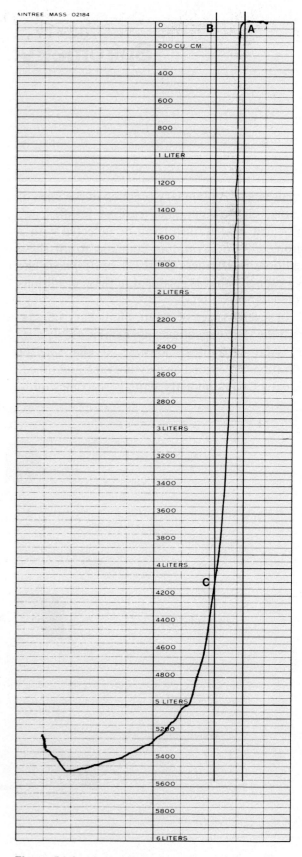

Figure 51.2   An expirogram.

the pen is dry, remove it and replace it with a new pen. Be sure the pen is oriented properly with respect to the drum. It can be rotated to the correct marking position. Also, check the vertical position of the pen to see that it is recording on the zero line. The pen can be adjusted vertically within its socket by simply pushing or pulling on it. Major adjustments should be made by loosening the set screw.

4. Plug the electric cord into the electrical outlet. *This apparatus must not be used in outlets that lack a ground circuit.* The three-pronged plug must fit into a three-holed receptacle. If an adapter is used, be sure to make the necessary ground hookup. Electricity and water can be lethal!

Performing the Test

1. Before exhaling into the mouthpiece, check out the following items to make sure everything is ready.
   - Bell is in lowered position so that the recording pen is exactly on 0.
   - A clean sheet of paper is on the kymograph drum.
   - A sterile mouthpiece has been inserted in the end of the breathing tube.
2. Apply a nose clip to the subject to prevent leakage through the nose. Allow the subject to hold the tube.
3. Before turning on the instrument explain to the subject that (1) he or she will be taking as deep a breath as possible, (2) he or she will expel as much air as possible into the mouthpiece, *as rapidly and completely as possible*, and (3) inhalation should take place *before* the mouthpiece is placed in his or her mouth.
4. Turn on the kymograph and tell the subject to perform the test as described. As the subject blows into the tube, encourage him or her to *push, push, push* to get all the air out.
5. Turn off the kymograph.
6. After recording as many tracings as desired, remove the paper from the drum.

Analysis of Tracing

To determine the FEV$_T$ from this expirogram, proceed as follows:

1. Draw a vertical line through the starting point of exhalation (see point A, figure 51.2).
2. Set a pair of dividers to the distance between two vertical lines. This distance represents one second.
3. Place one point of the dividers on point A and establish point B to the left of point A. Draw a parallel vertical line through point B. This line intersects the expiratory curve at point C.

   The one second volume (FEV$_1$) or one second timed vital capacity is represented by the distance B–C. This volume is read directly from the volume markings. In figure 51.2, it is 4,100 ml. The total vital capacity is also read directly from the volume markings. It is 5,500 ml.
4. Next, correct the recorded total vital capacity for body temperature, ambient pressure, and water saturation (BTPS). To do this, consult table III in Appendix A for the conversion factor and multiply the recorded vital capacity by this factor.

   Example: If the temperature in the spirometer is 25°C, the conversion factor is 1.075. Thus:

   5500 ml. $\times$ 1.075 = 5913 ml.

5. Also, convert the one second timed vital capacity (FEV$_1$) in the same manner:

   4100 ml. $\times$ 1.075 = 4408 ml.

6. Divide the FEV$_1$ by the total vital capacity:

   $$\frac{4408}{5913} = 75\%$$

7. Determine how the subject's vital capacity compares with the predicted (normal) values in tables IV and V of Appendix A.

   Example: Individual in the above test was a 26-year-old male, 6'3" (184 cm.) tall. From table V his predicted vital capacity is 4,545 ml.

   $$\frac{\text{Actual VC}}{\text{Predicted VC}} \times 100\% = \frac{5913}{4545} \times 100 = 131\%$$

Laboratory Report

Complete the last portion of Laboratory Report 50, 51.

Part 11 The Digestive System

Although this unit consists of only two exercises, both of them are quite comprehensive and complete in scope. The first portion of Exercise 52 relates primarily to anatomical discussion and should be completed prior to entering the laboratory for slide studies. The laboratory work involves the study of taste buds, salivary glands, and tissue studies of the stomach and intestines.

Exercise 53 is an in-depth study of the three basic hydrolytic reactions that occur in digestion; i.e., carbohydrate, protein, and fat hydrolysis. After removing the pancreas from a rat, the enzymes will be extracted and analyzed. The broad scope of this experiment will require that students work in teams to facilitate completion within a single laboratory period. Questions on the Laboratory Report will also review all hydrolytic reactions of the digestive system.

The digestion and absorption of food by the digestive system is accomplished by the alimentary tract, liver, and pancreas. In this exercise we will study the components of this system.

Alimentary Canal

Figure 52.1 is a simplified illustration of the digestive system. The alimentary canal, which is about thirty feet in length, has been foreshortened in the intestines for clarity. The following text which pertains to this illustration is related to the passage of food through its full length.

Materials:

manikin

When food is taken into the oral cavity, it is chewed and mixed with saliva that is secreted by many glands of the mouth. The most prominent of these glands are the parotid, submaxillary, and sublingual glands. The **parotid glands** are located in the cheeks in front of the ears, one on each side of the head. The **sublingual glands** are located under the tongue and are the most anterior ones of the three pairs. The **submandibular** glands are situated posterior to the sublinguals, just inside of the body of the mandible. (Figure 52.4 shows the position of these glands more precisely.)

After the food has been completely mixed with saliva it passes to the **stomach** by way of a long tube, the **esophagus.** The food is moved along the esophagus by wave-like constrictions called *peristaltic waves*. These constrictions originate in the **oropharynx,** which is the cavity at the top of the esophagus, posterior to the tongue. The upper opening of the stomach through which the food enters is the **cardiac valve.** The upper rounded portion, or **fundus**, of the stomach holds the bulk of the food to be digested. The lower portion, or **pyloric region,** is smaller in diameter, more active, and accomplishes most of the digestion that oc-

curs in the stomach. That region between the fundus and the pyloric portion is the **body.** After the food has been acted upon by the various enzymes of the gastric fluid, it is forced into the small intestine through the **pyloric valve** of the stomach.

The *small intestine* is approximately 23 feet long and consists of 3 parts: the duodenum, jejunum, and ileum. The first 10 to 12 inches make up the **duodenum.** The **jejunum** comprises the next 7 or 8 feet, and the last coiled portion is the **ileum.** Complete digestion and absorption of food takes place in the small intestines.

Indigestible food and water pass from the ileum into the large intestine, or **colon,** through the **ileocecal valve.** This valve is shown in a cutaway section. The large intestine has four sections: the ascending, transverse, descending, and sigmoid colons. The **ascending colon** is the portion of the large intestine that the ileum empties into. At its lower end is an enlarged compartment or pouch, the **cecum,** which has a narrow tube extending down from it which is the **appendix.** The ascending colon ascends on the right side of the abdomen until it reaches the under surface of the liver where it bends abruptly to the left, becoming the **transverse colon.** The **descending colon** passes down the left side of the abdomen where it changes direction again, becoming the **sigmoid colon.** The last portion of the alimentary canal is the **rectum,** which is about five inches long and terminates with an opening, the **anus.**

Leading downward from the inferior surface of the **liver** is the **hepatic duct.** This duct joins the **cystic duct,** which connects with the round sac-like **gallbladder.** Bile, which is produced in the liver, passes down the hepatic duct and up the cystic duct to the gallbladder where it is stored until needed. When the gallbladder contracts, bile is forced down the cystic duct into the **common bile duct,** which extends from the juncture of the cystic and hepatic ducts to the intestine.

Between the duodenum and the stomach lies

another gland, the **pancreas.** Its duct, the **pancreatic duct,** joins the common bile duct and empties into the duodenum.

Assignment:

Label figure 52.1. Disassemble the manikin and identify all of these structures.

Answer the questions on the Laboratory Report pertaining to this part of the exercise.

Oral Anatomy

A typical normal mouth is illustrated in figure 52.2. To provide maximum exposure of the oral structures, the lips (*labia*) have been retracted away from the teeth, and the cheeks (*buccae*) have been cut. The lips are flexible folds which meet laterally at the *angle* of the mouth where they are continuous with the cheeks. The lining of the lips,

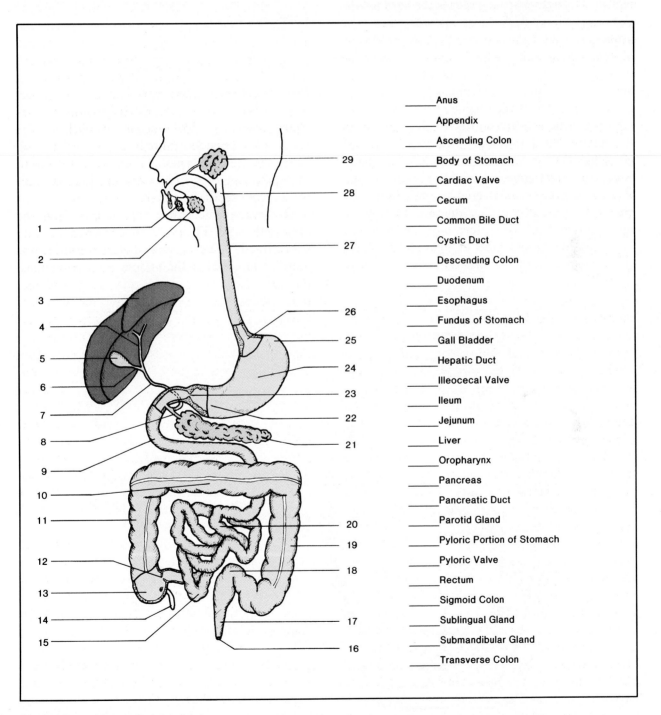

_____ Anus

_____ Appendix

_____ Ascending Colon

_____ Body of Stomach

_____ Cardiac Valve

_____ Cecum

_____ Common Bile Duct

_____ Cystic Duct

_____ Descending Colon

_____ Duodenum

_____ Esophagus

_____ Fundus of Stomach

_____ Gall Bladder

_____ Hepatic Duct

_____ Illeocecal Valve

_____ Ileum

_____ Jejunum

_____ Liver

_____ Oropharynx

_____ Pancreas

_____ Pancreatic Duct

_____ Parotid Gland

_____ Pyloric Portion of Stomach

_____ Pyloric Valve

_____ Rectum

_____ Sigmoid Colon

_____ Sublingual Gland

_____ Submandibular Gland

_____ Transverse Colon

Figure 52.1 The alimentary canal.

cheeks, and other oral surfaces consists of mucous membrane, the *mucosa*. Near the median line of the mouth on the inner surface of the lips, the mucosa is thickened to form folds, the **labial frenula,** or *frena*. Of the two frenula, the upper one is usually stronger. The delicate mucosa that covers the neck of each tooth is the **gingiva.**

The hard and soft palates are distinguishable as differently shaded areas on the roof of the mouth. The lighter shaded area is the **hard palate.** Posterior to it is a darker area, the **soft palate.** The **uvula** is a part of the soft palate. It is the finger-like projection which extends downward over the back of the tongue. It varies considerably in size and shape in different individuals.

On each side of the tongue at the back of the mouth are the **palatine tonsils.** Each tonsil lies in a recess bounded anteriorly by a membrane, the **glossopalatine arch,** and posteriorly by a membrane, the **pharyngopalatine arch.** The tonsils consist of lymphoid tissue covered by epithelium. The epithelial covering of these structures dips inward into the lymphoid tissue, forming gland-like pits called **tonsillar crypts** (label 11, figure 52.3). These crypts connect with channels that course through

the lymphoid tissue of the tonsil. If they are infected, however, their protective function is impaired and they may actually serve as foci of infection. Inflammation of the palatine tonsils is called *tonsillitis*. Enlargement of the tonsils tends to obstruct the throat cavity and interfere with the passage of air to the lungs.

Assignment:

Label figure 52.2.

The Tongue

Figure 52.3 shows the tongue and adjacent structures. It is a mobile mass of striated muscle completely covered with mucous membrane. The tongue is subdivided into three parts: the apex, body, and root. The **apex** of the tongue is the most anterior tip which rests against the inside surfaces of the front teeth. The **body** *(corpus)* is the bulk of the tongue which extends posteriorly from the apex to the root. The posterior border of the body is arbitrarily located somewhere anterior to the tonsillar material of the tongue. Extending down

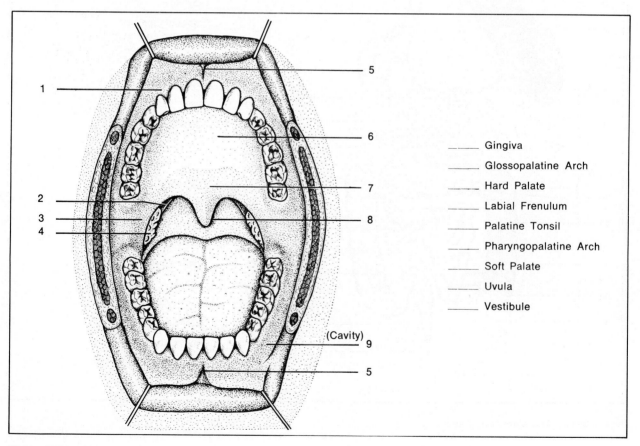

_____ Gingiva

_____ Glossopalatine Arch

_____ Hard Palate

_____ Labial Frenulum

_____ Palatine Tonsil

_____ Pharyngopalatine Arch

_____ Soft Palate

_____ Uvula

_____ Vestibule

Figure 52.2   The oral cavity.

the median line of the body is a groove, the **central sulcus.** The **root** of the tongue is the most posterior portion. The major portion of its surface is covered by the **lingual tonsil.**

Extending upward from the posterior margin of the root of the tongue is the **epiglottis.** On each side of the tongue are seen the oval **palatine tonsils.**

The dorsum of the tongue is covered with several kinds of projections called *papillae.* The cut-out section is an enlarged portion of its surface to reveal the anatomical differences between these papillae. The majority of the projections have tapered points and are called **filiform papillae.** These projections are sensitive to touch and give the dorsum a rough texture. This roughness provides friction for the handling of food. Scattered around among the filiform papillae are the larger rounded **fungiform papillae.** Two fungiform papillae are shown in the sectioned portion. A third type of papilla is the donut-shaped **vallate** (*circumvallate*) **papilla,** which is the largest of the three kinds. They are arranged in a "V" near the posterior margin of the dorsum. Although the exact number may vary in individuals, eight vallate papillae are shown.

The fungiform and vallate papillae contain **taste buds,** the receptors for taste. The circular furrow of the vallate papilla in the cut-out section reveals the presence of these receptors. Comparatively speaking, the vallate papillae contain many more taste buds than the fungiform papillae.

A fourth type of papilla is the **foliate papilla.** These projections exist as vertical rows of folds of mucosa on each side of the tongue, posteriorly. A few taste buds are also scattered among these papillae.

Assignment:

Label figure 52.3.

The Major Salivary Glands

The salivary glands are grouped according to size. The largest glands, of which there are three pairs, are referred to as the *major salivary glands.* Small glands, about 2–5 mm. in diameter, are called the *minor salivary glands;* they are embedded under

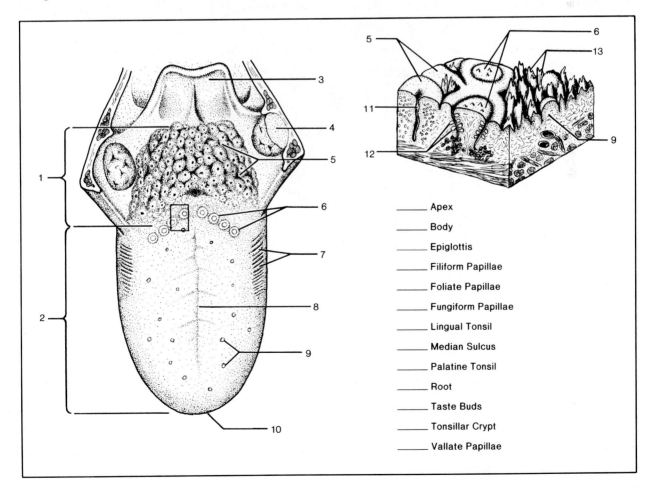

_____ Apex

_____ Body

_____ Epiglottis

_____ Filiform Papillae

_____ Foliate Papillae

_____ Fungiform Papillae

_____ Lingual Tonsil

_____ Median Sulcus

_____ Palatine Tonsil

_____ Root

_____ Taste Buds

_____ Tonsillar Crypt

_____ Vallate Papillae

Figure 52.3 Tongue anatomy.

the mucosa of the lips, palate, cheeks, and tongue. Our principal concern here will be with the major glands that are illustrated in figure 52.4.

Parotid Glands The parotid glands are the largest of the major salivary glands. Note in figure 52.4 that the parotid gland lies in front of the ear between the skin and the **masseter muscle.** Leading from this gland is **Stenson's duct,** which passes over the masseter and through the **buccinator muscle** to an outlet near the upper second molar. The secretion of the parotid glands is a clear watery fluid that contains the digestive enzyme *salivary amylase.* Acid (sour) substances in the mouth act as stimuli, causing the parotids to secrete.

Submandibular *(submaxillary)* **Glands** These glands lie within the body of the mandible. Note how the bone of the mandible has been cut away to reveal a portion of the left submandibular gland. Note, also, how the flat **mylohyoid** muscle extends

somewhat into the gland. The secretions of this gland empty into the oral cavity through **Wharton's duct,** which exits through the **opening of Wharton's duct** under the tongue near the lingual frenulum. The **lingual frenulum** (label 2, figure 52.4) is the mucosal fold on the median line between the tongue and the floor of the mouth.

The secretion of the submandibulars is a little more viscous than that of the parotids since it contains some *mucin.* Mucin in saliva helps to hold the food together in a bolus during swallowing. Bland substances such as bread and milk stimulate these glands.

Sublingual Glands The sublingual glands are located under the tongue. A single sublingual is seen in figure 52.4, with several short **ducts of Rivinus** protruding upward. The number of ducts is variable. The duct openings lie in a ridge of mucous membrane called the **sublingual fold.** Can you feel this fold with the tip of your tongue?

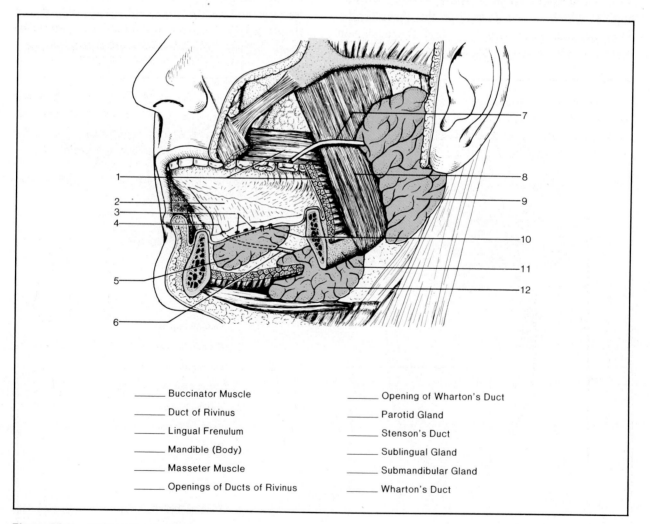

_____ Buccinator Muscle	_____ Opening of Wharton's Duct
_____ Duct of Rivinus	_____ Parotid Gland
_____ Lingual Frenulum	_____ Stenson's Duct
_____ Mandible (Body)	_____ Sublingual Gland
_____ Masseter Muscle	_____ Submandibular Gland
_____ Openings of Ducts of Rivinus	_____ Wharton's Duct

Figure 52.4 Major salivary glands.

The sublinguals differ from the other two pair in that the sublinguals are primarily mucous glands. The high mucin content of this secretion makes it quite "ropey."

Assignment:

Label figure 52.4.

The Teeth

Every individual develops two sets of teeth during the first twenty-one years of life. The first set, which begins to appear at approximately six months of age, are the *deciduous* teeth. Various other terms such as *primiry, baby,* or *milk* teeth are also used in reference to these teeth. The second set of teeth, known as *permanent,* or *secondary,* teeth, begin to appear when the child is about six years of age. As a result of normal growth from the sixth year on, the jaws enlarge, the secondary teeth begin to exert pressure on the primary teeth, and *exfoliation,* or shedding, of the deciduous dentition occurs.

The Deciduous Teeth

The deciduous teeth number twenty in all—five in each quadrant of the jaws. Normally, all twenty have erupted by the time the child is two years old. Starting with the first tooth at the median line,

they are named as follows: **central incisor, lateral incisor, cuspid, first molar,** and **second molar.** The naming of the lower teeth follows the same sequence.

Once the two year old child has all of the deciduous teeth, no visible change in the teeth occurs until around the sixth year. At this time the first permanent molars begin to erupt.

The Permanent Teeth

For approximately five years (7th to 12th year) the child will have a *mixed dentition,* consisting of both deciduous and permanent teeth. As the submerged permanent teeth enlarge in the tissues, the roots of the deciduous teeth undergo *resorption.* This removal of the underpinnings of the deciduous teeth results eventually in exfoliation.

Figure 52.5 illustrates the permanent dentition. Note that there are sixteen teeth in one half of the mouth—thus, a total of thirty-two teeth. Naming them in sequence from the median line of the mouth they are: **central incisor, lateral incisor, cuspid, first bicuspid, second bicuspid, first molar, second molar,** and **third molar.** The third molar is also called the *wisdom tooth.*

Comparisons of the teeth in figure 52.5 reveal that the anterior teeth have single roots and the posterior teeth may have several roots. Of the bicuspids, the only ones that have two roots *(bifur-*

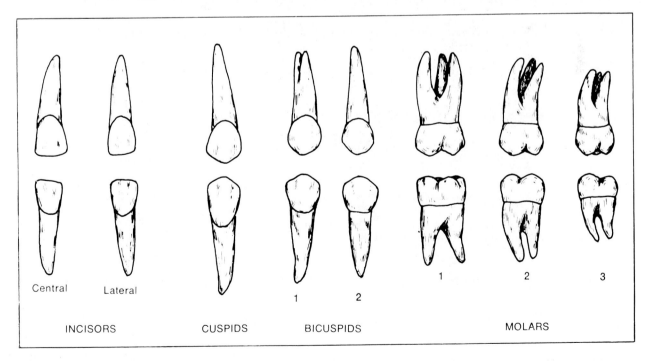

Central   Lateral

INCISORS   CUSPIDS   BICUSPIDS   MOLARS

Figure 52.5   The permanent teeth.

cated) are the maxillary first bicuspids. Although all of the maxillary molars are shown with three roots *(trifurcated)* there may be considerable variability, particularly with respect to the third molar. The roots of the mandibular molars are generally bifurcated. The longest roots are seen on the maxillary cuspids.

Assignment:

Identify the teeth of skulls that are available in the laboratory.

Tooth Anatomy

The anatomy of an individual tooth is shown in figure 52.6. Longitudinally, it is divided into two portions, the crown and root. The line where these two parts meet is the **cervical line** or *cemento-enamel juncture.*

The dentist sees the crown from two aspects: the anatomical and clinical crowns. The **anatomical crown** is the portion of the tooth that is covered with enamel. The **clinical crown,** on the other hand, is the portion of the crown that is exposed in the mouth. The structure and physical condi-

tion of the soft tissues around the neck of the tooth will determine the size of the clinical crown.

The tooth is composed of four tissues: the enamel, dentin, pulp, and cementum. **Enamel** is the most densely mineralized and hardest material in the body. Ninety-six percent of enamel is mineral. The remaining four percent is a carbohydrate-protein complex. Calcium and phosphorus make up over fifty percent of the chemical structure of enamel. Microscopically this tissue is made up of very fine rods or prisms that lie approximately perpendicular to the outer surface of the crown. It has been estimated that the upper molars may have as many as twelve million of these small prisms per tooth. The hardness of enamel enables the tooth to withstand the abrasive action of one tooth against another.

Dentin is the material that makes up the bulk of the tooth and lies beneath the enamel. It is not as hard or brittle as the enamel and resembles bone in composition and hardness. This tissue is produced by a layer of **odontoblasts** (label 15) which lie in the outer margin of the pulp cavity.

Cementum is the hard dental tissue that covers the anatomical root of the tooth. It is a modified

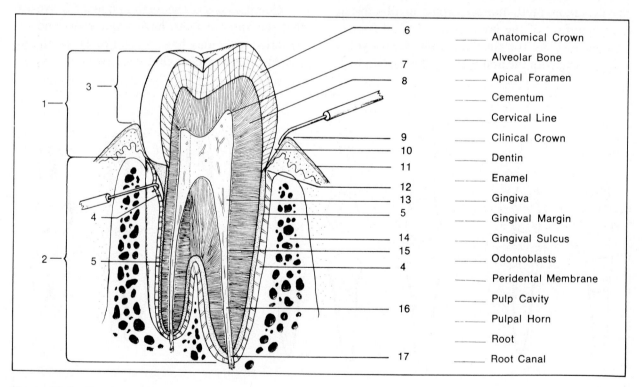

_____	Anatomical Crown
_____	Alveolar Bone
_____	Apical Foramen
_____	Cementum
_____	Cervical Line
_____	Clinical Crown
_____	Dentin
_____	Enamel
_____	Gingiva
_____	Gingival Margin
_____	Gingival Sulcus
_____	Odontoblasts
_____	Peridental Membrane
_____	Pulp Cavity
_____	Pulpal Horn
_____	Root
_____	Root Canal

Figure 52.6 Tooth anatomy.

bone tissue, somewhat resembling osseous tissue. It is produced by cells called *cementocytes*, which are very similar to osteocytes. The primary function of the cementum is to provide attachment of the tooth to surrounding tissues in the alveolus.

The cavity that occupies the central portion of the tooth is called the **pulp cavity.** Extending down into the roots, this cavity becomes the **root canals.** Where this chamber extends up into the cusps of the crown, it forms the **pulpal horns.** The tissue of the pulp is essentially loose connective tissue. It consists of fibroblasts, intercellular ground substance, and white fibers. Permeating this tissue are blood vessels, lymphatic vessels, and nerve fibers. The functions of the pulp are to (1) provide nourishment for the living cells of the tooth, (2) provide some sensation to the tooth, and (3) produce dentin. The blood vessels and nerve fibers that enter the pulp cavity do so through openings, the **apical foramina,** in the tips of the roots. Inflammation of the pulp, or *pulpitis,* may result in the destruction of the blood vessels and nerves producing a dead or *devitalized* tooth.

Between the cementum and the alveolar bone lies a vascular layer of connective tissue, the **peridental membrane.** On the left side of figure 52.6 it is shown pulled away from the root. One of the principal functions of this membrane is tooth retention. The membrane consists of bundles of fibers that extend from the cementum to the alveolar bone, firmly holding the tooth in place. Its presence also acts as a cushion to reduce the trauma of occlusal action. Another very important function of this membrane is tooth sensitivity to touch. This sensitivity is due to the presence of free nerve ending receptors in the membrane. The slightest touch at the surface of the tooth is transmitted to these nerve endings through the medium of the peridental membrane. Even if the apical parts of the membrane are removed, as in root tip resection, the sense of touch is not impaired.

To demonstrate the depth of the **gingival sulcus,** a probe is seen on the right side of figure 52.6. This crevice is frequently a site for bacterial putrefaction and the origin of *gingivitis.* The upper edge of the gingiva is called the **gingival margin.** The portion of the gingiva that lies over the alveolar bone is called the **alveolar mucosa.**

Assignment:

Label figure 52.6.

Microscopic Studies

1. **Taste Buds.** Examine a prepared microscope slide of a section through the vallate papillae of the tongue. Draw a few taste buds and surrounding tissue.
2. **Salivary Glands.** Examine a prepared microscope slide of the salivary gland tissue under high power. Draw a few cells.
3. **Stomach Wall.** Examine a prepared microscopic slide of the stomach wall. Examine first under low power to identify the inner lining (gastric mucosa), muscle layers, and peritoneum. Make a drawing that illustrates, diagrammatically, the arrangement of these layers.

    Next, examine the tissue under higher power (high-dry or oil immersion). Make a drawing of four or five cells of the inner lining.
4. **Intestinal Wall.** Examine a prepared slide of a section through the intestines under low and high powers. Do you see smooth muscle layers similar to those seen in the stomach? Draw a few cells of the inner lining.

Laboratory Report

Complete the Laboratory Report for this exercise.

53 The Chemistry of Hydrolysis

The basic nutrients of the body consist of small quantities of vitamins and minerals plus large amounts of carbohydrates, fats, and proteins. While vitamins and minerals play a vital role in all physiological activities, it is the carbohydrates, fats and proteins that provide the energy and raw materials for growth and tissue repair.

Food, as it is normally ingested, consists of large complex molecules of carbohydrates, fats, and proteins that cannot pass through the intestinal wall unless they are converted to smaller molecules by digestion in the stomach and intestines. The conversion of these large molecules to smaller ones is accomplished by digestive enzymes. All digestive enzymes split these molecules by hydrolysis.

Hydrolysis is a process whereby the larger molecules are split into smaller units by combining with water. Although hydrolysis without enzymes will occur automatically at body temperature, the process is extremely slow; the catalytic action of digestive enzymes, on the other hand, greatly hastens the process. The end result of this action is to reduce carbohydrates to monosaccharides, fats to fatty acids and glycerol, and proteins to amino acids. The large size of some food molecules requires that different enzymes work at different levels to produce intermediate size molecules before the end products are produced.

Digestive enzymes are produced by the salivary glands, the stomach wall, the intestinal wall, and the pancreas. Since the pancreas produces all three of the basic types of digestive enzymes (proteases, lipases, and carbohydrases), we will focus our attention here on this gland.

An enzyme extract from the pancreas of a freshly killed rat will be analyzed for the presence of the three types. Figure 53.1 reveals the steps that will be employed for extracting the pancreatic juice. Since enzymes are relatively unstable and easily denatured by certain adverse environmental conditions, it will be essential that chemical purity, cleanliness, and temperature control be maintained.

Once the pancreatic juice has been extracted, tests will be made for evidence of carbohydrase, lipase, and protease activity. Small amounts of the pancreatic juice will be added to substrates of starch, fat, and protein to observe the hydrolytic action of the various enzymes. Color changes that occur in each test will indicate the breaking of bonds within the large molecules to produce smaller molecules. Positive and negative test controls will be used for comparisons with the actual tests.

The class will be divided up into groups of four students. While two of the students are dissecting the rat, the other two will set up the necessary supplies and equipment for the extraction and assay procedures.

Extraction Procedure

Remove the pancreas from a freshly killed rat and produce the pancreatic extract using the following procedures.

Materials:

> freshly killed rat
> refrigerated centrifuge
> centrifuge tubes
> balance
> Sorvall blender
> beaker, 30 ml. size
> Erlenmeyer flask, 125 ml. size
> graduated cylinder, 100 ml. size
> crushed ice

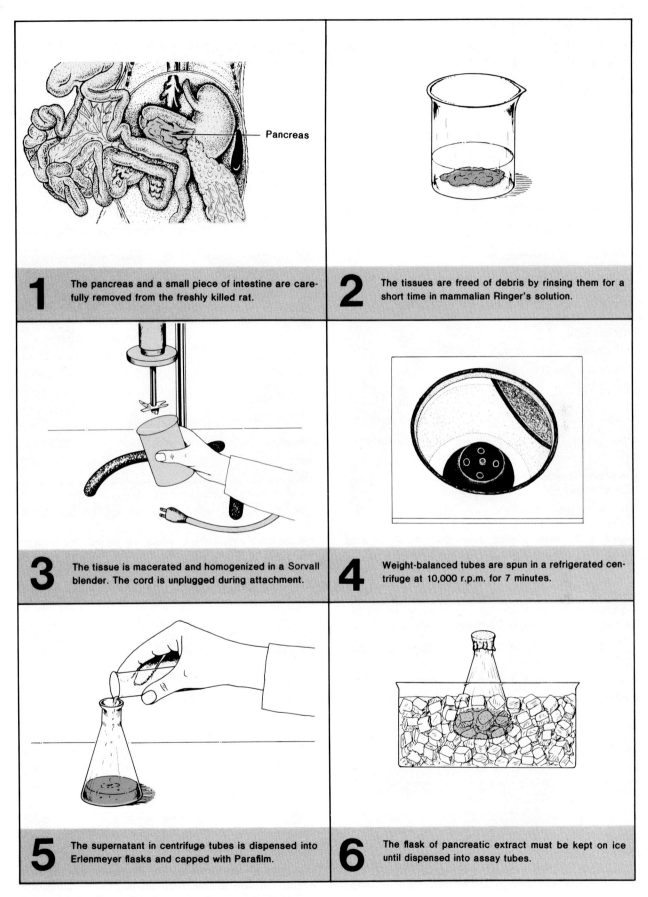

1 The pancreas and a small piece of intestine are carefully removed from the freshly killed rat.

2 The tissues are freed of debris by rinsing them for a short time in mammalian Ringer's solution.

3 The tissue is macerated and homogenized in a Sorvall blender. The cord is unplugged during attachment.

4 Weight-balanced tubes are spun in a refrigerated centrifuge at 10,000 r.p.m. for 7 minutes.

5 The supernatant in centrifuge tubes is dispensed into Erlenmeyer flasks and capped with Parafilm.

6 The flask of pancreatic extract must be kept on ice until dispensed into assay tubes.

Pancreas

Figure 53.1   Procedure for extracting pancreatic enzymes.

cold mammalian Ringer's solution
dissecting pan
dissecting instruments
dissecting pins
Parafilm

1. Follow the steps outlined in figures 53.2 through 53.5 to open up the abdominal cavity of a freshly killed rat and remove the pancreas. Place the gland into a beaker containing a small amount (about 20 ml.) of cold mammalian Ringer's solution. Snip off a small piece of the intestine and add that, too.

2. After washing the tissues in the Ringer's solution, macerate and homogenize them in a blender. The tissues of several groups should be blended together to get ample bulk.

3. Distribute the blended mixture into evenly balanced centrifuge tubes and spin the tubes in a refrigerated centrifuge (Beckman or other) for 7 minutes at 10,000 rpm. Tubes must be balanced by weighing carefully and adjusting the contents until they are equal.

4. Decant supernatant from the tubes into Erlenmeyer flasks for each group. The super-

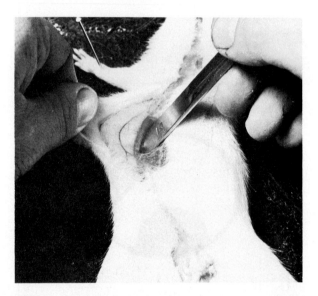

Figure 53.2 After pinning down the feet and cutting through the skin, separate the skin from the musculature with the scalpel handle.

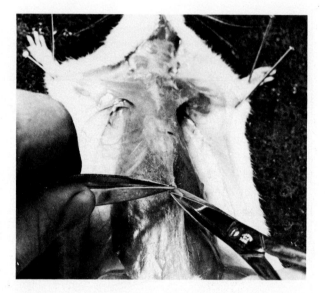

Figure 53.3 Hold the musculature up with the forceps while cutting through the wall. Be careful not to damage any organs.

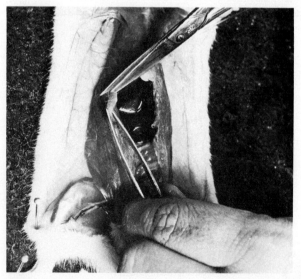

Figure 53.4 As the muscular body wall is opened, pull flaps of muscle wall laterally and pin them down to keep the cavity open.

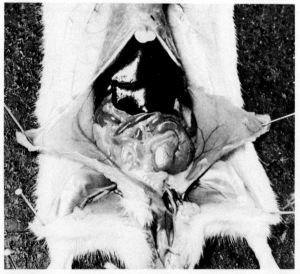

Figure 53.5 When the abdominal organs have been exposed, carefully expose the pancreas by lifting the liver with forceps.

natant will be a milky pink mixture of digestive enzymes.

5. Place the flask of pancreatic extract into a container of ice to prevent enzyme deterioration.

Enzyme Assay

Figure 53.6 illustrates the number of test tubes that will be set up to assay the pancreatic extract for carbohydrase, protease, and lipase. The detection of the presence of each enyzme will depend on a specific color test after the tubes have been

incubated in a 38°C. water bath for one hour. Tube 1 in each series contains pancreatic extract, a substrate, and necessary buffering agents. The last tube in each series is a **positive test control** that contains the enzyme we are testing for in tube 1. All other tubes in each series should give negative results because of certain deficiencies. Note that tube 2 in each series contains the same ingredients as tube 1 except for two significant variables: (1) the tube is placed on ice while the others are incubated and (2) the pancreatic extract is not

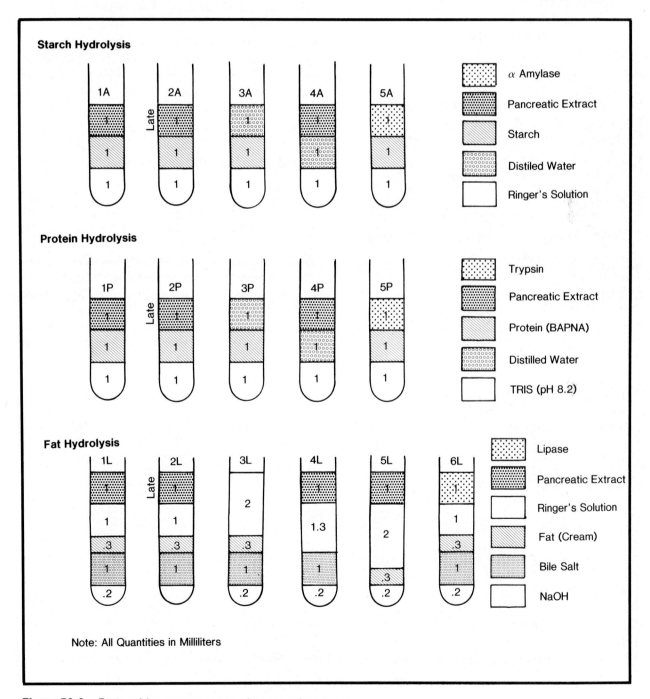

Figure 53.6 Protocol for enzyme assay of pancreatic extract.

added until after one hour on ice. This tube is designated as our **zero time control.**

All four members of each team will work co-operatively to set up these 16 tubes. After the tubes have been in the water baths for one hour, 1 ml. of pancreatic juice will be added to the three zero time tubes. Various tests will then be performed to detect the presence of hydrolysis.

Materials:

> water baths (38°C.)
> Vortex mixer
> pipetting devices
> test tube rack (wire type)
> test tubes (16 mm. × 150 mm.), 16 per
>   group
> hot plates
> pipettes, 1 ml., 5 ml., and 10 ml. sizes
> graduated cylinder (10 ml. size)
> bromthymol blue pH color standards
> china marking pencil
> photocolorimeter (Spectronic 20)
> spot plates

Solutions in Dropping Bottles

I-KI solution	2% sodium taurocholate;
Benedict's solution	a bile salt
Barfoed's solution	soluble starch, 0.1%
bromthymol blue	cream, unhomogenized
amylase, 1%	albumin, 0.1%
trypsin, 1%	NaCl, 0.2%
lipase, 1%	NaOH, 0.1N
	TRIS (ph 8.2)

Carbohydrase

Carbohydrate in plants is stored, primarily, in the form of starch. Starch is composed of 98% amylose and 2% amylopectin. Amylose is a straight chain polysaccharide of glucose units attached by **α–1–4 glucosidic bonds.** Amylopectin is a branching polysaccharide with side chains attached to the main chain by **α–1–6 glucosidic bonds.** These linkages are illustrated in figure 53.7.

The digestion of starch is catalyzed by *α amylase,* an enzyme present in saliva, pancreatic juice, and intestinal juice. This enzyme randomly attacks the α–1–4 bonds. An *α–1–6 glucosidase* is necessary to break the α–1–6 bond. The action of α amylase on starch results in the formation of molecules of maltose and isomaltose. These two molecules are converted to glucose by the enzymes *maltase* and *isomaltase* in the epithelial cells of the small intestines.

Set up the five tube test series for α amylase according to the following protocol:

1. With a china marking pencil, label the test tubes 1A through 5A (*A* for amylase).

2. With a 5 ml. or 10 ml. pipette, deliver 1 ml. of **mammalian Ringer's solution** to each of the five tubes. Use a mechanical pipetting device such as the one illustrated in figure 53.10.

3. Flush out the pipette with water and deliver 1 ml. of **distilled water** to tubes 3 and 4.

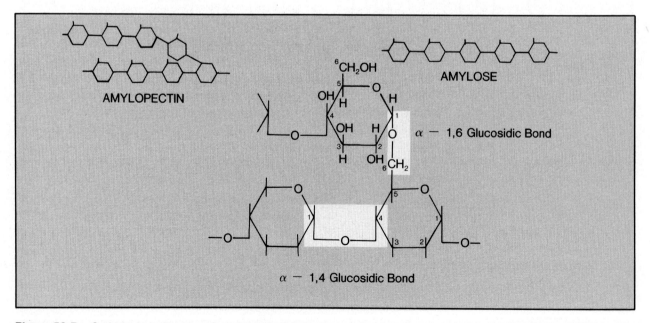

Figure 53.7 Carbohydrate linkage where hydrolysis occurs.

4. Using the same pipette, deliver 1 ml. of **starch solution** to tubes 1, 2, 3, and 5. Note that tube 4 does not get any starch.
5. With a fresh 5 ml. pipette, deliver 1 ml. of **pancreatic juice** to tubes 1 and 4.
6. With a fresh 1 ml. pipette, deliver 1 ml. of α **amylase** to tube 5.
7. Mix each tube on a Vortex mixer for a few seconds (see figure 53.11).
8. Remove tube 2 from the series, cover it with Parafilm, and place it in a rack that is immersed in ice water.
9. Cover the other four tubes with Parafilm and place the test tube rack with these tubes in a 38°C. water bath for **one hour.**

Protease

The dietary proteins are derived, primarily, from meats and vegetables. These proteins consist of long chains of amino acids that are connected to each other by **peptide linkages.** Note in figure 53.8 that when a peptide linkage is hydrolyzed, the carbon of one amino acid takes on OH^- to form a carboxyl group (COOH), and H^+ is added to the other amino acid to form an amine (NH_2) group.

Protein digestion begins in the stomach with the action of *pepsin* and is completed in the small intestine. Pancreatic juice contains *trypsin, chymotrypsin,* and *carboxypolypeptidase,* which convert proteoses, peptones, and polypeptides to polypeptides and amino acids. *Amino polypeptidase* and *dipeptidase* within the epithelial cells of

the small intestines hydrolyze the final peptide linkages as small polypeptides and dipeptides pass through the intestinal mucosa into the portal blood.

Since there are considerable physical differences between the linkages of different amino acids, there need be, and are, a multiplicity of proteolytic enzymes; no single enzyme can possibly digest protein all the way to its amino acids.

In our assay of pancreatic juice, we will use a synthetic peptide substrate that is designated as BAPNA (benzoyl D-L p-arginine nitroaniline). If our pancreatic extract contains trypsin, this protein will be hydrolyzed to release a yellow aniline dye, giving visual evidence of hydrolysis. Trypsin will be used in our positive test control.

1. With a china marking pencil, label five test tubes 1P through 5P (*P* for protease).
2. With a 5 ml. or 10 ml. pipette, deliver 1 ml. of **pH 8.2 buffered solution (TRIS)** to each of the five tubes.
3. Flush out the pipette with water and deliver 1 ml. of **distilled water** to tubes 3 and 4.
4. Using the same pipette, deliver 1 ml. of **BAPNA solution** to tubes 1, 2, 3, and 5. Tube 4 which lacks this protein, will be a negative test control.
5. With a fresh 5 ml. pipette, deliver 1 ml. of **pancreatic extract** to tubes 1 and 4.
6. With a fresh 1 ml. pipette, deliver 1 ml. of **trypsin** to tube 5.
7. Mix each tube on a Vortex mixer for a few seconds (see figure 53.11).

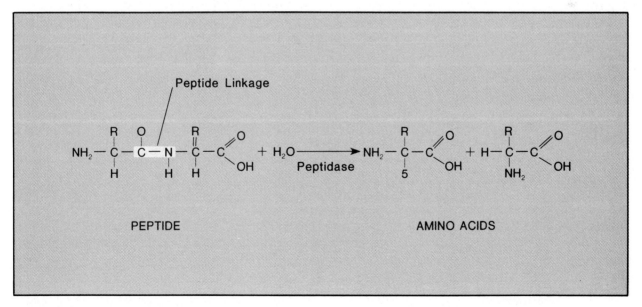

Figure 53.8 Protein hydrolysis.

8. Remove tube 2 from the series, cover it with Parafilm, and place it in a rack that is immersed in ice water.
9. Cover the other four tubes with Parafilm and place the test tube rack with these tubes in a 38°C. water bath for **one hour.**

Lipase

The hydrolysis of fats to glycerol and fatty acids is catalyzed by *lipase*. Although this enzyme is present in both gastric and pancreatic secretions, nearly all fat digestion occurs in the small intestine.

Lipase hydrolysis differs from carbohydrase and protease hydrolysis in that it requires an emulsifying agent to hasten the process. Bile salts, which are produced by the liver, act like detergents to break up large fat globules into smaller globules, greatly increasing the surface of fat particles. Since lipases are only able to work on the surface of fat globules, the increased surface area greatly enhances the speed of hydrolysis.

The most common fats of the diet are neutral fats, or **triglycerides.** As illustrated in figure 53.9 each triglyceride molecule is composed of a glycerol nucleus and three fatty acids. Although the final end products of triglyceride hydrolysis are fatty acids and glycerol, there are intermediate products of **monoglycerides** and **diglycerides.** Diagrammatically, the process looks like this:

TRIGLYCERIDE DIGLYCERIDE MONOGLYCERIDE GLYCEROL FATTY ACIDS

Set up six test tubes to assay pancreatic extract for the presence of lipase.

1. With a china marking pencil, label six test tubes 1L through 6L (*L* for lipase).
2. Using a 1 ml. pipette, deliver .2 ml. of **0.1N NaOH** to each of the tubes. Use a mechanical pipetting device such as the one in figure 53.10.
3. With a 5 ml. pipette, deliver 1 ml. of **mammalian Ringer's solution** to tubes 1, 2, and 6; 2 ml. to tubes 3 and 5; and 1.3 ml. to tube 4.
4. Flush out the pipette with water and deliver 1 ml. of **bile salt solution** to each of the tubes except tube 5.
5. Flush out the pipette with water again and deliver .3 ml. of **cream** to all tubes except tube 4.
6. With a fresh 5 ml. pipette, deliver 1 ml. of **pancreatic juice** to tubes 1, 4, and 5.
7. Deliver 1 ml. of **lipase** to tube 6, using a fresh 1 ml. pipette.
8. Mix each tube on a Vortex mixer for a few seconds.
9. Add 2 drops of **bromthymol blue** to each tube. Check the color against a bromthymol blue standard of pH 7.2. If the correct color does not develop, add NaOH, drop by drop, slowly, until the tube reaches the correct color of pH 7.2.
10. Remove tube 2 from the series, cover it with Parafilm, and place it in a rack that is immersed in ice water.

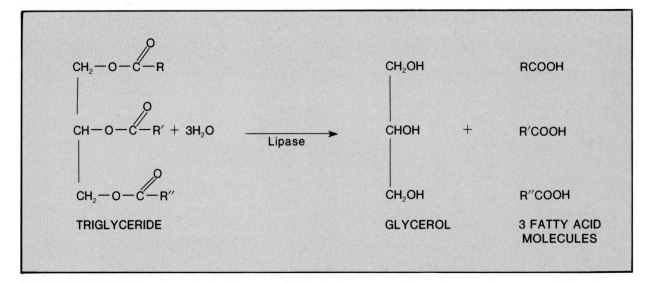

Figure 53.9 Fat hydrolysis.

11. Cover the other five tubes with Parafilm and place the test tube rack with these tubes in a 38°C. water bath for **one hour.**

Evaluation of Tests

Remove the tubes from both water baths after one hour and proceed as follows to complete the experiment:

1. To tubes 2 of each series add 1 ml. of pancreatic juice and mix for a few seconds on a Vortex mixer. Place each #2 tube in place within its series.

2. **Carbohydrase Test:** Remove a sample of test solution from each of the five tubes (1A through 5A) and place in separate depressions of a spot plate. Add 2 drops of IKI solution to each of the solutions on the plate. If starch is present due to lack of hydrolysis, the spot solution will be blue. If hydrolysis has occurred, test remaining solution in tube with Barfoed's and Benedict's tests to determine degree of digestion. See Appendix C for these tests. Record your results on the Laboratory Report.

3. **Protease Test:** Since hydrolysis produces a yellow color in the test tube, record the color of each tube on the Laboratory Report. If a Spectronic 20 photocolorimeter is set up and calibrated at 410 mμ, take a reading for each tube and record the Optical Densities (OD) in the table on the Laboratory Report. Be sure to use appropriate cuvettes for the contents of each tube. (Before taking any readings, the photocolorimeter should be standardized with a blank made up of 3 ml. of BAPNA and .1 ml. of 0.001 M HCl.)

4. **Lipase Test:** Since fat hydrolysis results in a lowering of pH due to the formation of fatty acids, the bromthymol blue in the tubes will change from blue to yellow. Compare the six tubes with bromthymol blue color standard to determine the pH changes that have occurred.

Laboratory Report

Complete the Laboratory Report for this exercise.

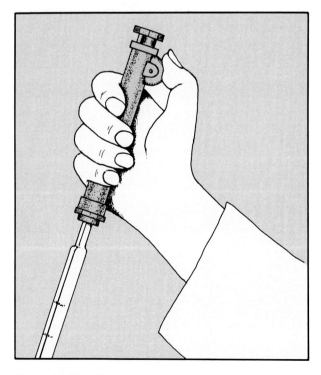

Figure 53.10 The volume of deliveries with pipetting device is controlled with the thumb.

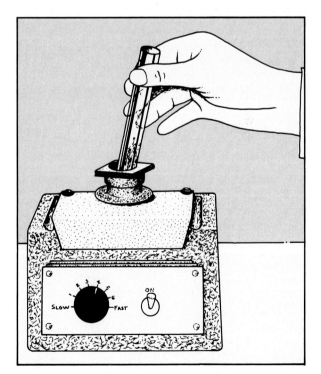

Figure 53.11 Each tube in the assay procedure must be mixed on the Vortex mixer.

Part 12 The Excretory System

Although various organs, such as the intestines, lungs, and skin, play secondary roles in the excretion of metabolic wastes, this unit will concentrate primarily on the work of the kidneys. In Exercise 54 the anatomy and physiology of the urinary system will be studied. A sheep kidney will be used for dissection and prepared microscope slides will be used for tissue studies. Prior to doing the laboratory work all illustrations should be labeled.

Exercise 55 consists of approximately twelve clinical tests of urine as performed in a routine urine analysis. Reagent test strip as well as classic methods for urine analysis are given. The methods to be used will be indicated by the laboratory instructor.

Anatomy of the Urinary System

This study of the urinary system consists of three parts: labeling of illustrations, sheep kidney dissection, and microscopic studies. The sequence of events will occur in that order. It is assumed that most of the illustrations will be labeled prior to entering the laboratory.

Organs of the Urinary System

Figure 54.1 illustrates the components of the urinary system with its principal blood supply. It consists of two kidneys, two ureters, the urinary bladder, and a urethra. The **kidneys** are somewhat bean-shaped, of dark brown color, and located behind the peritoneum. The right kidney is usually positioned somewhat lower than the left one, probably because of its displacement by the liver. Each kidney is supplied blood through a **renal artery** that is a branch off the **abdominal aorta.** Blood leaves each kidney through a **renal vein** that empties into the **inferior vena cava.**

Urine passes from each kidney to the urinary bladder through a **ureter.** The upper end of each ureter is enlarged to form a funnel-like **pelvis.** The lower end of each ureter enters the posterior surface of the bladder.

Leading from the urinary bladder to the exterior is a short tube, the **urethra.** In the male the urethra is about 20 centimeters long; in females it is approximately 4 centimeters in length.

The exit of urine from the bladder is called *micturition.* The passage of urine from the bladder is controlled by two sphincter muscles, the sphincter vesicae and the sphincter urethrae. The **sphincter vesicae** is a smooth muscle sphincter that is near the exit of the bladder. When approximately 300 ml. of urine has accumulated in the bladder, the muscular walls of the bladder are stretched sufficiently to initiate a parasympathetic reflex that causes the bladder wall to contract. These contractions force urine past the sphincter vesicae into the urethra above the **sphincter ure-**

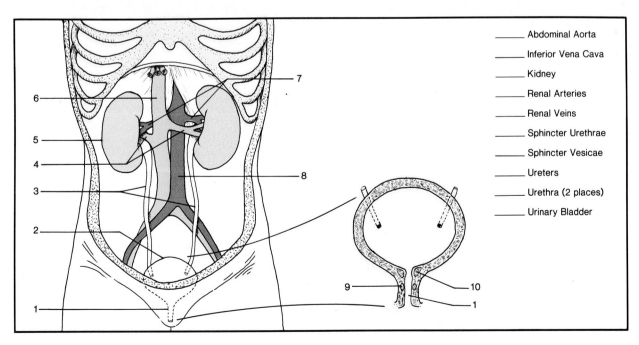

Figure 54.1 The urinary system.

thrae. This second sphincter, which is located approximately 1–3 centimeters below the sphincter vesicae, consists of skeletal muscle fibers and is voluntarily controlled. The presence of urine in the urethra above this sphincter creates the desire to micturate; however, since the valve is under voluntary control, micturition can be inhibited. When both sphincters are relaxed, urine passes from the body.

Assignment:

Label figure 54.1.

Kidney Anatomy

Figure 54.2 reveals a frontal section of a human kidney. Its outer surface is covered with a thin fibrous **renal capsule** (label 8). In addition to this thin covering, the kidney is provided support and

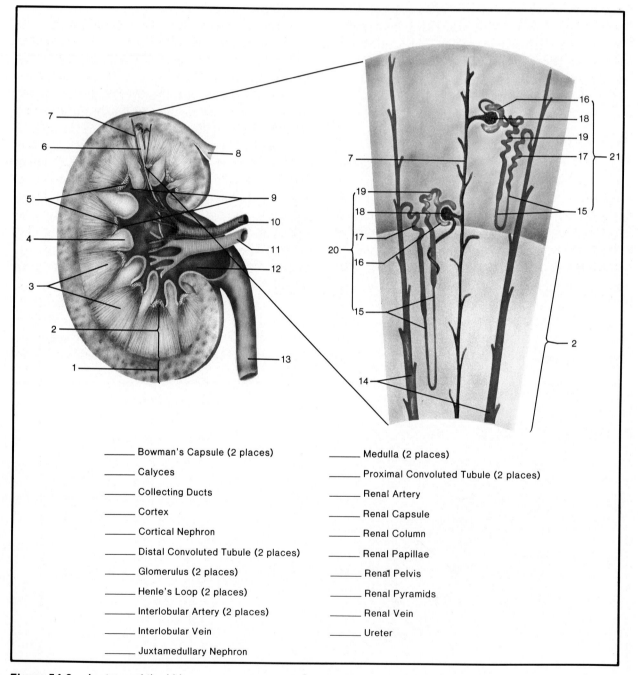

_____ Bowman's Capsule (2 places)

_____ Calyces

_____ Collecting Ducts

_____ Cortex

_____ Cortical Nephron

_____ Distal Convoluted Tubule (2 places)

_____ Glomerulus (2 places)

_____ Henle's Loop (2 places)

_____ Interlobular Artery (2 places)

_____ Interlobular Vein

_____ Juxtamedullary Nephron

_____ Medulla (2 places)

_____ Proximal Convoluted Tubule (2 places)

_____ Renal Artery

_____ Renal Capsule

_____ Renal Column

_____ Renal Papillae

_____ Renal Pelvis

_____ Renal Pyramids

_____ Renal Vein

_____ Ureter

Figure 54.2 Anatomy of the kidney.

protection by a fatty capsule which completely encases it. The latter is not shown in figures 54.1 or 54.2.

Immediately under the capsule is the **cortex** of the kidney. The cortex is reddish-brown due to its great blood supply. The lighter inner portion is called the **medulla.** The medulla is divided into cone-shaped **renal pyramids.** Nine of these pyramids are seen in figure 54.2. Cortical tissue, in the form of **renal columns,** extend down between the pyramids. Each renal pyramid terminates as a **renal papilla,** which projects into a calyx. The **calyces** are short tubes that receive urine from the renal papillae; they empty into the large funnel-like **renal pelvis.**

Nephrons

The basic functioning unit of the kidney is the *nephron.* The enlarged section in figure 54.2 reveals two nephrons. Figure 54.3 illustrates a single nephron in greater detail. It has been estimated that there are around one million nephrons in each kidney. Approximately 80% of the nephrons are located in the cortex; these are designated as **cor-**tical nephrons. The remainder, or **juxtamedullary nephrons** (label 20, figure 54.2), are located partially in the cortex and partially in the medulla.

The formation of urine by the nephron results from three physiological activities that occur in different regions of the nephron: (1) filtration, (2) reabsorption, and (3) secretion. Because of the differing functions of various regions of the nephron, the fluid that first forms at the beginning of the nephron is quite unlike the urine that enters the calyces of the kidney.

Note that each nephron consists of an enlarged end, the **renal corpuscle,** and a long tubule which empties eventually into the calyx of the kidney. Each renal corpuscle has two parts: an inner tuft of capillaries, the **glomerulus,** and an outer double-walled cap-like structure, the **glomerular** *(Bowman's)* **capsule.** Blood that enters the kidney through the **renal artery** reaches each nephron through an **interlobular artery** (label 2, figure 54.3). A short **afferent arteriole** conveys blood into the glomerulus from the interlobular artery. High intraglomerular blood pressure causes a considerable quantity of blood fluid to pass from the glomerulus into the glomerular capsule. Only blood cells and large protein molecules fail to pass

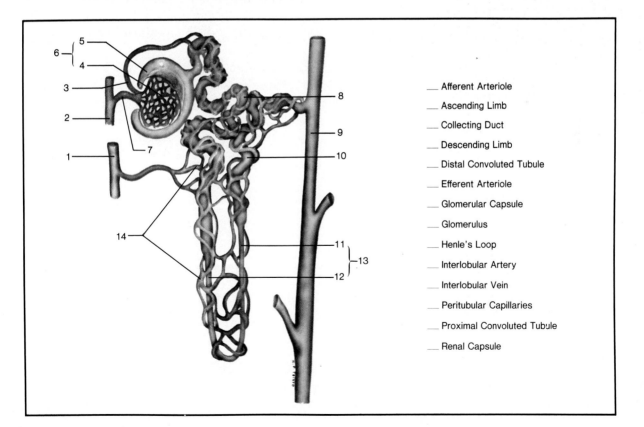

___ Afferent Arteriole

___ Ascending Limb

___ Collecting Duct

___ Descending Limb

___ Distal Convoluted Tubule

___ Efferent Arteriole

___ Glomerular Capsule

___ Glomerulus

___ Henle's Loop

___ Interlobular Artery

___ Interlobular Vein

___ Peritubular Capillaries

___ Proximal Convoluted Tubule

___ Renal Capsule

Figure 54.3   The nephron.

through. This fluid, *glomerular filtrate,* contains glucose, amino acids, urea, salts, and a great deal of water.

As the glomerular filtrate passes down the long tubule, reabsorption by the cells of the tubule removes water, glucose, amino acids, and some of the salts and urea. Without reabsorption of water from the filtrate, dehydration would soon occur. Glucose and amino acids, of course, are needed by the cells of the body and must be conserved. The tubule consists of three distinct areas: the proximal convoluted tubule, Henle's loop, and the distal convoluted tubule.

Henle's loop is a deep downward fold in the tubule that greatly increases the absorptive area of the nephron. That part of the loop nearest the glomerulus is the **descending limb;** the other part is its **ascending limb.** The **proximal convoluted tubule** is the twisted portion of the tubule that is between the renal corpuscle and Henle's loop. About 80% of the water is reabsorbed here. The **distal convoluted tubule** lies between Henle's loop and the **collecting tube** (label 9, figure 54.3). Reabsorption of water in this region of the nephron is facilitated by the *antidiuretic hormone* of the posterior lobe of the pituitary gland. A collecting tube receives urine from many nephrons. Note that the **peritubular capillaries** that surround the nephron tubule originate from the **efferent arteriole** (label 3) of the glomerulus; thus, the very blood that gives up its water, glucose, and amino acids in the renal capsule is restocked with these essentials. In addition to reabsorption, the tubule cells affect the composition of the urine by secreting ammonia, uric acid, and other substances into the lumen of the tubule.

Assignment:

Label figures 54.2 and 54.3.

Microscopic Study

Examine a prepared slide of kidney tissue that has been made from the renal cortex. Locate a renal corpuscle under low power. Examine it under high power and make a drawing of it on the Laboratory Report.

Sheep Kidney Dissection

The structure of the human kidney is well represented in the sheep. Either fresh or preserved material may be used. Fresh kidneys reveal the tissues much as they appear in real life. Preserved kidneys that are injected are best for observing the circulatory system of the organ.

Materials:

dissecting instruments
long knife
fresh sheep kidneys
dissecting trays

1. If the kidneys are still encased in fat, peel it off carefully. As you lift the fat away from the kidney, look carefully for the **adrenal gland,** which should be embedded in the fat near one end of the kidney. Remove the adrenal gland from the fat and cut it in half. Note that the gland has a distinct outer **cortex** and inner **medulla.**
2. Probe into the surface of the kidney with a sharp dissecting needle to see if you can differentiate the **capsule** from the underlying tissue.
3. With a long knife, slice the kidney longitudinally to produce a frontal section similar to figure 54.2. Wash out the cut halves with running water.
4. Identify all the structures seen in the kidney section of figure 54.2.

Laboratory Report

Complete the Laboratory Report for this exercise.

Urine: Composition and Tests

Approximately 150 liters of plasma are purified each day by glomerular filtration and tubular absorption to produce .6 to 2.5 liters of urine. The amount of urine produced is influenced by environmental temperature, amount of fluid intake, time of day, emotional state, and many other factors.

The composition of urine reveals much about the manner in which the body is functioning. Metabolic waste products such as carbon dioxide, urea, uric acid, creatinine, sodium chloride, and ammonia are normally present and have no particular pathological significance. The presence of albumin, glucose, ketones, and various other substances, however, may indicate malfunction of the kidneys or some other organ in the body.

In this exercise, we will have an opportunity to do some of the more routine tests that are performed in the analysis of a urine sample. For many of the tests, a *Test Strip Method* will also be included which utilizes specially prepared reagent test strips. These convenient test strips, manufactured by the Ames Company, are designed primarily for patient use and doctor's office laboratories. The larger clinical labs will generally use other methods for reasons of economy.

Following a discussion of the normal constituents of urine will be a series of tests to detect the presence of abnormal substances. This series may be performed as a demonstration by the instructor in which positive control samples are used along with a supposedly normal sample; or, the tests may be performed by each student on his or her own urine sample.

Normal Constituents

Normal urine is actually a highly complex aqueous solution of organic and inorganic substances. The majority of the constituents are either waste products of cellular metabolism or products derived directly from certain foods that are eaten. The total amount of solids in a 24 hour urine sample will average around 60 grams. Of this total, 35 grams would be organic and 25 grams inorganic.

The most important organic substances are urea, uric acid, and creatinine. **Urea** is a product formed by the liver from ammonia and carbon dioxide. Ninety-five percent of the nitrogen content of urine is in the form of this substance. **Uric acid** is an end product of the oxidation of purines in the body. By weight, there is normally about sixty times as much urea as uric acid in urine. **Creatinine** is a hydrated form of creatine. There may be twice as much creatinine as uric acid in the urine.

The principal inorganic constituents of urine are chlorides, phosphates, sulfates, and ammonia. **Sodium chloride** is the predominant chloride and makes up about half of the inorganic substances. Since **ammonia** is toxic to the body and lacking in plasma, there is very little of it normally present in fresh urine. The small amount that is present is probably secreted in the nephron by the nephron tubule. Urine that is allowed to stand at room temperature for 24 hours or longer may give off an odor of ammonia due to the breakdown of urea by bacterial action.

Because of the efficient absorptive properties of the cells of the nephron tubule there should be no appreciable amounts of glucose or amino acids in urine. About .3 to 1.0 gram of **glucose** in a 24 hour volume of urine would be normal excretion. Occasionally, higher percentages may result in individuals during emotional stress.

Abnormal Constituents
(Test Procedure)

In determining the presence or absence of pathology through urine analysis, it is necessary to perform both physical and chemical tests. Of the various physical tests that are available, only the appearance of the urine and specific gravity will be observed. The chemical tests will be for pH, protein, mucin, glucose, ketones, hemoglobin, and bilirubin. The significance of each abnormality will accompany the specific test.

Collection of Specimen

The manner of collecting a urine sample is determined by the type of tests to be performed. If a quantitative analysis is to be performed, a 24 hour collection is necessary. When making qualitative tests, random sampling is satisfactory. When looking for pathological substances, it is best to collect urine 3 hours after a meal. The first urine collected in the morning is least likely to contain pathological evidence and is usually discarded in 24 hour specimens. Urine should be collected in a clean container and stored in a cool place until tested. In qualitative testing it should be tested 1–2 hours after voiding. Catheterization is necessary only in bacteriological examinations.

Appearance

The first thing to consider in a routine urine analysis is the visual appearance of the urine. Its color and turbidity can provide clues as to evidence of pathology.

Color Normal urine will vary from light straw to amber color. The color of normal urine is due to a pigment called *urochrome,* which is the end-product of hemoglobin breakdown:

Hemoglobin → Hematin → Bilirubin

Urochrome ← Urochromogen ↙

Deviations from normal color that have pathological implications are:

Milky: pus, bacteria, fat or chyle.

Reddish Amber: urobilinogen or porphyrin. Urobilinogen is produced in the intestine by the action of bacteria on bile pigment. Porphyrin may be evidence of liver cir-

rhosis, jaundice, Addison's disease, and other conditions.

Brownish Yellow or Green: bile pigments. Yellow foam is definite evidence of bile pigments.

Red to Smoky Brown: blood and blood pigments.

Carrots, beets, rhubarb, and certain drugs may color the urine, yet have no pathological significance. Carrots may cause increased yellow color due to carotene, beets cause reddening, and rhubarb may cause urine to become brown.

Evaluate your urine sample according to the above criteria and record the information on the Laboratory Report.

Transparency (Cloudiness) A fresh sample of normal urine should be clear, but may become cloudy after standing a while. Cloudy urine may be evidence of phosphates, urates, pus, mucus, bacteria, epithelial cells, fat, and chyle. Phosphates disappear with the addition of dilute acetic acid and urates dissipate with heat. Other causes of turbidity can be analyzed by microscopic examination.

After shaking your sample, determine the degree of cloudiness and record it on the Laboratory Report.

Specific Gravity

The specific gravity of a twenty-four hour specimen of normal urine will be between 1.015 to 1.025. Single urine specimens may range from 1.002 to 1.030. The more solids in solution, the higher will be the specific gravity. The greater the volume of urine in a twenty-four-hour specimen, the lower will be the specific gravity. A low specific gravity will be present in chronic nephritis and diabetes insipidus. A high specific gravity may indicate diabetes mellitis, fever, and acute nephritis.

To determine the specific gravity of each sample do as follows:

Materials:

urinometer (cylinder and hydrometer)
thermometer
filter paper

1. Fill the urinometer cylinder three-fourths full of well-mixed urine. Remove any foam on the

surface with a piece of filter paper.

2. Insert the hydrometer into the urine and read the graduation on the stem at the level of the bottom of the miniscus.

3. Take the temperature of the urine and record the *adjusted specific gravity*. Hydrometers are graduated for a specific temperature, usually 25°C. If the temperature of the urine is greater than this, it is necessary to add .001 for each 3°C. The same amount is subtracted for each 3°C. below 25°C.

4. Wash the cylinder and hydrometer with soap and water after completing this test.

Hydrogen Ion Concentration

Although freshly voided urine is usually acid (around pH 6), the normal range is between 4.8 and 7.5. The pH will vary with the time of day and diet. Twenty-four hour specimens are less acid than fresh specimens and may become alkaline after standing, due to bacterial decomposition of urea. High acidity is present in acidosis, fevers, and high protein diets. Excess alkalinity may be due to urine retention in the bladder, chronic cystitis, anemia, obstructing gastric ulcers, and alkaline therapy. The simplest way to determine pH is to use pH indicator paper strips.

Materials:

pH indicator paper strip (*pHydrion* or nitrazine papers)

1. Dip a strip of pH paper into the urine three consecutive times and shake off the excess liquid.

2. After one minute, compare the color with color chart. Record observations on the Laboratory Report.

Protein

Although the large size of protein molecules normally prevents their presence in normal urine, certain conditions can allow them to filter through. Excessive muscular exertion, prolonged cold baths, and excessive ingestion of protein may result in *physiologic albuminurea. Pathologic albuminurea,* on the other hand, exists when albumin of the urine is due to kidney congestion, toxemia of pregnancy, febrile diseases, and anemias.

Many tests are used for the detection of albumin. They all function in the same way—to precipitate the albumin by chemicals or heat. Albumin is soluble, and when heated or treated with certain chemicals it becomes insoluble, forming a visible precipitate.

Exton's method, which utilizes sulfosalicylic acid, will be used for albumin detection.

Materials:

Exton's reagent
test tubes, one for each urine sample
test tube holder
Bunsen burner
pipettes (5 ml. size) or graduates

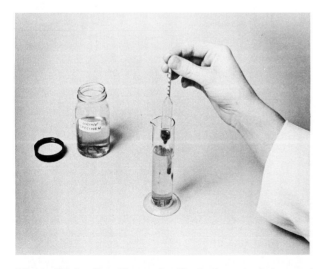

Figure 55.1 Specific gravity. The hydrometer is lowered into the urine after the foam has been removed with a piece of filter paper. The reading is taken from the bottom of the meniscus.

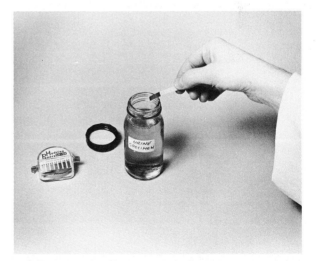

Figure 55.2 Determination of pH. Only fresh urine samples should be tested because pH tends to rise with aging of urine. Read the pH one minute after paper has been dipped.

1. Pour or pipette equal volumes of urine and Exton's reagent into a test tube. Approximately 3 ml. of each should suffice.
2. Shake the tube from side to side or roll between the palms to mix thoroughly.
3. Look for precipitation. If no cloudiness occurs, albumin is absent. Record results on Laboratory Report.

Test Strip Method

Shake the sample of urine and dip the test portion of an *Albustix* test strip into the urine. Touch the tip of the strip against the edge of the urine container to remove excess urine. Immediately, compare the test area with the color chart on the bottle. Note that the color scale runs from yellow (negative) to turquoise (+ + + +). Record your results on the Laboratory Report.

Mucin

Inflammation of the mucous membranes of the urinary tract and vagina may result in large amounts of mucin being present in urine. Since it can be confused with albumin, its presence should be confirmed. Mucin and mucoid are glycoproteins that will reduce Benedict's reagent in the presence of acid or alkali. Glacial acetic acid will be used here for the detection of mucin.

Materials:

    test tubes
    test tube holder

Bunsen burner
acetic acid (glacial)
sodium hydroxide (10%)
pipettes (5 ml. size) or graduates
filter paper
funnel
ring stand
small graduate

1. If urine was positive for albumin, remove the albumin by boiling 5 ml. of urine for a few minutes and filtering while hot.
2. After the urine has cooled, add 6 ml. of **distilled water** to 2 ml. of the urine to prevent precipitation of urates.
3. Add a few drops of **glacial acetic acid.** If mucin is present, the urine will become turbid.
4. To further prove that the precipitate is mucin, add a few drops of 10% **sodium hydroxide.** The precipitate will disappear if it is mucin.

Glucose

As stated above, only a small amount of glucose is normally present in urine (.01 to .03 gm. per 100 ml. urine). When urine contains glucose amounts greater than this, *glycosurea* exists. This is usually an indication of diabetes mellitus. Lack of insulin production by the pancreas is the cause of this disease. Insulin is necessary for the conversion of excess glucose to glycogen in the liver and muscles. It is also essential in the oxidation of glucose by the cells. A deficiency of insulin, thus,

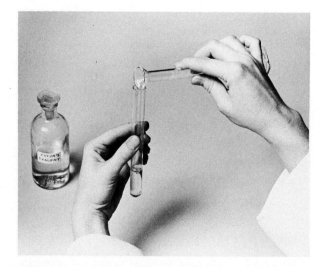

Figure 55.3 Protein. Exton's reagent, which contains sulfosalicylic acid, is added to the urine to detect the presence of albumin.

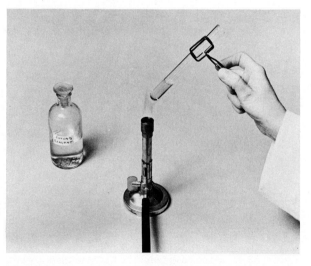

Figure 55.4 Protein. If precipitate forms after adding Exton's reagent and gently heating it, the presence of albumin is confirmed.

will result in high blood levels of glucose.

The renal threshold of glucose is around 160 mg. per 100 ml. Glycosurea indicates that blood levels of glucose exceed this amount and the kidneys are unable to accomplish 100% reabsorption of this carbohydrate.

The detection of glucose in urine is usually performed with **Benedict's reagent.** In the presence of glucose a precipitate ranging from yellowish-green to red will form. The color of the precipitate will indicate the percentage of glucose present.

Materials:

    test tubes
    electric hot plate
    beaker (250 ml. size)
    Benedict's qualitative reagent
    pipettes (5 ml. size) or graduates

1. Label one tube for each sample to be tested. One tube should be for a known positive sample.
2. Pour 5 ml. of Benedict's reagent into each tube.
3. Add 8 drops (0.5 ml.) of urine to each tube of Benedict's reagent. Use separate clean pipettes for each urine sample.
4. Place the tubes in a beaker of warm water and bring to boil for 5 minutes.
5. Determine the amount of glucose present by color:

Negative	= clear blue to cloudy green
+	= yellowish green (.5 to 1 gm.%)
+ +	= greenish yellow (1 to 1.5 gm.%)
+ + +	= yellow (1.5 to 2.5 gm.%)
+ + + +	= orange (2.5 to 4 gm.%) red (4 gm.% and over)

Test Strip Method

Shake the sample of urine and dip the test portion of a *Clinistix* test strip into the urine. **Ten seconds** after wetting, compare the color of the test area with the color chart on the label of the bottle. Note that there are three degrees of positivity. The light intensity generally indicates .25% or less glucose. The dark intensity indicates .5% or more glucose. The medium intensity has no quantitative significance. Record your results on the Laboratory Report.

Ketones

Normal catabolism of fats produces carbon dioxide and water as final end products. When there is inadequate carbohydrate in the diet, or when there is a defect in carbohydrate metabolism, the body begins to utilize an increasing amount of fatty acids. When this increased fat metabolism reaches a certain point, fatty acid utilization becomes incomplete, and intermediary products of fat metabolism occur in the blood and urine. These

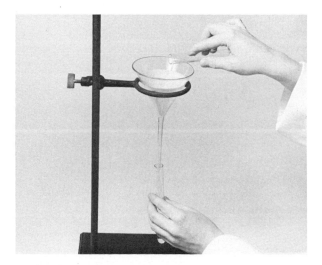

Figure 55.5 Mucin. When testing for mucin in urine, it is necessary to remove the albumin first by filtering the boiled urine.

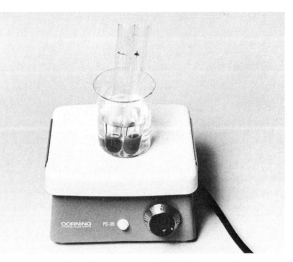

Figure 55.6 Glucose. Positive evidence of glycosuria is determined by boiling urine in Benedict's reagent for five minutes.

intermediary substances are the three ketone bodies: acetoacetic acid (diacetic acid), acetone, and beta hydroxybutyric acid. The presence of these substances in urine is called *ketonuria*.

Diabetes mellitus is the most important disorder in which ketonuria occurs. Progressive diabetic ketosis is the cause of diabetic acidosis, which can eventually lead to coma or death. It is for this reason that the detection of ketonuria in diabetics is of great significance.

The method we will use for detection of ketones is *Rothera's Test*.

Materials:

Rothera's reagent
test tubes
ammonium hydroxide (concentrated)

1. Add about 1 gm. of Rothera's reagent to 5 ml. of urine in a test tube.
2. Layer over the urine 1 ml. to 2 ml. of concentrated ammonium hydroxide by allowing it to flow gently down the side of the inclined test tube.
3. If a **pink-purple ring** develops at the interface ketones are present. No ring or a brown ring is negative.

Test Strip Method

As with the above rapid methods, dip the test portion of a *Ketostix* test strip into the urine sample, and tap dry on the edge of the urine container. **Fifteen seconds** after wetting, compare the color of the test strip with the color chart on the label of the bottle. Record your results on the Laboratory Report.

Hemoglobin

When red blood cells disintegrate (hemolysis) in the body, hemoglobin is released into the surrounding fluid. If the hemolysis occurs in the blood vessels, the hemoglobin becomes a constituent of plasma. Some of it will be excreted by the kidneys into urine. If the red blood cells enter the urinary tract due to disease or trauma, the cells will hemolyze in the urine. The presence of hemoglobin in urine is called *hemoglobinuria*.

Hemoglobinuria may be evidence of hemolytic anemia, transfusion reactions, yellow fever,

smallpox, malaria, hepatitis, mushroom poisoning, renal infarction, burns, etc. The simplest way to test for hemoglobin is to use *Hemastix* test strips as follows:

1. Shake the sample of urine and dip the test portion of a *Hemastix* test strip into the urine.
2. Tap the edge of the strip against the edge of the urine container and let dry for **30 seconds.**
3. Compare the color of the test strip with the color chart on the bottle.
4. Record your results on the Laboratory Report.

Bilirubin

As indicated earlier in this exercise, bilirubin is a product of hemoglobin breakdown. It is the second stage in hemoglobin degradation, forming from hematin. It is normally present in the urine in very small quantities. When present in large amounts, however, it usually indicates a disorder of the liver caused by infection of hepatoxic agents. The presence of significant amounts of bilirubin is designated as *bilirubinuria*.

The simplest way to test for bilirubin is to use *Bili-Labstix*. The procedure is as follows:

1. Shake the sample of urine and dip the test portion of a *Bili-Labstix* test strip into the urine.
2. Tap the edge of the strip against the edge of the urine container and let dry for **20 seconds.**
3. Compare the color of the test strip with the color chart on the bottle. The results are interpreted as negative, small (+), moderate (+ +), and large (+ + +) amounts of bilirubin.
4. Record your results on the Laboratory Report.

Microscopic Study

This is the most important phase of urine analysis, yet the scope of its implications goes considerably beyond the limitations of this course. A complete microscopic examination will include not only an analysis of the sediment, but also a bacteriological determination.

Normal urine will contain an occasional leukocyte, some epithelial cells, mucus, bacteria, and crystals of various kinds. The experienced technologist has to determine when these substances exist in excess amounts and be able to identify the various types of casts, cells, and crystals that predominate.

Figure 55.7 illustrates only a few of the elements that might be encountered in urine. No attempt shall be made here to identify all particulate matter in urine.

Materials:

> microscope slides and cover glasses
> capillary pipettes
> centrifuge
> centrifuge tubes (conical tipped)
> pipettes (5 ml. size) or graduates
> wire loop

1. Pour 5 ml. of urine into a centrifuge tube after shaking the urine sample to re-suspend the sediment. Be sure to balance the centrifuge with an even number of loaded tubes.
2. Centrifuge the tubes for 5 minutes at a slow speed (1500 r.p.m.).
3. Pour off all urine and allow the sediment on the side of the tube to settle down into the bottom of the tube.
4. With a capillary pipette or flamed wire loop, transfer a small amount of the sediment to a microscope slide and cover with a cover glass.
5. Examine under the microscope with low and high power objectives. Reduce the lighting by adjusting the diaphragm. Refer to figure 55.7 to identify structures.

Laboratory Report

Complete the Laboratory Report for this exercise.

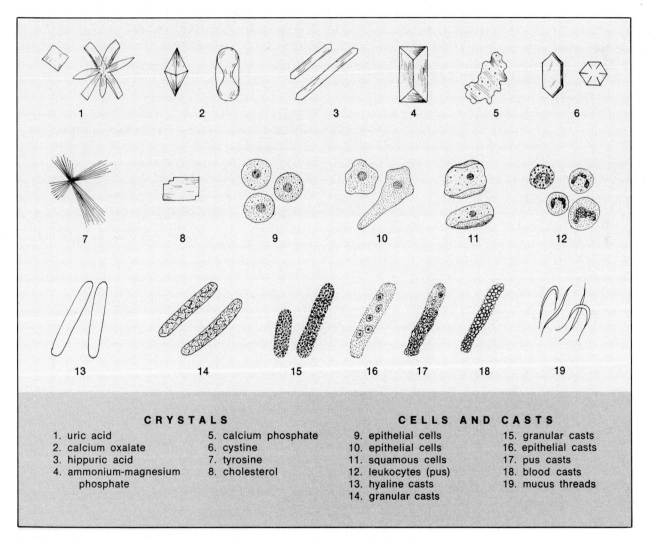

CRYSTALS

1. uric acid	5. calcium phosphate
2. calcium oxalate	6. cystine
3. hippuric acid	7. tyrosine
4. ammonium-magnesium phosphate	8. cholesterol

CELLS AND CASTS

9. epithelial cells	15. granular casts
10. epithelial cells	16. epithelial casts
11. squamous cells	17. pus casts
12. leukocytes (pus)	18. blood casts
13. hyaline casts	19. mucus threads
14. granular casts	

Figure 55.7 Microscopic elements in urine.

Part 13 The Endocrine and Reproductive Systems

The regulatory action of the endocrine glands in the reproductive process logically places these two systems together. Exercise 56, which pertains to the endocrine glands, is primarily descriptive in nature. The laboratory work consists essentially of microscopic studies of the various glands. The principal value of this exercise is that it is a summarization of the functions of the various hormones. Questions on the Laboratory Report will test your understanding of hormonal action.

Exercise 57 pertains to the anatomy of the male and female reproductive organs. Spermatogenesis and oogenesis are also studied.

56 The Endocrine Glands

In previous exercises, many of the endocrine glands have been mentioned in connection with various physiological activities. In this exercise, all the glands will be studied as a unified system to summarize the functions of the various hormones. Endocrinology is a vast subject; no assumption should be made here that this review is comprehensive. The main purpose of this exercise is to tie together in a unified whole a brief summary of the entire endocrine system.

This exercise consists of two parts: (1) a discussion of the structure and function of the glands as related to figures 56.1 and 56.2, and (2) microscopic studies of the various glands. It is suggested that the first portion be completed prior to doing slide studies.

The Thyroid Gland

The thyroid gland consists of two lateral lobes with a connecting isthmus. It is located just below the larynx and anterior to the trachea. Microscopically it consists of large numbers of spherical sacs called **follicles** that are filled with a colloidal suspension. The major constituent of this colloid is a glycoprotein, **thyroglobulin.** The principal hormones produced by the thyroid gland are thyroxine, triiodothyronine, and calcitonin.

The Thyroid Hormone

The thyroid hormone consists of thyroxine, triiodothyronine, and small quantities of closely related iodinated hormones. These hormones form within the thyroglobulin molecule, emerge from the molecule, and are absorbed by blood vessels in the gland. Once in the blood the hormones combine with blood proteins and are carried to all tissues of the body for utilization. Excess hormones are stored in thyroglobulin.

The function of the thyroid hormone is to increase metabolic activity of most tissues of the body. About 90% of the thyroid hormone is thyroxine; 10% is triiodothyronine. Although triiodothyronine is four times as potent as thyroxine, the effectiveness of thyroxine lasts about four times as long.

Since the thyroid hormone increases the metabolism of all cells of the body, the following changes occur when its levels are increased: (1) bone growth in children is accelerated, and rapid maturation and ossification at the epiphyses occurs early; (2) carbohydrate metabolism is enhanced by rapid glucose uptake, increased glycolysis, accelerated gluconeogenesis, and increased insulin production; (3) body weight is decreased; (4) cardiac output and heart rate is increased; (5) the respiratory rate is increased; and (6) mental activity is increased.

Exactly how the thyroid hormone increases metabolic activity is not known. However, we do know that the following changes occur: (1) protein synthesis is stepped up through increased synthesis of RNA by DNA; (2) the quantity of cellular enzymes is increased; and (3) the size and number of mitochondria in cells is increased.

Hypothyroidism (thyroid hormone deficiency) may result in a condition called *cretinism* in children. This condition may be due to a congenitally small or absent thyroid gland, a genetically deficient thyroid gland, or a lack of iodine in the diet. Children with this condition are mentally and physically retarded. Adults who have a severe long-term thyroid hormone deficiency develop a condition called *myxedema.*

Hyperthyroidism (thyroid hormone excess) is characterized by weight loss, rapid pulse, intolerance to heat, nervousness, extreme fatigue, inability to sleep, and tremor of the hands. In most cases, hyperthyroidism causes some degree of eyeball protrusion, or *exophthalmos.*

An enlarged thyroid gland is referred to as a "goiter". Goiters may be present in both hypothyroidism and hyperthyroidism. Goiters due to

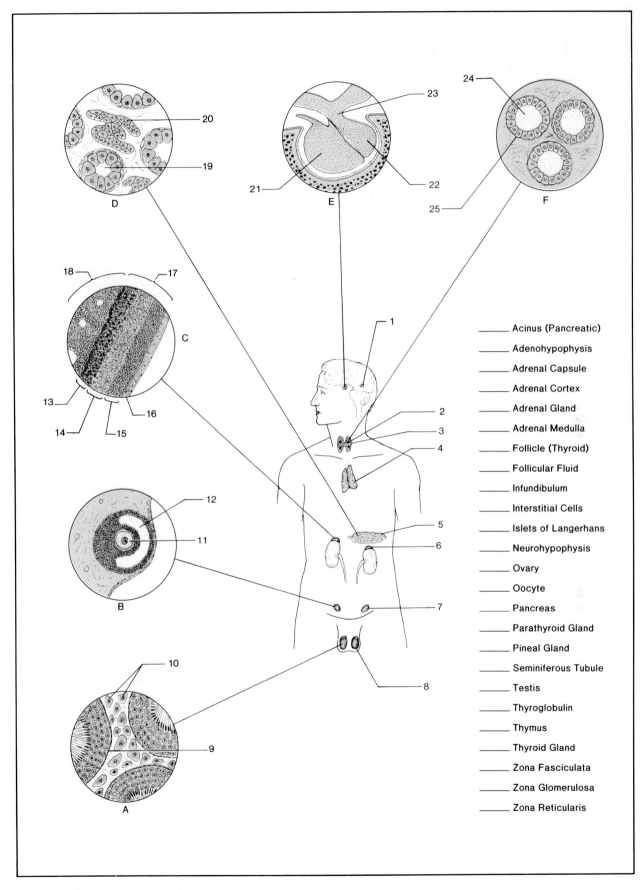

Acinus (Pancreatic)

Adenohypophysis

Adrenal Capsule

Adrenal Cortex

Adrenal Gland

Adrenal Medulla

Follicle (Thyroid)

Follicular Fluid

Infundibulum

Interstitial Cells

Islets of Langerhans

Neurohypophysis

Ovary

Oocyte

Pancreas

Parathyroid Gland

Pineal Gland

Seminiferous Tubule

Testis

Thyroglobulin

Thymus

Thyroid Gland

Zona Fasciculata

Zona Glomerulosa

Zona Reticularis

Figure 56.1 The endocrine glands.

iodine deficiency are called *endemic goiters.* Hypothyroid goiters which are not caused by iodine deficiency are called *idiopathic nontoxic goiters.* Some idiopathic non-toxic goiters, incidentally, produce normal quantities of the thyroid hormone. The exact cause of this type of goiter is unknown. Individuals with severe hyperthyroidism have a *toxic goiter.* The gland in this case may be increased by three or four times.

Calcitonin

Calcitonin is a large polypeptide produced by the parafollicular cells in the interstitium of the thyroid gland. If levels of this hormone are increased experimentally, the calcium level in the blood is rapidly decreased. This is caused by the inhibiting effect of calcitonin on osteoclasts and the stimulation of osteoblastic activity. Over a period of time, however, osteoblastic activity is actually depressed and there is no significant effect of calcitonin on the plasma calcium level. It appears that calcitonin actually has only a weak effect on the calcium levels in blood of adults. In children it plays an important role in bone remodeling during growth.

The Parathyroid Glands

Four small pea-sized parathyroid glands lie embedded in the posterior surface of the thyroid gland. Occasionally, parathyroid tissue is formed outside of the thyroid gland. As many as a dozen small individual parathyroid glands may be distributed throughout the neck region.

These glands produce the **parathyroid hormone,** which regulates the calcium-phosphorous ratio in the blood and tissues. The hormone is produced by *chief cells* in the gland. When this hormone is injected into an animal the plasma calcium level is elevated and the phosphorous level is lowered. This is accomplished by the action of the parathyroid hormone on osteoclasts and kidney tissue. The hormone mobilizes calcium from the bone by (1) activating all osteoclasts to remove calcium from bone; (2) promoting the formation of new osteoclasts from mesenchymal stem cells; and (3) slowing down the conversion of osteoclasts to osteoblasts. It removes phosphorous from the blood by impeding the reabsorption of phos-

phorous ions in the renal tubules. To prevent calcium loss in the kidneys, the parathyroid hormone facilitates calcium reabsorption in the renal tubules. This increased reabsorption of calcium prevents complete depletion of this mineral from the body. Removal of all parathyroid tissue results in *tetany* and death.

Production of the parathyroid hormone by the parathyroid glands is regulated by the level of calcium in the blood. Excess calcium levels inhibit hormone production; a deficiency of calcium causes the gland to produce more of the hormone.

The Thymus Gland

This gland consists of two long lobes and lies in the upper chest region above the heart. The cortex contains a meshwork of reticular connective tissue similar to lymphoid tissue. Within the reticular framework are thousands of lymphocytes. The medulla contains structures called **Hassal's corpuscles.** These structures are concentrically arranged flattened epithelial cells. In addition, the medulla contains some lymphocytes.

It appears that the principal role of the thymus is to process lymphocytes into T-lymphocytes, as described on pages 243 and 244. It is speculated that this gland produces one or more hormones that function in T-lymphocyte formation. **Thymosin** and several other polypeptide extracts from this gland have been under study.

The Pancreas

The pancreas consists of two principal types of secretory tissues: (1) sac-like structures called **acini,** which secrete digestive enzymes that pass to the duodenum and (2) the **islets of Langerhans,** which are clusters of cells between the acini that secrete directly into the blood stream.

The islets of Langerhans produce three hormones: insulin, glucagon, and somatostatin. All of these hormones except insulin are also secreted by the mucosa of the gastrointestinal tract. Three different kinds of cells in the islets of Langerhans account for the production of these hormones: alpha, beta, and delta cells. Glucagon is produced by the alpha cells; insulin by beta cells; and somatostatin is produced by delta cells.

Insulin is an anabolic hormone which promotes the storage of glucose, fatty acids, and amino acids. **Glucagon** is catabolic in that it mobilizes glucose, fatty acids, and amino acids. These two hormones are, thus, reciprocal in their action. Insulin excess causes *hypoglycemia* (low blood sugar), which causes convulsions and coma. Glucagon deficiency can also cause hypoglycemia. Insulin deficiency results in *diabetes mellitus,* a serious debilitating disease in which *hyperglycemia* (high blood sugar) is present. An excess of glucagon can further aggravate diabetes. In addition to its catabolic effects on carbohydrates, fats, and proteins, glucagon stimulates the production of the growth hormone, insulin, and pancreatic somatostatin.

Somatostatin is a growth-inhibiting hormone in that it inhibits the release of the growth hormone by the anterior pituitary gland. In addition, it inhibits the secretion of insulin and glucagon. Patients with somatostatin-secreting tumors develop hyperglycemia and other diabetes-like symptoms. The symptoms disappear when the tumors are removed.

Insulin production is influenced by a variety of stimulatory and inhibitory factors; however, feedback control by the blood glucose level directly on the beta cells is the major controlling factor. As glucose levels become elevated, insulin production is stepped up; when the glucose level is normal or low, the rate of insulin secretion is low.

The Adrenal Glands

Illustration C, figure 56.1, illustrates the histological nature of this gland. Note that the gland has an outer **cortex** (label 17) and an inner **medulla.** The entire gland is enclosed in a **capsule** of fibrous connective tissue. Note, also, that the cortex consists of three distinct layers: an outer **zona glomerulosa,** an inner **zona reticularis,** and an intermediate **zona fasciculata.**

The embryological origins of the two portions of the adrenal gland explain their differences in function. The cells of the medulla originate from the neural crest of the embryo. This embryonic tissue also gives rise to ganglionic cells of the sympathetic nervous system; thus, we could expect a close kinship in function of the adrenal medulla and the sympathetic nervous system. Cells of the adrenal cortex, on the other hand, arise from embryonic tissue associated with the gonads. Hormones having some relationship to the reproductive organs might be expected from the cortex.

Adrenal Medulla

The hormones produced by the adrenal medulla are the catecholamines **norepinephrine** and **epinephrine.** Separate cells in the medulla produce each of these hormones. In the human, 80% of the catecholamine output of the adrenal medulla is epinephrine; 20% is norepinephrine. Norepinephrine is also produced by the postganglionic nerve fibers of the autonomic nervous system.

These circulating hormones have almost the same effect on different organs as direct sympathetic stimulation, except that the effects of epinephrine and norepinephrine last about ten times as long. The longer duration of hormonal effects is due to their slow removal from the blood.

Production of these hormones is initiated when the adrenal medulla is stimulated by sympathetic nerve fibers. The effects of norepinephrine and epinephrine on different tissues in the body depend on the type of receptors that exist in those tissues. There are two classes of adrenergic receptors: alpha and beta. The alpha receptors are subdivided into α_1 and α_2; the beta receptors into β_1 and β_2.

Both norepinephrine and epinephrine increase the force and rate of contraction of the isolated heart. These reactions are mediated by β_1 receptors. Norepinephrine causes vasoconstriction in all organs through α_1 receptors. Epinephrine causes vasoconstriction everywhere except in the muscles and liver, where β receptors bring about vasodilation. Increased mental alertness is also induced by both catecholamines, which may be induced by the increased blood pressure.

Blood glucose levels are increased by both norepinephrine and epinephrine. This is accomplished by glycogenolysis (glycogen $\rightarrow$ glucose) in the liver. β_2 receptors account for this reaction.

Norepinephrine and epinephrine are equally potent in mobilizing free fatty acids through β receptors. The metabolic rate is also increased by these hormones; how this takes place is not precisely understood at this time.

Adrenal Cortex

The adrenal cortex secretes many different hormones, all of which belong to a group of substances called *steroids*. Collectively, they are referred to as **corticosteroids.** All of them are synthesized from cholesterol and have similar molecular structures. While destruction of the adrenal medulla is more or less inconsequential to survival, obliteration of the adrenal cortex results in death.

The corticosteroids fall into three groups: the mineralcorticoids, the glucocorticoids, and the androgenic hormones. The **mineralcorticoids** are hormones that control the excretion of Na^+ and K^+, which affects the electrolyte balance. The **glucocorticoids** affect primarily the metabolism of glucose and protein. The **androgenic hormones** produce some of the same effects on the body as the male sex hormone testosterone. Although over 30 corticosteroids have been isolated, only 5 are produced in significant amounts. They are the mineralcorticoids *aldosterone* and *deoxycorticosterone*, the glucocorticoids *cortisol* and *corticosterone*, and the androgen *dehydroepiandrosterone*. The two most important corticosteroids are aldosterone and cortisol.

Aldosterone Any condition which impairs the production of this mineralcorticoid will result in death within two weeks if salt replacement or mineralcorticoid therapy is not provided. In the absence of aldosterone the K^+ concentration in extracellular fluids rises, Na^+ and Cl^- concentrations decrease, and the total volume of body fluids becomes greatly reduced. These conditions cause reduced cardiac output, shock, and death.

The most important function of aldosterone is to increase the rate of renal tubular absorption of sodium. The reabsorption process occurs primarily in the ascending limb of Henle's loop.

Aldosterone is produced by cells of the zona glomerulosa. The mechanisms that regulate its production are complex. Some of the controlling factors are (1) ACTH, (2) potassium ion concentration of extracellular fluid, (3) the renin-angiotensin system, and (4) the quantity of body sodium.

Cortisol Approximately 95% of glucocorticoid activity results from the action of cortisol (hydro-

cortisone, compound F). When an excess of cortisol exists, the following events occur:

1. The rate of gluconeogenesis in the liver is facilitated. Gluconeogenesis is the conversion of amino acids and glycerol to glucose.
2. The rate of glucose utilization by cells is decreased. This decrease in glucose utilization plus gluconeogenesis results in hyperglycemia. This condition is also referred to as *adrenal diabetes.*
3. Protein catabolism in all cells except the liver is increased. This raises the level of plasma proteins.
4. Amino acid transport to muscle cells is decreased.
5. Fatty acids are mobilized from adipose tissue.
6. Some mineralcorticoid activity takes place.

Cushing's syndrome, a condition in which there is an excess of glucocorticoids induced by pituitary tumor, is characterized by many symptoms caused by the above physiological reactions. Protein depletion causes these patients to have poorly developed muscles and wounds that heal poorly. Hair is thin and scraggly. Bone dissolution takes place due to protein depletion and the anti-vitamin D action of the glucocortocoids. Hyperglycemia exists. Mineralcorticoid action causes salt and water retention. Total glucocorticoid insufficiency caused by cancer or tuberculosis of the adrenals results in *Addison's disease.*

Cortisol is produced primarily by cells in the zona fasciculata and to some extent by cells in the zona reticularis. Although several substances such as vasopressin, serotonin, and angiotensin II stimulate the adrenal cortex, it is primarily the action of ACTH produced by the anterior pituitary that induces the cells of the adrenal cortex to produce cortisol. ACTH production is initiated by stress through the hypothalamus. Stress induces the hypothalamus to produce CRF, which passes to the anterior pituitary causing the latter to produce ACTH.

When large doses of glucocorticoids are injected inflammation in tissues is inhibited. Glucocorticoids also suppress manifestations of allergies caused by histamine. Inhibition of production of the growth hormone by the anterior pituitary can also be produced with large doses. None of these effects are observed from normal physiologic doses.

The Pineal Gland

The pineal gland is situated on the roof of the third ventricle under the posterior end of the corpus callosum. It is supported by a stalk which contains postganglionic sympathetic nerve fibers that do not seem to extend into the gland. The gland consists of neuroglial and parenchymal cells that suggest a secretory function.

In young animals and infants the gland is large and more gland-like; cells tend to be arranged in alveoli. Just before puberty the gland begins to regress and small concretions of calcium carbonate form that are called *pineal sand*. There is some evidence, though not conclusive, that the gland contains some gonadotropin peptides. Most attention, however, is on the production of melatonin by the gland.

Melatonin is an indole which is synthesized from serotonin. Its production appears to be regulated by daylight (circadian rhythm). During daylight its production is suppressed. At night the gland becomes active and produces considerable quantities of melatonin. Regulation occurs through the eyes. Light striking the retina sends messages to the pineal gland through a retina-hypothalmic pathway. Norepinephrine reaches the cells from the postganglionic sympathetic nerve endings in the pineal stalk. Beta adrenergic receptors in the cells are mediators in the inhibition of melatonin production. Although considerable speculation exists that melatonin inhibits the estrus cycle in the human, as it probably does in lower animals, there is no definitive proof at this time that this is the case.

The Testes

A section through the testis reveals that it consists of coils of **seminiferous tubules** where spermatozoa are produced, and **interstitial cells** where the male sex hormone, **testosterone,** is secreted. Illustration A, figure 56.1, shows portions of three seminiferous tubules with interstitial cells between them.

Testosterone is produced in considerable quantities during embryological development, but from early childhood to puberty very little of the hormone is produced. With the onset of puberty at the age of 10 or 11 years the testes begin to produce large quantities of testosterone as the development of sexual maturity takes place.

During embryological development, testosterone is responsible for the development of the male sex organs. If the testes are removed from a fetus at an early stage, the fetus will develop a clitoris and vagina instead of a penis and scrotum, even though the fetus is a male. The hormone also controls the development of the prostate gland, seminal vesicles, and male ducts while suppressing the formation of female genitals.

Embryological development of the testes occurs within the body cavity. During the last two months of development the testes descend into the scrotum through the inguinal canals. The descent of the testes is controlled by testosterone.

During puberty the testes become larger and produce a great deal of testosterone, which causes considerable enlargement of the penis and scrotum. At the same time, the male secondary sexual characteristics develop. These characteristics are (1) the appearance of hair on the face, axillae, chest and pubic region; (2) the enlargement of the larynx accompanied by a voice change; (3) increased skin thickness; (4) increased muscular development; (5) increased bone thickness and roughness; and (6) baldness.

Shortly after secretion, testosterone enters a cell, is converted to dihydrotestosterone, and forms a bond with a protein receptor. This new compound migrates to the nucleus and induces DNA to produce RNA. After several days DNA production increases, also. Excess testosterone that is not assimilated by cells within 15 minutes after secretion is degraded by the liver to androgens and excreted.

During embryological development the production of testosterone is regulated by **chorionic gonadotropin,** a hormone produced by the placenta. At the onset of puberty the testes are stimulated to produce testosterone by the **luteinizing hormone (LH),** which is produced by the anterior lobe of the pituitary gland. This hormone stimulates the interstital cells to produce the hormone. Maturation of spermatozoa in the testes is controlled by the **follicle stimulating hormone (FSH),** which is also produced by the anterior pituitary gland. Testosterone assists FSH in this process, also.

The Ovaries

In addition to producing ova, the ovaries produce *estrogens* and *progesterone*. During fetal development small groups of cells move inward from the germinal epithelium of the ovary and develop into *primoidial follicles.* The ovaries of a young female at the onset of puberty are believed to contain between 100,000 and 400,000 of these immature follicles. During all the reproductive years of a woman only about 400 of these follicles will reach maturity and expel their ova.

Before puberty each follicle contains a single primary oocyte of 46 chromosomes. After puberty the follicles enlarge, meiosis takes place and the chromosome number is reduced to 23 chromosomes, converting the primary oocyte to a secondary oocyte (see figure 57.6). As this reduction division occurs in the maturing follicle (**Graafian follicle**), estrogens are produced within the follicle.

Estrogens

The principal function of estrogens is to promote the growth of specific cells in the body and to control the development of female secondary sex characteristics. In addition to being produced in the Graafian follicles, estrogens are produced by the corpus luteum, placenta, adrenal cortex, and testes.

Of the six or seven estrogens that have been isolated from the plasma of women, β **estradiol, estrone,** and **estriol** are the most abundant ones produced. The most potent of all the estrogens is β **estradiol.**

During puberty when the estrogens are produced in large quantities the following changes occur: (1) the female sex organs become fully developed; (2) the vaginal epithelium changes from cuboidal to stratified epithelium; (3) the uterine lining (endometrium) becomes more glandular in preparation for implantation; (4) the breasts form; (5) osteoblastic activity increases with more rapid bone growth; (6) calcification of epiphyses in long bones is hastened; (7) pelvic bones enlarge and change shape, increasing the size of the pelvic outlet; (8) fat deposition under the skin, in the hips, buttocks, and thighs is increased; and (9) skin vascularization is increased.

Estrogens enter target cells much in the same way that testosterone does. After combining with a protein receptor, the bound estrogen stimulates DNA to produce RNA, and protein synthesis is accelerated. The production of estrogens at puberty is initiated by the secretion of FSH.

Progesterone

Once ovulation takes place in the ovary the Graafian follicle is replaced by a **corpus luteum.** Progesterone is the principal hormone produced by this body. The ovary in figure 56.2 shows the development of Graafian follicles and the corpus luteum. The most important function of this hormone is to promote secretory changes in the endometrium in preparation for implantation of the fertilized ovum. In addition, the hormone causes mucosal changes in the fallopian tubes and promotes the proliferation of alveolar cells in the breasts in preparation for milk production. When the plasma level of progesterone falls due to regression of the corpus luteum, deterioration of the endometrium takes place and menstruation occurs. If pregnancy occurs the corpus luteum enlarges and produces additional quantities of progesterone. The conversion of a Graafian follicle into a corpus luteum is completely dependent on LH production.

The Pituitary Gland

The pituitary gland, or hypophysis, consists of an anterior portion, the *adenohypophysis,* and a posterior portion, the *neurohypophysis.* It is attached to the brain by a stalk, the infundibulum.

Adenohypophysis

This portion of the pituitary gland forms as an outpocketing of the foregut in the embryo. Its proximity to the hypothalamus enables various releasing factors that are produced in the hypothalamus to pass to the adenohypophysis through a vascular connection called the *hypophyseal portal system.* Six hormones are produced by six different kinds of cells in the gland. Some of the cells are acidophilic; others are basophilic. Except for the growth hormone, all hormones produced by the adenohypophysis are targeted specifically on glands.

Somatotropin (SH) This hormone, also called the *growth hormone, (GH),* increases the growth rate of all cells in the body by enhancing amino acid uptake and protein synthesis. Excess production of the hormone during the growing years causes *gigantism* due to the stimulation of growth in the epiphyses of the long bones. An adult with excess SH production develops *acromegaly,* which is

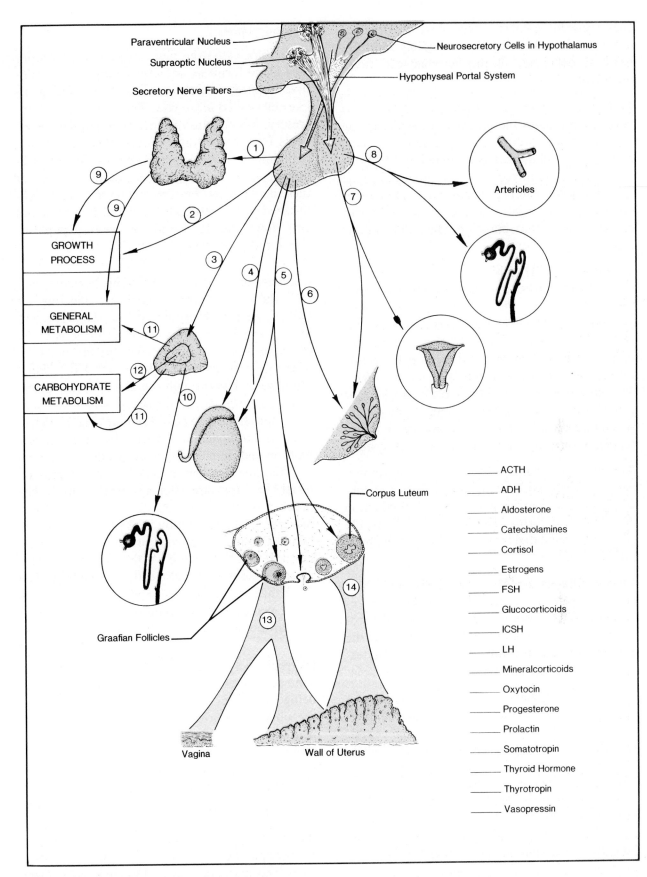

Figure 56.2 The pituitary regulatory mechanism.

ACTH
ADH
Aldosterone
Catecholamines
Cortisol
Estrogens
FSH
Glucocorticoids
ICSH
LH
Mineralcorticoids
Oxytocin
Progesterone
Prolactin
Somatotropin
Thyroid Hormone
Thyrotropin
Vasopressin

characterized by enlargement of the small bones of the hands and feet and the mandible and forehead. A deficiency of the hormone can cause *dwarfism*.

Prolactin *(Luteotropic Hormone, LTH)* This hormone promotes the production of milk in the breasts after childbirth. The release of this hormone prior to childbirth is inhibited by PIF *(prolactin inhibiting factor)*, which is produced in the hypothalamus. The presence of large amounts of estrogens and progesterone prior to childbirth causes the hypothalamus to produce this inhibiting factor.

Thyrotropin (TSH) This hormone regulates the rate of iodine uptake and the synthesis of the thyroid hormone by the thyroid gland. Excess amounts of TSH will cause hyperthyroidism. Hypothyroidism results from a deficiency of the hormone. TSH production is regulated by the hypothalmic *thyrotropin-releasing factor (TRE)*. This hormone is secreted by nerve endings in the hypothalamus and passes through the hypophyseal portal system. Thyrotropin production is also regulated by a thyroid hormone feedback system.

Adrenocorticotropin (ACTH) This hormone, which is produced by basophilic cells of the adenohypophysis, controls glucocorticoid production of the adrenal cortex. The production of ACTH is partially regulated by a glucocorticoid feedback mechanism.

Follicle Stimulating Hormone (FSH) This hormone promotes the development of Graafian follicles in the ovaries and the maturation of spermatozoa in the testes. FSH production by the adenohypophysis is regulated by estrogen and testosterone feedback mechanisms.

Luteinizing Hormone (LH) Maturation of the Graafian follicles and ovulation during the menstrual cycle are dependent on this hormone. Although FSH contributes much to early follicle development, ovulation is entirely dependent on LH. The production of estrogens and progesterone is dependent on LH also. In addition, this hormone stimulates the interstitial cells to produce testosterone; thus, it has been referred to as the **interstitial cell stimulating hormone (ICSH)**. LH production by the adenohypophysis is regulated by estrogen and progesterone feedback mechanisms.

Neurohypophysis

Embryologically, the posterior lobe of the pituitary gland forms as an outpouching of the primitive brain. It is composed mainly of neuroglial-like cells called *pituicytes*. These cells are nonsecretory, but function primarily to support nerve fibers from nerve tracts that originate in the hypothalamus. Two hormones, the *antidiuretic hormone* and *oxytocin*, are liberated by the ends of these nerve tracts. The hormones are actually synthesized in the cell bodies of the nerve cells and pass down the fibers into the neurohypophysis where they are released.

Antidiuretic Hormone (ADH) This hormone controls the permeability of nephron-collecting tubules to water reabsorption. When ADH is present in even minute amounts, water in urine is easily reabsorbed back into the blood through the walls of the collecting tubules. When ADH is lacking, rapid water loss through the kidneys takes place. *Diabetes insipidis* is a disease in which the neurohypophysis fails to provide enough ADH. Without this hormone the osmotic balance of body fluids cannot remain stable very long.

Another function of ADH is to increase arterial blood pressure when a severe blood loss has reduced blood volume. A 25% loss of blood will increase ADH production by 25 to 50 times normal. ADH increases blood pressure by constricting arterioles. Because ADH has this potent pressor capability it is also called **vasopressin.** The mechanism for increased ADH production during blood loss lies in baroreceptors that are located in the carotid, aortic, and pulmonary regions.

Oxytocin Any substance that will cause uterine contractions during pregnancy is designated as being "oxytocic". During childbirth, pressure of the unborn child on the uterine cervix causes a neurogenic reflex to the neurohypophysis, stimulating it to produce the hormone oxytocin. This hormone, being oxytocic, increases the strength of uterine contractions to facilitate childbirth.

Oxytocin also plays an important role in milk ejection during nursing. When the infant suckles the breast, sensory impulses to the hypothalamus via the spinal cord causes the production of oxytocin. This hormone is carried to the breasts where it causes alveoli in the breast to contract, forcing out milk.

Laboratory Assignment

Illustrations

Label figures 56.1 and 56.2. When labeling figure 56.2, keep in mind two things: (1) some of the hormones are not under direct pituitary control, and (2) synonyms for certain hormones are listed (for example, oxytocin and vasopressin).

Microscopic Studies

Examine prepared slides of available endocrine glands and make drawings of each type of tissue. Identify as many structures as possible that are unique about each specific gland.

Laboratory Report

Complete the Laboratory Report by answering all the questions.

57 The Reproductive Organs

This exercise on the reproductive system will include both gross and microscopic studies of the human reproductive organs. In addition, the cytological nature and significance of spermatogenesis and oogenesis will be studied.

Materials:

> models of human reproductive organs
> prepared slides of human testis
> ocular micrometer
> cat and dissection instruments

The Male Organs

Figure 57.1 is a sagittal section of the male reproductive system. The primary sex organs are the paired oval **testes** *(testicles),* which lie enclosed in a sac, the **scrotum.** The external nature of the testes provides a slightly lower temperature (94–95° F.), which favors the development of mature spermatozoa.

Lying over the superior and posterior surfaces of each testis is an elongated, flattened body, the **epididymis,** where immature sperm are stored after they leave the testis. The cutaway section of the testis in figure 57.1 reveals the relationship of the epididymis to the testis. Each testis is divided up into several chambers by partitions, or **septa,** that contain coiled up **seminiferous tubules.** A single seminiferous tubule is shown held out of one of the compartments. All the seminiferous tubules anastomose to form a network of tubules called the **rete testis.** Cilia within the rete testis move the immature spermatozoa through this maze of tubules into 10 or 15 **vasa efferentia** that lead directly into the epididymis. The outer wall, or **tunica albuginea,** that surrounds each testis consists of fibrous connective tissue.

In the act of ejaculation the spermatozoa leave the epididymis by way of the **vas deferens.** Tracing this duct upward, we see that it passes over the pubic bone and bladder into the pelvic cavity. The terminus of the vas deferens is enlarged to form the **ampulla of the vas deferens.** An accessory sex gland, the **seminal vesicle,** and the ampulla empty into the **common ejaculatory duct.** This latter duct passes through the **prostate gland** and empties into the **prostatic urethra.** From here the spermatozoa continue through the **penile urethra** out of the body.

The prostate gland, seminal vesicles, and Cowper's glands contribute alkaline secretions to the seminal fluid, which stimulates sperm motility. The prostate gland is the largest of these secondary sex glands. **Cowper's** *(bulbourethral)* **glands** are the smallest glands of the three. These pea-sized glands have ducts about one inch long that empty into the urethra at the base of the penis. The secretion of these glands is a clear mucoid fluid that lubricates the end of the penis and prepares the urethra for seminal fluid.

The **penis** consists of three cylinders of erectile tissue: two corpora cavernosa and one corpus spongiosum. The **corpus spongiosum** is a cylinder of tissue that surrounds the penile urethra. The two **corpora cavernosa** are located in the dorsal part of the organ and are separated on the midline by a **septum.** The distal end of the corpus spongiosum is enlarged to form a cone-shaped **glans penis.** The enlarged portion of the urethra within the glans is the **navicular fossa.** Erection of the penis occurs when the sponge-like tissue of the three corpora fills with blood.

Over the end of the glans penis lies a circular fold of skin, the **prepuce.** Around the neck of the glans, and on the inner surface of the prepuce, are scattered small *preputial glands.* These sebaceous glands produce a secretion of peculiar odor that readily undergoes decomposition to form a whitish substance called *smegma. Circumcision* is a surgical procedure that involves the removal of the prepuce to facilitate sanitation.

Assignment:

Label figure 57.1.

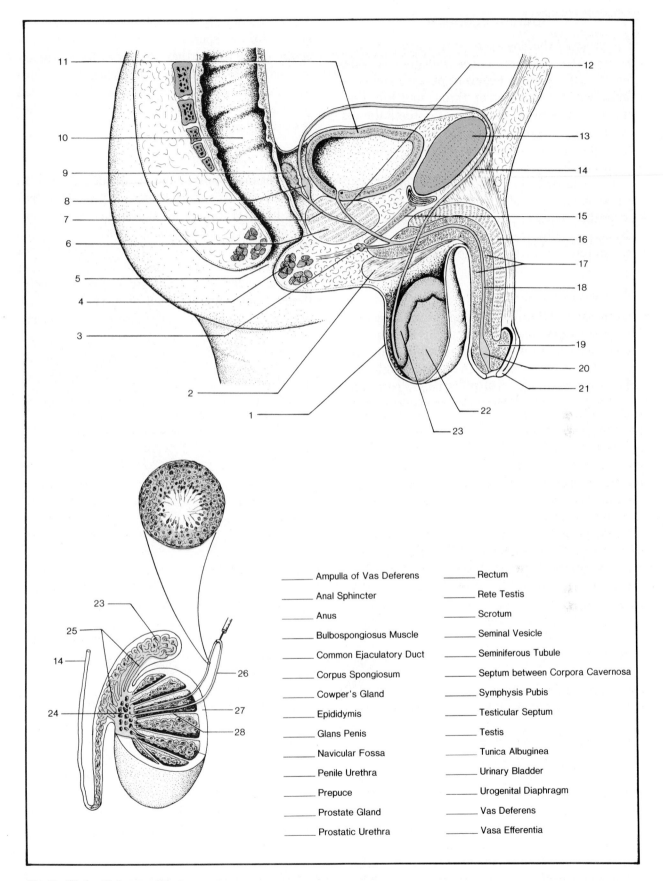

Figure 57.1 Male reproductive organs.

_____ Ampulla of Vas Deferens

_____ Anal Sphincter

_____ Anus

_____ Bulbospongiosus Muscle

_____ Common Ejaculatory Duct

_____ Corpus Spongiosum

_____ Cowper's Gland

_____ Epididymis

_____ Glans Penis

_____ Navicular Fossa

_____ Penile Urethra

_____ Prepuce

_____ Prostate Gland

_____ Prostatic Urethra

_____ Rectum

_____ Rete Testis

_____ Scrotum

_____ Seminal Vesicle

_____ Seminiferous Tubule

_____ Septum between Corpora Cavernosa

_____ Symphysis Pubis

_____ Testicular Septum

_____ Testis

_____ Tunica Albuginea

_____ Urinary Bladder

_____ Urogenital Diaphragm

_____ Vas Deferens

_____ Vasa Efferentia

Spermatogenesis

The process whereby spermatozoa are produced in the testes is known as *spermatogenesis*. This function of the testes begins at puberty and continues without interruption throughout life. Figure 57.2 illustrates a section of a seminiferous tubule and a diagram of the various stages in the development of mature spermatozoa. The formation of these germ cells takes place as the result of both mitosis and meiosis. In this study of spermatogenesis, slides of human testis will be studied under the microscope. Before examining the slides, however, familiarize yourself with the characteristic differences between meiosis and mitosis.

Mitosis

All spermatozoa originate from **spermatogonia** of the **primary germinal epithelium.** This layer of cells is located at the periphery of the seminiferous tubule. These cells contain the same number of chromosomes as other body cells; i.e., 23 pairs, and are said to be *diploid*. Mitotic division of these cells

occurs constantly, producing other spermatogonia. As the spermatogonia move toward the center of the tubule they enlarge to form **primary spermatocytes.** In this stage the homologous chromosomes unite to form chromosomal units called *tetrads*. This union of homologous chromosomes is called *synapsis*.

Meiosis

Each primary spermatocyte divides further to produce two secondary spermatocytes. This division, however, occurs by meiosis instead of mitosis. *Meiosis,* or reduction division, results in the distribution of a *haploid* number of chromosomes to each **secondary spermatocyte.** The result is that each secondary spermatocyte has only twenty-three chromosomes instead of forty-six. The individual chromosomes of the secondary spermatocytes are formed by the splitting of the tetrads along the line of previous conjugation to form chromosomes called *dyads*. The second meiotic division occurs when each secondary spermatocyte divides to produce two haploid **spermatids.** In

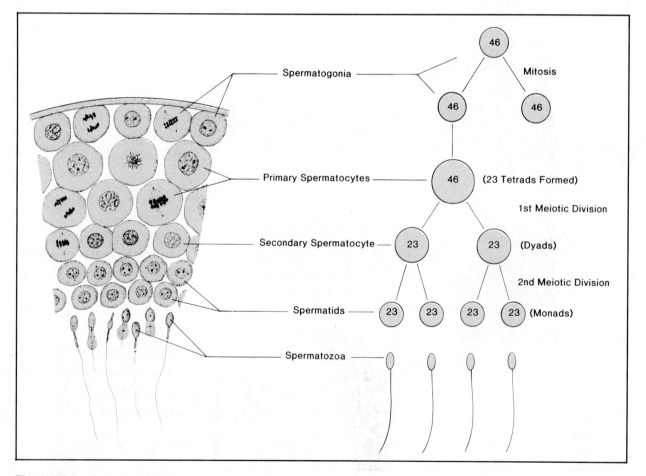

Figure 57.2 Spermatogenesis.

this division the dyads split to form single chromosomes, or *monads*. Each spermatid metamorphoses directly into a mature sperm cell. Note that four spermatozoa form from each spermatogonium.

Assignment:

Examine a slide of the human testis under low and high power magnifications. Identify the seminiferous tubules. Identify the various types of cells by referring to figure 57.2.

The Female Organs

Figures 57.3, 57.4, and 57.5 illustrate the female reproductive organs. Models and wall charts will be helpful in identifying all the structures.

External Genitalia

The **vulva** of the external female reproductive organs includes the mons pubis, labia majora, labia minora, and hymen. All of these structures are shown in figure 57.3. The **mons pubis** *(mons veneris)* is the most anterior portion and consists of a firm cushion-like elevation over the symphysis pubis. It is covered with hair.

Two folds of skin on each side of the vaginal orifice (label 11) lie over the opening. The larger exterior folds are the **labia majora.** Their exterior surfaces are covered with hair; their inner surfaces are smooth and moist. The **labia majora** are homologous (of similar embryological origin) to the scrotum of the male. Medial to the labia majora are the smaller **labia minora.** These folds meet anteriorly on the median line to form a fold of skin, the **prepuce of the clitoris.** The **clitoris** is a small protruberance of erectile tissue under the prepuce that is homologous to the penis on the male. It is highly sensitive to sexual excitation. The fold of skin that extends from the clitoris to each labium minora is called the **frenulum of the clitoris.** Posteriorly, the labia minora join to form a transverse fold of skin, the **posterior commissure,** or **fourchette.** Between the anus and the posterior commissure is the **central tendinous point of the perineum.** The **perineum** corresponds to the outlet of the pelvis.

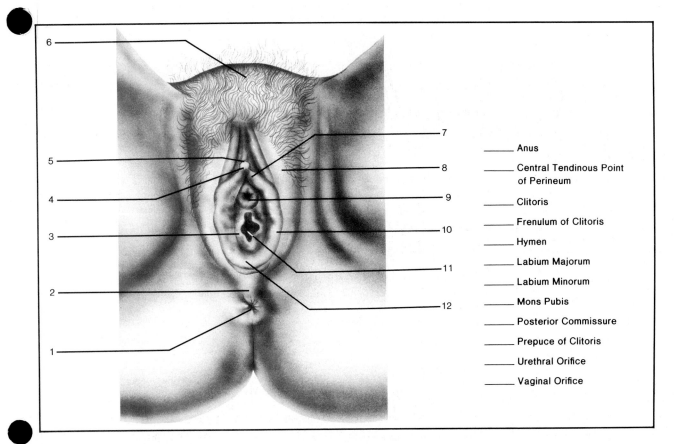

_____ Anus

_____ Central Tendinous Point of Perineum

_____ Clitoris

_____ Frenulum of Clitoris

_____ Hymen

_____ Labium Majorum

_____ Labium Minorum

_____ Mons Pubis

_____ Posterior Commissure

_____ Prepuce of Clitoris

_____ Urethral Orifice

_____ Vaginal Orifice

Figure 57.3 Female genitalia.

The *vestibule* is the area between the labia minora, extending from the clitoris to the fourchet. Situated within the vestibule are the vaginal orifice, hymen, urethral orifice, and openings of the vestibular glands. The **urethral orifice** lies about 2–3 centimeters posterior to the clitoris. Many small **paraurethral glands** surround this opening. They are homologous to the prostate glands of the male. On either side of the vaginal orifice are openings from the two **greater vestibular (Bartholin's) glands.** These glands are homologous to Cowper's glands. They provide mucous secretion for vaginal lubrication. The **hymen** is a thin fold of mucous membrane that separates the vagina from the vestibule. It may be completely absent or cover the vaginal orifice partially or completely. Its condition or absence is not a determinant of virginity.

Assignment:

Label figure 57.3.

Internal Organs

Figure 57.4 is a posterior view of the female reproductive organs which shows the relationship of the vagina, uterus, ovaries, and uterine tubes. It also reveals many of the principal supporting ligaments.

The Uterus The uterus is a pear-shaped thick-walled hollow organ which consists of a fundus, corpus, and cervix. The **fundus** is the dome-shaped portion at the large end. The body, or **corpus,** extends from the fundus to the cervix. The **cervix,** or neck, of the uterus is the narrowest portion; it is about one inch long. Within the cervix is an **endocervical canal** that has an outer opening, the **external os,** that opens into the **vagina;** an inner opening, the **internal os,** opens into the cavity of the corpus.

The thick muscular portion of the uterine wall is called the **myometrium.** The inner lining, or **endometrium,** consists of mucosal tissue. The anterior and posterior surfaces of the corpus, as well as the fundus, are covered with peritoneum.

The Uterine *(Fallopian)* **Tubes** Extending laterally from each side of the uterus is a uterine tube. The distal end of each tube is enlarged to form a funnel-shaped fimbriated **infundibulum** that surrounds an **ovary.** Although the infundibulum

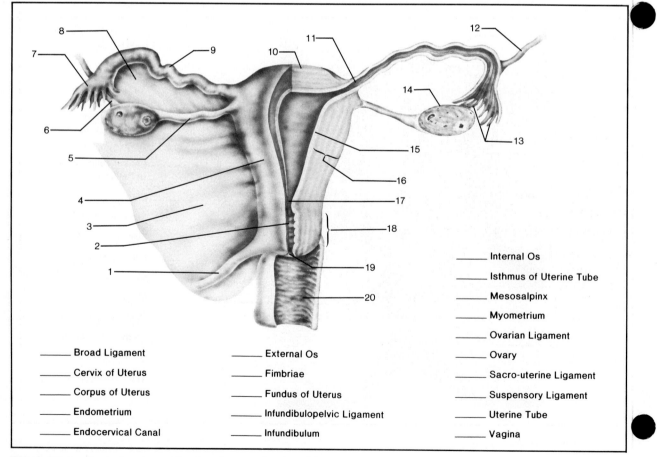

_____ Broad Ligament

_____ Cervix of Uterus

_____ Corpus of Uterus

_____ Endometrium

_____ Endocervical Canal

_____ External Os

_____ Fimbriae

_____ Fundus of Uterus

_____ Infundibulopelvic Ligament

_____ Infundibulum

_____ Internal Os

_____ Isthmus of Uterine Tube

_____ Mesosalpinx

_____ Myometrium

_____ Ovarian Ligament

_____ Ovary

_____ Sacro-uterine Ligament

_____ Suspensory Ligament

_____ Uterine Tube

_____ Vagina

Figure 57.4 Posterior view of the female reproductive organs.

doesn't usually contact the ovary, one or more of the finger-like **fimbriae** on the edge of the infundibulum usually do contact it. The infundibula and fimbriae receive oocytes produced by the ovary.

The uterine tubes are lined with ciliated columnar cells which assist in transporting egg cells from the ovary into the uterus. The tubes also have a muscular wall which produces peristaltic movements that help to propel the egg cells. Fertilization usually occurs somewhere along the uterine tube. Note that each uterine tube narrows down to form an **isthmus** near the uterus.

The Urinary Bladder Figure 57.5 reveals the relationship of the urinary bladder to the reproductive organs. Note that it lies between the uterus and the **symphysis pubis.** The base of the bladder is in direct contact with the anterior vaginal wall. The **urethra** is positioned between the vagina and the symphysis pubis.

The superior surface of the bladder is covered with peritoneum which is continuous with the peritoneum on the anterior face of the uterus. The small cavity lined with peritoneum between the bladder and uterus is known as the **anterior cul-de-sac.** The space between the uterus and rectum is the **posterior cul-de-sac of Douglas.**

Ligaments The uterus, ovaries, and uterine tubes are held in place and supported by various ligaments formed from the peritoneum and connective and muscular tissue. Figure 57.4 reveals the majority of them. A few are shown in figure 57.5.

The largest supporting structure is the **broad ligament** (label 3, figure 57.4). It is an extension of the peritoneal layers that cover the fundus and corpus of the uterus. It extends up over the fundus and corpus of the uterus. It extends up over the uterine tubes to form a mesentery, the **mesosalpinx,** on each side of the uterus. This portion of the broad ligament, which is shown in figure 57.4 between the ovary and uterine tube, contains blood vessels that supply nutrients to the uterine tube.

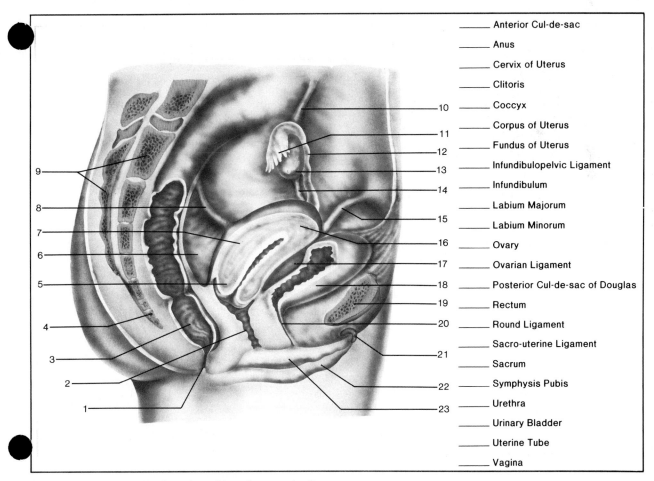

_____ Anterior Cul-de-sac
_____ Anus
_____ Cervix of Uterus
_____ Clitoris
_____ Coccyx
_____ Corpus of Uterus
_____ Fundus of Uterus
_____ Infundibulopelvic Ligament
_____ Infundibulum
_____ Labium Majorum
_____ Labium Minorum
_____ Ovary
_____ Ovarian Ligament
_____ Posterior Cul-de-sac of Douglas
_____ Rectum
_____ Round Ligament
_____ Sacro-uterine Ligament
_____ Sacrum
_____ Symphysis Pubis
_____ Urethra
_____ Urinary Bladder
_____ Uterine Tube
_____ Vagina

Figure 57.5 Midsagittal section of female reproductive organs.

Attached to the cervical region of the uterus are two **sacro-uterine ligaments** that anchor the uterus to the sacral wall of the pelvic cavity. A **round ligament** (label 15, figure 57.5) extends from each side of the uterus to the body wall. This ligament is not shown in figure 57.4, because it is anterior in position.

Each ovary is held in place by ovarian and suspensory ligaments. The **ovarian ligament** extends from the medial surface of the ovary to the uterus. The **suspensory ligament** is a peritoneal fold on the other side of the ovary that attaches the ovary to the uterine tube.

Each uterine tube is held in place by an **infundibulopelvic ligament** that is attached to the posterior surface of the infundibulum.

Assignment:

Label figures 57.4 and 57.5.

Oogenesis

The development of egg cells in the ovary is called *oogenesis*. It was pointed out on page 304 that as many as 400,000 immature follicles develop in the ovaries from the germinal epithelium. At the onset of puberty all the potential ova in these follicles contain forty-six chromosomes and are designated as **primary oocytes.** Only about 400 of these oocytes will mature during the reproductive lifetime of a woman.

Note in figure 57.6 that the primary oocyte, which forms from an oogonium of the germinal epithelium during embryological development, has 46 chromosomes. In the prophase stage of a dividing primary oocyte, the homologous pairs of chromosomes unite by synapsis to form 23 chromosomes. Each new chromosome is made up of 4 chromatids; thus, it is called a *tetrad*. When the primary oocyte finally divides in the first meiotic division, each tetrad is divided into two chromosomes with two chromatids each. These chromosomes are called *dyads*. The first meiotic division produces a **secondary oocyte** with all the yolk of the primary oocyte and a **polar body** with no yolk. Both the secondary oocyte and polar body each contain 23 dyads.

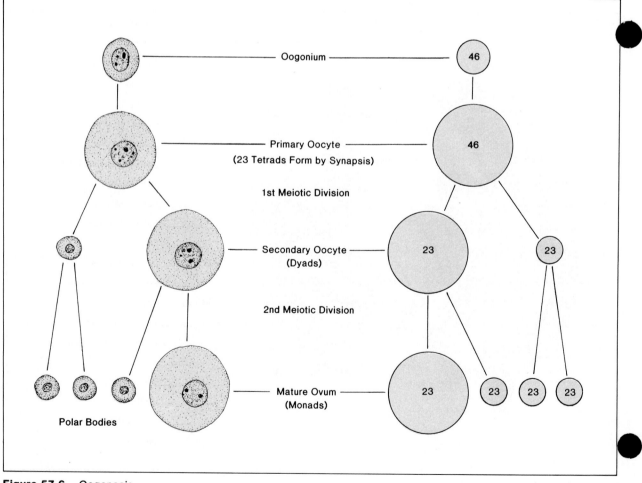

Figure 57.6 Oogenesis.

314

The second meiotic division produces a **mature ovum** with 23 chromosomes that are *monads;* that is, each chromosome consists of a single chromatid. Another polar body is also formed in this division. The first polar body divides also to produce two new polar bodies with monads. The end result is that one primary oocyte produces one mature ovum and three polar bodies. The polar bodies never serve any direct function in the fertilization process.

When ovulation takes place the "ovum" is a secondary oocyte. Usually, it is necessary for the sperm to penetrate the cell to trigger the second meiotic division. When the 23 chromosomes in the sperm unite with the chromosomes of the mature ovum, a **zygote** of 46 chromosomes is formed.

Assignment:

Study figure 57.6 thoroughly to familiarize yourself with the terminology and kinds of divisions. Several questions on the Laboratory Report pertain to this phenomenon.

Laboratory Report

Complete the Laboratory Report for this exercise.

LABORATORY REPORT	**1**	Student: _____
		Desk No: _____ Section: _____

Anatomical Terminology, Body Cavities, and Membranes

A. Illustration Labels

Record the label numbers for all the illustrations in the appropriate places in the answer columns.

B. Terminology

From this list of positions, sections, and membranes, select those that are applicable to the following questions. In some cases more than one answer may apply.

Relative Positions		*Sections*	*Membranes*
anterior—1	lateral—7	frontal—13	mesentery—17
caudad—2	medial—8	midsagittal—14	pericardium, parietal—18
cephalad—3	posterior—9	sagittal—15	pericardium, visceral—19
distal—4	proximal—10	transverse—16	peritoneum, parietal—20
dorsal—5	superior—11		peritoneum, visceral—21
inferior—6	ventral—12		pleura, parietal—22
			pleura, pulmonary—23

1. Type of section that divides the body into unequal right and left sides.
2. Type of section that divides the body into equal right and left halves.
3. Type of section that divides the body into front and back portions.
4. Two types of sections that will reveal both of the lungs and the heart in each section.
5. Three sections that are classified as being longitudinal sections.
6. A hat is worn on the _____ surface of the head.
7. The tongue is _____ to the palate.
8. The cheeks are _____ to the tongue.
9. The shoulder of a dog is _____ to its hip.
10. The shoulder of the human is _____ to his or her hip.
11. The fingertips are _____ to all structures of the hand.
12. The shoulder is the _____ *(proximal, distal)* portion of the arm.
13. Double-layered membrane that surrounds the heart.
14. Double-layered membrane that holds abdominal organs in place.
15. Serous membrane attached to the thoracic wall.
16. Serous membrane attached to the surface of the lung.

Select the surfaces on which the following are located:

17. Adam's apple.
18. Palm of the hand.
19. Ear.
20. Kneecap.

Answers

Terms	Fig. 1.2
1. _____	_____
2. _____	_____
3. _____	_____
4. _____	_____
5. _____	_____
6. _____	_____
7. _____	_____
8. _____	_____
9. _____	_____
10. _____	_____
11. _____	_____
12. _____	_____
13. _____	_____
14. _____	_____
15. _____	_____
16. _____	_____
17. _____	_____
18. _____	
19. _____	**Fig. 1.3**
20. _____	_____

Fig. 1.1	
_____	_____
_____	_____
_____	_____
Fig. 1.4	_____
_____	_____
_____	_____
_____	_____
_____	_____
_____	_____
_____	_____
_____	_____

C. Localized Areas

Identify the specific areas of the body described by the following statements.

antebrachium—1   calf—6   glutea—11   iliac—16
antecubital—2   costal—7   groin—12   lumbar—17
axilla—3   cubital—8   ham—13   plantar—18
brachium—4   epigastric—9 hypochondriac—14   pectoral—19
buttocks—5   flank—10   hypogastric—15   popliteal—20
                                                   umbilical—21

1. The underarm area.
2. The upper arm.
3. The forearm.
4. The "rump" area.
5. The sole of the foot.
6. The elbow area.
7. Anterior surface of the antebrachium.
8. Posterior portion of lower leg.
9. Depression on back of leg behind knee.
10. Area over the ribs on the dorsum.
11. Upper chest region.
12. Back portion of abdominal body wall that extends from lower edge of rib cage to hipbone.
13. Side of abdomen between lower edge of rib cage and upper edge of hipbone.
14. Abdominal area that surrounds the navel.
15. Area of the abdominal wall that covers the stomach.
16. Abdominal area which is lateral to the pubic region.
17. Abdominal area which is lateral to the epigastric area.
18. Abdominal area which is lateral to the umbilical area.
19. Area of arm on opposite side of elbow.
20. Abdominal areas between the transpyloric and transtubercular planes.

D. Body Cavities

From the following list of cavities select those applicable to the following statements.

abdominal—1   dorsal—4   spinal—8
abdominopelvic—2   pelvic—5   thoracic—9
cranial—3   pericardial—6   ventral—10
                   pleural—7

1. Cavity that consists of cranial and spinal cavities.
2. Most inferior portion of the abdominopelvic cavity.
3. Cavity separated into two major divisions by the diaphragm.
4. Cavity inferior to the diaphragm that consists of two parts.

Select the cavity in which the following organs are located:

5. Kidneys.   9. Rectum.   13. Spleen.
6. Duodenum.   10. Pancreas.   14. Spinal cord.
7. Brain.   11. Lungs and heart. 15. Urinary bladder.
8. Heart.   12. Liver.   16. Stomach.

Answers

Areas	Fig. 1.5
1. _____	_____
2. _____	_____
3. _____	_____
4. _____	_____
5. _____	_____
6. _____	_____
7. _____	_____
8. _____	_____
9. _____	_____
10. _____	_____
11. _____	_____
12. _____	_____
13. _____	_____
14. _____	_____
15. _____	_____
16. _____	**Fig. 1.6**
17. _____	_____
18. _____	_____
19. _____	_____
20. _____	_____

Cavities	
1. _____	_____
2. _____	_____
3. _____	_____
4. _____	_____
5. _____	**Fig. 1.7**
6. _____	_____
7. _____	_____
8. _____	_____
9. _____	_____
10. _____	_____
11. _____	
12. _____	
13. _____	
14. _____	
15. _____	
16. _____	

Gross Structures: Organ Systems

A. System Functions

Select the system (or systems) that perform the following functions in the body. Record the numbers in the answer column. More than one answer may apply.

circulatory—1	lymphatic—5	respiratory—9
digestive—2	muscular—6	reticuloendothelial—10
endocrine—3	nervous—7	skeletal—11
integumentary—4	reproductive—8	urinary—12

1. Carries heat from the muscles to the surface of the body for dissipation.
2. Provides rigid support for muscles.
3. Movement of ribs in breathing.
4. Transportation of food from the intestines to all parts of body.
5. Provides a shield of protection for vital organs such as the brain and heart.
6. Disposes of urea, uric acid, and other cellular wastes.
7. Removes carbon dioxide from the blood.
8. Destroys microorganisms once they break through the skin.
9. Phagocytic cells of this system remove bacteria from lymph.
10. Helps to maintain normal body temperature by dissipating excess heat.
11. Carries messages from receptors to centers of interpretation.
12. Movement of arms and legs.
13. Produces various kinds of blood cells.
14. Enables the body to adjust to changing internal and external environmental conditions.
15. Returns tissue fluid to the blood.
16. Controls the rate of growth.
17. Acts as a shield against invasion by bacteria.
18. Carries fats from the intestines to the blood.
19. Carries hormones from glands to all tissues.
20. Eliminates excess water from body.
21. Ensures continuity of the species.
22. Breaks food particles down into small molecules.
23. Controls the development of the ovaries and testes.
24. Production of spermatozoa and ova.
25. Coordination of all body movements.

Answers
System Functions
1. _____
2. _____
3. _____
4. _____
5. _____
6. _____
7. _____
8. _____
9. _____
10. _____
11. _____
12. _____
13. _____
14. _____
15. _____
16. _____
17. _____
18. _____
19. _____
20. _____
21. _____
22. _____
23. _____
24. _____
25. _____

B. Organ Placement

Match the system at the right to the following organs.

1. Bones	circulatory—1		**Answers**	
2. Pancreas	digestive—2		**Organ Placement**	
3. Bronchi	endocrine—3			
4. Brain	integumentary—4			
5. Lungs	lymphatic—5			

Organ	System
1. Bones	circulatory—1
2. Pancreas	digestive—2
3. Bronchi	endocrine—3
4. Brain	integumentary—4
5. Lungs	lymphatic—5
6. Kidneys	muscular—6
7. Heart	nervous—7
8. Hair	reproductive—8
9. Esophagus	respiratory—9
10. Dermis	skeletal—10
11. Teeth	urinary—11
12. Veins	
13. Uterus	
14. Ureters	
15. Trachea	
16. Tonsils	
17. Toenails	
18. Stomach	
19. Spleen	
20. Thyroid	

Answers

Organ Placement

1. _____
2. _____
3. _____
4. _____
5. _____
6. _____
7. _____
8. _____
9. _____
10. _____
11. _____
12. _____
13. _____
14. _____
15. _____
16. _____
17. _____
18. _____
19. _____
20. _____

3

Student: _____

Desk No: _____ Section: _____

The Microscope

A. Completion Questions

Record the answers to the following questions in the column at the right.

1. What effect *(increase* or *decrease)* does closing the diaphragm have on the following?
   a. Image brightness
   b. Image contrast
   c. Resolution
2. In general, at what position should the condenser be kept?
3. List three fluids that may be used for cleaning lenses.
4. Express the maximum resolution of the compound microscope in terms of micrometers (μm).
5. What is the magnification of objects observed through a $100\times$ oil immersion objective with a $7.5\times$ eyepiece?
6. If you are getting $225\times$ magnification with a $45\times$ high-dry objective, what would be the power of the eyepiece?
7. Immersion oil must have the same refractive index as _____ to be of any value.
8. Substage filters should be of a _____ color to get the maximum resolution of the optical system.
9. How can one greatly increase the bulb life on a microscope lamp if voltage is variable?
10. What characteristic of a microscope enables one to switch from one objective to another without appreciably altering the focus?

B. True-False

Record these statements as True or False in the answer column.

1. The safest way to use the oil immersion lens is to start with the low power objective and then switch to oil immersion.
2. The higher powered objectives have shorter working distances than the $10\times$ objective.
3. The resolution of a microscope can be greatly increased by using a $20\times$ ocular instead of one that is only $10\times$.
4. Eyepieces are of such simple construction that almost anyone can safely disassemble them for cleaning.
5. When swinging the oil immersion into position after using high-dry, one should always raise the objective a little first to avoid damaging the oil immersion lens.
6. Lens tissue is the only safe cleaning tissue to use on microscope lenses.

Answers
Completion
1a. _____
b. _____
c. _____
2. _____
3a. _____
b. _____
c. _____
4. _____
5. _____
6. _____
7. _____
8. _____
9. _____
10. _____
True-False
1. _____
2. _____
3. _____
4. _____
5. _____
6. _____

C. Multiple Choice

Select the best answer for the following statements.

1. The resolution of a microscope is increased by
   1. using blue light
   2. stopping down the diaphragm
   3. lowering the condenser
   4. raising the condenser to its highest point
   5. both 1 and 4

2. Microscope lenses may be cleaned with
   1. lens tissue and optically safe boxed tissues
   2. a soft linen handkerchief
   3. an air syringe
   4. both 1 and 3
   5. 1, 2, and 3

3. When changing from low power to high power, it is generally necessary to
   1. lower the condenser
   2. open the diaphragm
   3. close the diaphragm
   4. both 1 and 2

4. The most commonly used ocular is
   1. 5×
   2. 10×
   3. 15×
   4. 20×

5. The magnification of an object seen through the 10× objective with a 10× ocular is
   1. ten times
   2. twenty times
   3. 1000 times
   4. none of these

Answers
Multiple-Choice
1. _____
2. _____
3. _____
4. _____
5. _____

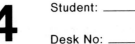

Student: _____

Desk No: _____ Section: _____

Cytology: Basic Cell Structure

A. Figure 4.1
Record the labels for figure 4.1 in the answer column.

B. Cell Drawings
Sketch in the space below, one or two epithelial cells as seen under high-dry magnification. Label the **nucleus, cytoplasm,** and **cell membrane.**

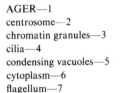

C. Cell Structure
From the list of structures below, select those that are described by the following statements. More than one structure may apply to some of the statements.

AGER—1	Golgi apparatus—8	nucleolus—15
centrosome—2	lysosome—9	nucleus—16
chromatin granules—3	microvilli—10	plasma membrane—17
cilia—4	mitochondria—11	polyribosomes—18
condensing vacuoles—5	microfilaments—12	RER—19
cytoplasm—6	microtubules—13	ribosomes—20
flagellum—7	nuclear envelope—14	secretory granules—21

1. Trilaminar unit membrane structure.
2. Double-layered unit membrane structure.
3. Prominent body in nucleus that produces ribosomes.
4. ER that lacks ribosomes.
5. A nonmembranous organelle consisting of two bundles of microtubules.
6. Bodies within the nucleus that represent chromosomal material.
7. Outer membrane of the cell.
8. All protoplasmic material between the plasma membrane and nucleus.
9. Extensive system of tubules, vesicles, and sacs within the cytoplasm.
10. Small bodies of RNA and protein attached to surfaces of the RER.
11. Sacs in the cytoplasm that contain digestive enzymes.
12. A unit membrane organelle in the cytoplasm that contains cristae.
13. Short, hair-like appendages on the surface of some cells.
14. ER that is covered by ribosomes.
15. Bodies produced by Golgi and exocytosed by the cell.
16. Surface protuberances that form a delicate brush border.
17. A layered concave structure in the cytoplasm that is similar to the basic structure of AGER.
18. Extranuclear bodies that possess DNA and are able to replicate themselves.
19. Unit membrane structure that is continuous with the plasma and nuclear membranes.
20. Chains and rosettes of small bodies scattered throughout the cytoplasm.

Answers

Cell Structure	Fig. 4.1
1. _____	_____
2. _____	_____
3. _____	_____
4. _____	_____
5. _____	_____
6. _____	_____
7. _____	_____
8. _____	_____
9. _____	_____
10. _____	_____
11. _____	_____
12. _____	_____
13. _____	_____
14. _____	_____
15. _____	_____
16. _____	_____
17. _____	_____
18. _____	_____
19. _____	_____
20. _____	_____

D. Organelle Functions

Select the organelles that perform the following functions. More than one structure may apply to some statements.

AGER—1   microfilaments—6   nucleus—11
centrosome—2   microtubules—7   plasma membrane—12
flagellum—3   microvilli—8   polyribosomes—13
Golgi apparatus—4 mitochondrion—9   RER—14
lysosome—5   nucleolus—10

1. The microcirculatory system of the cell.
2. Contains genetic code of cell.
3. Produces ribonucleoprotein for ribosomes.
4. Forms asters during mitosis.
5. Place where ATP is synthesized.
6. Brings about rapid hydrolysis (digestion) of cell after death.
7. Regulates the flow of materials into and out of cell (selectively permeable).
8. Synthesize protein for endogenous use.
9. Synthesizes protein for extracellular distribution.
10. Place where oxidative respiration occurs.
11. Synthesis, packaging, and transportation of substances secreted by cell.
12. Plays a role in flagellar growth.
13. Increases absorptive surface area of the cell.
14. Disposal organelles of cytoplasm, digesting fragments of mitochondria, red blood cells, etc.
15. Accomplishes some motile function for some cells.
16. Imparts the characteristic of contractility to muscle cells.

E. True-False

Evaluate the validity of the following statements concerning cell structure and function.

1. Practically all molecules that pass through the plasma membrane are actively assisted by the membrane.
2. The inner and outer layers of the plasma membrane are essentially lipoidal in nature with scattered molecules of protein and glycoprotein.
3. Ribosomes are 40% RNA and 60% protein.
4. Ribosomes are found in the cytoplasm, mitochondria, RER, and AGER.
5. Mitochondria are able to replicate themselves because they contain RNA.
6. Secretory cells have well-developed Golgi apparatus and RER.
7. Most molecules that pass through the plasma membrane enter through large pores in that membrane.
8. Intestinal and kidney cells characteristically have microvilli.
9. Cilia originate from basal bodies that are derived from centrioles.
10. Ribosomes are primarily involved in oxidative respiration.

Answers

Functions

1. _____
2. _____
3. _____
4. _____
5. _____
6. _____
7. _____
8. _____
9. _____
10. _____
11. _____
12. _____
13. _____
14. _____
15. _____
16. _____

True-False

1. _____
2. _____
3. _____
4. _____
5. _____
6. _____
7. _____
8. _____
9. _____
10. _____

Student: _____

Desk No: _____ Section: _____

Histology: Epithelial, Connective, and Supportive Tissues

A. Microscopic Study

If any drawings of tissues are to be made, make them on separate sheets of paper and label all identifiable structures.

B. Figure 5.3

Record the labels for this illustration in the answer column.

C. Tissue Structure

From the list of tissues select those that are described by the following statements. More than one tissue may apply in some cases.

1. Reinforced by white fibers only.
2. Reinforced by both white and yellow fibers.
3. Reinforced by yellow fibers only.
4. Consists of a single layer of flat cells.
5. Flattened cells in layers.
6. Cubical shaped cells.
7. Cells with large fat vacuoles.
8. Sponge-like bone tissue.
9. Dense bone tissue.
10. Tissue with trabeculae.
11. Cells contained in lacunae.
12. Elongated epithelial cells lacking cilia.
13. Elongated cells with hair-like projections.
14. Cells surrounded by a matrix of calcium salts.
15. Cartilage lacking in readily visible fibers.
16. Cells in rows surrounded by white fibers.
17. Contains fibers that often mature into collagenous fibers.

adipose—1
areolar—2
cancellous bone—3
ciliated columnar—4
compact bone—5
cuboidal—6
elastic cartilage—7
fibrocartilage—8
fibrous connective tissue—9
hyaline cartilage—10
osteocyte—11
plain columnar—12
reticular—13
simple squamous—14
stratified squamous—15

D. Tissue Location

From the above list of tissues select the ones that would be found in the following places.

1. Ligaments and tendons.
2. Lining of mouth.
3. Lining of glands.
4. Lining of trachea.
5. Lining of stomach.
6. Skull.
7. Reinforcement of pericardium.
8. Cartilage of external ear.
9. Fat within mesenteries.
10. Covering on ends of long bones.
11. Between skin and muscles.
12. Between vertebrae (disk).
13. Nasal septum cartilage.
14. Interspersed among lymphoblasts and macrophages of lymph nodes.

E. Tissue Function

From the above list of tissues select those that perform the following functions.

1. Secretion of hormones.
2. Provide rigid support.
3. Provide flexible support.
4. Storage of energy.
5. Absorption of food.
6. Hold organs together.
7. Reinforce lymph nodes.
8. Produce fibers for basement membrane of some epithelial cells.

Answers

Tissue Structure	Tissue Location
1. _____	1. _____
2. _____	2. _____
3. _____	3. _____
4. _____	4. _____
5. _____	5. _____
6. _____	6. _____
7. _____	7. _____
8. _____	8. _____
9. _____	9. _____
10. _____	10. _____
11. _____	11. _____
12. _____	12. _____
13. _____	13. _____
14. _____	14. _____

Tissue Structure	Tissue Function
15. _____	
16. _____	1. _____
17. _____	2. _____
Fig. 5.3	3. _____
_____	4. _____
_____	5. _____
_____	6. _____
_____	7. _____
_____	8. _____
_____	
_____	
_____	

The Integument (Ex. 6)

A. Figure 6.1

Record the labels for this illustration in the answer column.

B. Microscopic Study

After examining a section of skin under low and high power, make a drawing on a separate sheet of paper of a section as seen under high power. Label as many structures as are recognizable.

C. Questions

Select the answer that completes the following statements. Only one answer applies.

1. The epidermis consists of the following number of distinct layers:
   (1) three   (2) four   (3) five   (4) six
2. The papillary layer is a part of the
   (1) dermis   (2) epidermis   (3) stratum spinosum
3. Meissner's corpuscles are sensitive to
   (1) temperature   (2) pressure   (3) touch
4. Pacinian corpuscles are sensitive to
   (1) temperature   (2) pressure   (3) touch
5. Granules of the stratum granulosum consist of
   (1) melanin   (2) keratin   (3) eleidin
6. New cells of the epidermis originate in the
   (1) dermis   (2) stratum germinativum
   (3) stratum corneum   (4) corium
7. The outermost layer of the epidermis is the
   (1) stratum corneum   (2) stratum germinativum
   (3) stratum lucidum   (4) stratum spinosum
8. Melanin granules are produced by the
   (1) stratum granulosum   (2) melanocytes
   (3) corium   (4) stratum corneum
9. The hair follicle is
   (1) a shaft of hair   (2) the root of a hair
   (3) a tube in the skin   (4) none of these
10. Arrector pili assist in
    (1) maintaining skin tonus   (2) limiting excessive sweating
    (3) forcing out sebum   (4) none of these
11. Pacinian corpuscles are located in the
    (1) papillary layer   (2) reticular layer   (3) hypodermis
12. Meissner's corpuscles are located in the
    (1) papillary layer   (2) reticular layer   (3) hypodermis
13. Sweat is produced by
    (1) sebaceous glands   (2) eccrine glands
    (3) cerumenous glands   (4) none of these
14. Keratin's function is primarily to
    (1) destroy bacteria   (2) provide waterproofing   (3) cool the body
15. The coiled up portion of an apocrine gland is located in the
    (1) hair follicle   (2) dermis   (3) epidermis   (4) hypodermis
16. Sebum is secreted by
    (1) sweat glands   (2) eccrine glands
    (3) apocrine glands   (4) none of these
17. Apocrine glands are located
    (1) in the axillae   (2) on the scrotum
    (3) in the ear canal   (4) all of these
18. The secretions of the apocrine glands usually empty
    (1) into a hair follicle   (2) directly out through skin surface
    (3) both 1 and 2   (4) none of these

Answers

Fig. 6.1

Questions

1.
2.
3.
4.
5.
6.
7.
8.
9.
10.
11.
12.
13.
14.
15.
16.
17.
18.

7

Student: _____

Desk No: _____ Section: _____

Osmosis and Cell Membrane Integrity

A. Molecular Movement

1. Brownian Movement

a. Were you able to see any movement of carbon particles under the microscope? _____ .

b. What forces propel the movement of these particles? _____

_____

_____

_____

2. Diffusion Plot the distance of diffusion of molecules for the two crystals on the following graphs.

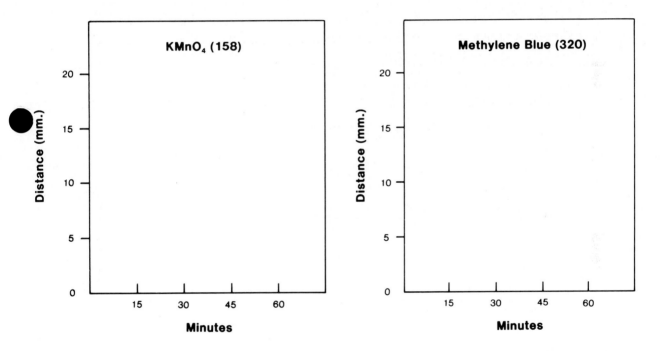

3. Conclusion What is the relationship of molecular weight to the rate of diffusion? _____

_____

_____

_____

_____

Osmosis and Cell Membrane Integrity

B. Osmotic Effects

As you examine each tube macroscopically and microscopically, record your results in the following table:

Solution	Macroscopic Appearance (lysis or no lysis)	Microscopic Appearance (crenation, lysis, or isotonic)
.15M Sodium Chloride		
.30M Sodium Chloride		
.28M Glucose $C_6H_{12}O_6$		
.30M Glycerine $C_3H_5(OH)_3$		
.30M Urea $CO(NH_2)_2$		

1. Are any of the above solutions isotonic for red blood cells? _____

   If so, which ones? _____

2. With the aid of the atomic weight table in the appendix, calculate the percentages of solutes in those solutions that proved to be isotonic. _____

   _____

   _____

Student: _____

Desk No: _____ Section: _____

The Skeletal Plan

A. Illustrations

Record the labels for figures 8.1, 8.2, and 8.3 in the answer columns.

B. Bone Terminology

Identify the terms described by the following statements.

1. A narrow slit.
2. A round bone.
3. A groove or furrow.
4. A cavity in a bone.
5. Shaft portion of bone.
6. Hollow chamber in shaft of bone.
7. Process to which muscle is attached.
8. An opening in a bone through which nerves and blood vessels pass.
9. An enlarged end of a bone.
10. Type of bone tissue in shaft of bone.
11. Tough fibrous covering of shaft of bone.
12. A sharp or long slender process.
13. Small rounded process on bone.
14. A rounded knuckle-like articulating process.
15. Linear growth area of a long bone.
16. Smooth gristle covering end of long bone.
17. Type of marrow in ends of bone.
18. A long tube-like passageway.
19. A shallow depression.
20. Lining of medullary canal.
21. A large roughened process on a bone.
22. Type of bone tissue in ends of bone.

articular cartilage—1
cancellous bone—2
compact bone—3
condyle—4
crest—5
diaphysis—6
endosteum—7
epiphyseal disk—8
epiphysis—9
fissure—10
foramen—11
fossa—12
head—13
meatus—14
medullary canal—15
periosteum—16
red marrow—17
sesamoid—18
sinus—19
spine—20
sulcus—21
trochanter—22
tubercle—23
tuberosity—24
yellow marrow—25

C. Bone Identification

Select the structures on the right that match the statements on the left.

1. Kneecap.
2. Shinbone.
3. Breastbone.
4. Collarbone.
5. Shoulder blade.
6. Upper arm bone.
7. Bones of spine.
8. Thigh bone.
9. Lateral bone of forearm.
10. Horse-shoe shaped bone.
11. Joint between os coxae.
12. Bones of shoulder girdle.
13. One half of pelvic girdle.
14. Medial bone of forearm.
15. Thin bone paralleling tibia (calfbone).

clavicle—1
femur—2
fibula—3
humerus—4
hyoid—5
os coxa—6
patella—7
radius—8
scapula—9
sternum—10
symphysis pubis—11
tibia—12
ulna—13
vertebrae—14

Answers

Terms	Bones
1. _____	1. _____
2. _____	2. _____
3. _____	3. _____
4. _____	4. _____
5. _____	5. _____
6. _____	6. _____
7. _____	7. _____
8. _____	8. _____
9. _____	9. _____
10. _____	10. _____
11. _____	11. _____
12. _____	12. _____
13. _____	13. _____
14. _____	14. _____
15. _____	15. _____
16. _____	Fig. 8.2
17. _____	_____
18. _____	_____
19. _____	_____
20. _____	_____
21. _____	_____
22. _____	_____
Fig. 8.1	_____
_____	_____
_____	_____
_____	_____
_____	_____
_____	_____
_____	_____
_____	_____
_____	_____
_____	_____

D. Medical

Select the condition that is described by the following statements. Consult your lecture text or medical dictionary.

closed reduction—1
Colle's fracture—2
comminuted fracture—3
compacted fracture—4
compound fracture—5
fissured fracture—6
greenstick fracture—7
open reduction—8

osteomalacia—9
osteomyelitis—10
osteoporosis—11
pathological fracture—12
Pott's fracture—13
rickets—14
simple fracture—15
none of these—16

1. Procedure used to set broken bones without using surgery.

2. Fracture in which skin is not broken.

3. Fracture caused by torsional forces.

4. Fracture caused by severe vertical forces.

5. Fracture due to weak bone structure, not trauma.

6. Skeletal softness in adults.

7. Fracture in which a piece of bone is broken out of shaft.

8. Fracture characterized by two or more fragments.

9. Fracture which occurs at an angle to the longitudinal axis of a bone.

10. Procedure used to set broken bones, utilizing surgery.

11. Bone condition in which increased porosity occurs due to widening of Haversian canals.

12. Incomplete bone fracture in which fracture is apparent only on convex surface.

13. Skeletal softness in children due to vitamin D deficiency.

14. Outward displacement of foot due to fracture of lower part of fibula and malleolus.

15. Displacement of hand backward and outward due to fracture at lower end of radius.

16. Infection of bone marrow.

17. Bone fracture extends only partially through a bone; incomplete fracture.

Answers

Medical

1. _____
2. _____
3. _____
4. _____
5. _____
6. _____
7. _____
8. _____
9. _____
10. _____
11. _____
12. _____
13. _____
14. _____
15. _____
16. _____
17. _____

Fig. 8.3

_____
_____
_____
_____
_____
_____
_____
_____
_____
_____

Student: _____

Desk No: _____ Section: _____

The Skull

A. Illustrations

Record the labels for all the illustrations of this exercise in the answer columns.

B. Bone Names

Select the bones that are described by the following statements. More than one answer may apply.

1. Cheekbone.
2. Lower jaw.
3. Upper jaw.
4. Sides of skull.
5. Forehead bone.
6. Bones of hard palate (2 pair).
7. Unpaired bones (usually).
8. Back and bottom of skull.
9. On walls of nasal fossa.
10. On median line in nasal cavity.
11. External bridge of nose.
12. Contain sinuses (4 bones).

ethmoid—1
frontal—2
lacrimal—3
mandible—4
maxilla—5
nasal—6
nasal conchae—7
occipital—8
palatine—9
parietal—10
sphenoid—11
temporal—12
vomer—13
zygomatic—14

C. Terminology

From the above list of bones select those which have the following structures.

1. Incisive fossa.
2. Cribriform plate.
3. Angle.
4. Inferior temporal line.
5. Crista galli.
6. Alveolar process.
7. Clinoid processes.
8. Mandibular condyle.
9. External acoustic meatus.
10. Coronoid process.
11. Antrum of Highmore.
12. Internal acoustic meatus.
13. Ramus.
14. Superior nasal concha.
15. Superior temporal line.
16. Mandibular fossa.
17. Zygomatic arch.
18. Mastoid process.
19. Symphysis.
20. Tear duct.
21. Sella turcica.
22. Mylohyoid line.
23. Styloid process.
24. Occipital condyle.

D. Sutures

Identify the sutures described by the following statements.

1. Between parietal and occipital.
2. Between two maxillae on median line.
3. Between the two parietals on the median line.
4. Between parietal and temporal.
5. Between maxillary and palatine.
6. Between frontal and parietal.

coronal—1
lambdoidal—2
median palatine—3
sagittal—4
squamosal—5
transverse palatine—6

Answers

Bones	Fig. 9.1	Fig. 9.2
1. ___	_____	_____
2. ___	_____	_____
3. ___	_____	_____
4. ___	_____	_____
5. ___	_____	_____
6. ___	_____	_____
7. ___	_____	_____
8. ___	_____	_____
9. ___	_____	_____
10. ___	_____	_____
11. ___	_____	_____
12. ___	_____	_____

Terms		
1. ___	_____	_____
2. ___	_____	_____
3. ___	_____	_____
4. ___	_____	
5. ___	_____	**Sutures**
6. ___	_____	1. ___
7. ___	_____	2. ___
8. ___	_____	3. ___
9. ___	_____	4. ___
10. ___	**Fig. 9.5**	5. ___
11. ___	_____	6. ___
12. ___	_____	
13. ___	_____	
14. ___	_____	
15. ___	_____	
16. ___	_____	
17. ___	_____	
18. ___	_____	
19. ___	_____	
20. ___	_____	
21. ___	_____	
22. ___	_____	
23. ___	_____	
24. ___		

E. Foramina Location

By referring to the illustrations select the bones on which the following foramina or canals are located.

1. Foramen magnum.
2. Optic canal.
3. Carotid canal.
4. Internal acoustic meatus.
5. Zygomaticofacial foramen.
6. Mandibular foramen.
7. Hypoglossal canal.
8. Anterior palatine foramen.
9. Foramen cecum.
10. Supraorbital foramen.
11. Infraorbital foramen.
12. Foramen ovale.
13. Mental foramen.

frontal—1
mandible—2
maxilla—3
occipital—4
palatine—5
parietal—6
sphenoid—7
temporal—8
zygomatic—9

F. Foramen Function

Select the cranial nerves and other structures that pass through the following foramina, canals and fissures. Note that the number following each cranial nerve corresponds to the number designation of the particular cranial nerve. (Example: the abducens nerve is also known as the 6th cranial nerve.) Also, note that the 5th cranial nerve (trigeminal) has three divisions.

1. Foramen ovale.
2. Foramen magnum.
3. Hypoglossal canal.
4. Optic canal.
5. Carotid canal.
6. Cribriform plate foramina.
7. Internal acoustic meatus.

Cranial Nerves
olfactory—1
optic—2
oculomotor—3
trochlear—4
ophthalmic—5.1
maxillary—5.2
mandibular—5.3
abducens—6
facial—7
statoacoustic—8
spinal accessory—11
hypoglossal—12
brain stem—13
internal carotid a.—14

G. Completion

Provide the name of the structures described by the following statements.

1. Provides anchorage for sternocleidomastoideus muscle.
2. Allow sound waves to enter the skull.
3. Provides attachment site for falx cerebri.
4. Allow compression of skull bones during birth.
5. Lighten the skull.
6. Point of articulation between mandible and skull.
7. Points of articulation between skull and first vertebra.
8. Point of attachment for some tongue muscles.

Answers

Location	Fig. 9.3	Fig.
1. ___	___	___
2. ___	___	___
3. ___	___	___
4. ___	___	___
5. ___	___	___
6. ___	___	___
7. ___	___	___
8. ___	___	___
9. ___	___	___
10. ___	___	___
11. ___	___	___
12. ___	___	___
13. ___	___	___

Function		
1. ___	___	___
2. ___	___	___
3. ___	___	___
4. ___	___	___
5. ___	___	___
6. ___	___	___
7. ___	___	___

Fig 9.7

Fig 9.6	
___	___
___	___
___	___
___	___
	___
	___
	___
	___

Completion	Fig. 9.8
1. ___	___
2. ___	___
3. ___	___
4. ___	___
5. ___	___
6. ___	___
7. ___	___
8. ___	___

Student: _____

Desk No: _____ Section: _____

The Vertebral Column and Thorax

A. Illustrations

Record the labels for figures 10.1 and 10.2 in the answer column.

B. Bone Names

Select the proper anatomical terminology for the following bones.

1. True rib.
2. Floating rib.
3. False rib.
4. Breastbone.
5. Tailbone.
6. First cervical vertebra.
7. Second cervical vertebra.
8. Inferior portion of breastbone.
9. Mid-portion of breastbone.
10. Superior portion of breastbone.

atlas—1
axis—2
coccyx—3
gladiolus—4
manubrium—5
sacrum—6
sternum—7
vertebral ribs—8
vertebrochondral ribs—9
vertebrosternal ribs—10
xiphoid—11

C. Structures

Select the structures or foramina that perform the following functions.

body—1
intervertebral disks—2
intervertebral foramina—3
odontoid process—4
rib cage—5
spinal curves—6
spinous process—7
sternum—8
transverse process—9
transverse foramina—10

1. Protects vital thoracic organs.
2. Portion of vertebra that supports body weight.
3. Openings through which spinal nerves exit.
4. Provides anterior attachment for vertebrosternal ribs.
5. Openings through which vertebral artery and vein pass.
6. Acts as a pivot for the atlas to move around.
7. Part of thoracic vertebra which provides attachment for rib.
8. Impart springiness to vertebral column (two things).

Answers

Bone Names	Fig. 10.1
1. _____	_____
2. _____	_____
3. _____	_____
4. _____	_____
5. _____	_____
6. _____	_____
7. _____	_____
8. _____	_____
9. _____	_____
10. _____	_____

Structures	
1. _____	_____
2. _____	_____
3. _____	_____
4. _____	_____
5. _____	_____
6. _____	_____
7. _____	_____
8. _____	_____

D. *Vertebral Differences*

Select the vertebrae from the column on the right that is unique for the following characteristics.

1. Odontoid process.
2. Large massive body.
3. Downward sloping spinous process.
4. Transverse foramina.
5. Rib facets on transverse process.

atlas—1
axis—2
cervical vertebrae—3
thoracic vertebrae—4
lumbar vertebrae—5

E. *Numbers*

Indicate the number of components that constitute the following.

1. Cervical vertebrae.
2. Thoracic vertebrae.
3. Lumbar vertebrae.
4. Sacrum (fused vertebrae).
5. Coccyx (rudimentary vertebrae).
6. Pairs of ribs.
7. Spinal curvatures.
8. Pairs of vertebrochondral ribs.
9. Pairs of vertebrosternal ribs.
10. Pairs of vertebral ribs.

F. *Medical*

Select the medical terms at the right that are described by the following statements. Consult your lecture text or medical dictionary.

1. Exaggerated thoracic curvature.
2. "Slipped disk."
3. Exaggerated lumbar curvature.
4. Lateral curvature of the spine.

herniation—1
kyphosis—2
lordosis—3
scoliosis—4

Answers

Vertebral Differences	Medical
1. _____	1. _____
2. _____	2. _____
3. _____	3. _____
4. _____	4. _____
5. _____	

Fig. 10.2

Numbers	
1. _____	_____
2. _____	_____
3. _____	_____
4. _____	_____
5. _____	_____
6. _____	_____
7. _____	_____
8. _____	
9. _____	
10. _____	

The Appendicular Skeleton and Articulations

A. Labels
Record the labels for all the illustrations of Exercise 11 in the appropriate columns.

B. Bone Names
Select the names of the bones described in the following statements.

1. Bones of palm of hand.
2. Upper arm bone.
3. Medial forearm bone.
4. Lateral forearm bone.
5. Wrist bones.
6. Fingers.
7. Components of shoulder girdle.

carpals—1
clavicle—2
humerus—3
metacarpals—4
phalanges—5
radius—6
scapula—7
tarsals—8
ulna—9

C. Landmarks (Upper Extremities)
Identify the depressions and processes of the arm and shoulder described by the following statements.

acromion process—1
capitulum—2
coracoid process—3
coronoid fossa—4
coronoid process—5
deltoid tuberosity—6
glenoid cavity—7
greater tubercle—8
head—9
lateral epicondyle—10

lesser tubercle—11
medial epicondyle—12
olecranon fossa—13
olecranon process—14
radial notch—15
radial tuberosity—16
semi-lunar notch—17
styloid process—18
trochlea—19

1. Distal medial condyle of humerus.
2. Distal lateral condyle of humerus.
3. Fossa on distal posterior surface of humerus.
4. Epicondyle of humerus adjacent to trochlea.
5. Depression on scapula that articulates with humerus.
6. Depression on ulna which is in contact with head of radius.
7. Process on scapula that articulates with clavicle.
8. Process on scapula that lies anterior and superior to glenoid cavity.
9. Process on anterior surface of ulna just below semi-lunar notch.
10. Process on humerus to which deltoid muscle is attached.
11. Condyle of humerus that articulates with ulna.
12. Condyle of humerus that articulates with radius.
13. Part of humerus that articulates with scapula.
14. Epicondyle adjacent to capitulum.
15. Large process near head of humerus.
16. Small process just above surgical neck of humerus.
17. Small distal medial process of ulna.
18. Depression on proximal end of ulna that articulates with trochlea of humerus.
19. Proximal process of ulna, sometimes called the "funny bone."
20. Fossa on distal anterior surface of humerus.

Answers

Bone Names	Fig. 11.1
1. _____	_____
2. _____	_____
3. _____	_____
4. _____	_____
5. _____	_____
6. _____	_____
7. _____	_____

Landmarks	
1. _____	_____
2. _____	Fig. 11.2
3. _____	
4. _____	_____
5. _____	_____
6. _____	_____
7. _____	_____
8. _____	_____
9. _____	_____
10. _____	_____
11. _____	_____
12. _____	_____
13. _____	_____
14. _____	_____
15. _____	_____
16. _____	_____
17. _____	_____
18. _____	_____
19. _____	_____
20. _____	_____

D. Bones and Joints (Lower Extremities)

Identify the bones and joints of the lower extremities that are described by the following statements.

1. Heel bone.
2. Bones of fingers.
3. Bones of toes.
4. Bones surrounding the obturator foramen.
5. One half of pelvic girdle.
6. Pelvic girdle, sacrum and coccyx.
7. Lower posterior bone of os coxa.
8. Upper bone of os coxa.
9. Anterior bone of os coxa.
10. Strongest bone of lower leg.
11. Longest, heaviest bone of leg.
12. Thin lateral bone of lower leg.
13. Bone adjacent to the sacroiliac joint.
14. Five elongated bones that form instep of foot.
15. Tarsal bone that articulates with tibia.
16. Largest tarsal bone.

calcaneous—1
femur—2
fibula—3
ilium—4
ischium—5
metatarsals—6
os coxa—7
patella—8
pelvis—9
phalanges—10
pubis—11
talus—12
tibia—13

E. Numbers

Supply the correct numbers for the following statements.

1. Number of phalanges in the large toe.
2. Number of phalanges in each of the other toes.
3. Number of phalanges in thumb.
4. Number of phalanges on each of the other four fingers.
5. Number designation for the small toe metatarsal.
6. Number designation for thumb metacarpal.
7. Number designation for little finger metacarpal.
8. Total number of metatarsals in each foot.
9. Total number of tarsal bones in each foot.
10. Total number of carpals in each wrist.

F. Comparisons

Compare the male and female pelves side-by-side to determine the validity (True or False) of these statements.

1. The greater pelvis (concavity of the expanded iliac bones above the pelvic brim) of the male is narrower than in the female.
2. The male sacrum and coccyx is more curved.
3. The muscular impressions on the female pelvis are less distinct (bone is smoother).
4. The sacrum of the female pelvis is shorter and wider.
5. The obturator foramina of the female pelvis are rounder and larger than in the male.
6. The iliac crests of the female pelvis protrude outward more than in the male.
7. The aperture of the female pelvis is larger and more oval than the male pelvis.
8. The acetabula of the female face more anteriorly.
9. The aperture of the female pelvis is heart-shaped.
10. The obturator foramina of the female pelvis tend toward being triangular and smaller than in the male.

Answers

Bones and Joints	Fig. 11.3
1. _____	_____
2. _____	_____
3. _____	_____
4. _____	_____
5. _____	_____
6. _____	_____
7. _____	_____
8. _____	_____
9. _____	_____
10. _____	_____
11. _____	_____
12. _____	_____
13. _____	
	Fig. 11.4
14. _____	
15. _____	_____
16. _____	_____

Numbers	
1. _____	_____
2. _____	_____
3. _____	_____
4. _____	_____
5. _____	_____
6. _____	_____
7. _____	_____
8. _____	_____
9. _____	_____
10. _____	_____

Comparisons	
1. _____	_____
2. _____	_____
3. _____	_____
4. _____	_____
5. _____	_____
6. _____	_____
7. _____	_____
8. _____	_____
9. _____	_____
10. _____	_____

G. Landmarks (Lower Extremities)

Identify the following processes, fossae and other structures of bones of the lower extremities.

acetabulum—1
greater trochanter—2
head of femur—3
head of fibula—4
iliac crest—5
intertrochanteric crest—6
intertrochanteric line—7
ischial spine—8
lateral condyle—9

lateral malleolus—10
lesser trochanter—11
medial condyle—12
medial malleolus—13
obturator foramen—14
sacroiliac—15
symphysis pubis—16
tibial tuberosity—17
tuberosity of ischium—18

1. Large prominent process of ischium.
2. Large process lateral to neck of femur.
3. Process just inferior to neck of femur on medial surface.
4. Proximal lateral process of tibia that articulates with femur.
5. Proximal medial process of tibia that articulates with femur.
6. Distal medial process of femur.
7. Distal lateral process of femur.
8. Fossa on os coxa that articulates with femur.
9. Joint between sacrum and ilium of os coxa.
10. Joint between pubic bones of os coxae.
11. Uppermost process on posterior edge of ischium.
12. Uppermost ridge of os coxa.
13. Large opening in os coxa surrounded by pubis and ischium.
14. Part of femur that fits into depression of os coxa.
15. Oblique ridge between trochanters on posterior surface of femur.
16. Oblique ridge between trochanters on anterior surface of femur.
17. Distal lateral process of fibula that forms outer ankle bone.
18. Distal medial process of tibia that forms the inner prominence of the ankle.
19. Proximal anterior process of tibia which protrudes below the knee cap.
20. End of fibula that articulates with upper end of tibia.

H. Articulation Illustrations

Record all the labels for the illustrations of Exercise 12 in the answer columns.

I. Joint Characteristics

Select the joints from the list on the right that have the following characteristics. More than one answer may apply in some cases.

1. Freely movable in one plane only.
2. Freely movable in two planes.
3. Slightly movable.
4. Immovable.
5. Movement in all directions.
6. Rotational movement only.
7. Fused joint of hyaline cartilage.
8. Contains fibrocartilaginous pad.
9. Condyle end in elliptical cavity.
10. Held together with interosseous ligament.
11. Bone ends are flat or slightly convex.
12. Each bone end is concave in one direction and convex in another direction.
13. Bone ends joined by fibrous connective tissue.

ball and socket—1
condyloid—2
gliding—3
hinge—4
pivot—5
saddle—6
sutures—7
symphysis—8
synchondrosis—9
syndesmosis—10

Answers	
Landmarks	**Fig. 12.1**
1. _____	_____
2. _____	_____
3. _____	_____
4. _____	_____
5. _____	_____
6. _____	_____
7. _____	_____
8. _____	_____
9. _____	_____
10. _____	_____
11. _____	**Fig. 12.2**
12. _____	_____
13. _____	_____
14. _____	_____
15. _____	_____
16. _____	_____
17. _____	_____
18. _____	_____
19. _____	_____
20. _____	_____
Chars.	
1. _____	**Fig. 12.3**
2. _____	_____
3. _____	_____
4. _____	_____
5. _____	_____
6. _____	_____
7. _____	_____
8. _____	_____
9. _____	_____
10. _____	_____
11. _____	_____
12. _____	_____
13. _____	_____

J. Joint Classification

By examining and manipulating the bones of a skeleton, classify the following joints. Two terms apply to each joint.

1. Hip joint.
2. Shoulder joint.
3. Intervertebral joint.
4. Interlocking joint between cranial bones.
5. Phalanges to metacarpals.
6. Phalanges to metatarsals.
7. Knee joint.
8. Elbow joint.
9. Metaphysis of long bone of child.
10. Symphysis pubis.
11. Articulation between tibia and fibula at distal ends.
12. Wrist (carpals to carpals).
13. Ankle (talus to tibia).
14. Ankle (tarsals to tarsals).
15. Axis-atlas rotation.
16. Finger joint (between phalanges).
17. Toe joints (between phalanges).
18. Talus to navicular.

Basic Types
slightly movable—1
freely movable—2
immovable—3

Sub-types
ball and socket—4
condyloid—5
gliding—6
hinge—7
pivot—8
saddle—9
suture—10
symphysis—11
synchondrosis—12
syndesmosis—13

K. Medical

Consult your lecture text, medical dictionary or other source for the following.

ankylosis—1
bursitis—2
chronic fibrositis—3
dislocation—4
gouty arthritis—5
osteoarthritis—6
rheumatism—7
rheumatoid arthritis—8
sprain—9
strain—10
subluxation—11
tenosynovitis—12

1. Arthritis resulting from degenerative changes in joint (wear and tear).
2. Arthritis of confusing etiology, characterized by deformity and immobility of joints; often treated with cortisone.
3. Arthritis due to excess levels of uric acid in blood.
4. General term applied to soreness and stiffness in muscles and joints.
5. Inflammation of tendon and its sheath.
6. Inflammation of fibrous connective tissue causing tenderness and stiffness.
7. Inflammation of synovial bursa.
8. Injury to joint by displacement, resulting in swelling and discoloration.
9. Injury to joint in which ligaments are stretched without swelling or discoloration.
10. Fixation or fusion of a joint; usually preceded by infection.
11. Displacement of bones from normal orientation in a joint.
12. Partial or incomplete dislocation.

Answers	
Fig. 12.4	**Classification**
_____	1. _____
_____	2. _____
_____	3. _____
_____	4. _____
_____	5. _____
_____	6. _____
_____	7. _____
_____	8. _____
_____	9. _____
_____	10. _____
_____	11. _____
_____	12. _____
_____	13. _____
_____	14. _____
_____	15. _____
_____	16. _____
_____	17. _____
_____	18. _____

Medical
1. _____
2. _____
3. _____
4. _____
5. _____
6. _____
7. _____
8. _____
9. _____
10. _____
11. _____
12. _____

Histology: Nerve and Muscle Tissues

A. Cell Drawings
On a separate sheet of paper, sketch representative cells of the various types of neurons and muscle cells from slides that are available. Staple the drawings to this report when it is handed in for grading.

B. Figure 13.1
Record the label numbers for this illustration in the answer column.

C. Nerve Cell Types
From the list of cells below, select those that are described by the following statements. More than one cell may apply in some cases.

1. Afferent neurons.
2. Efferent neurons.
3. Neuroglial cells.
4. Specialized cells of the retina.
5. Phagocytes of the CNS.
6. Pseudounipolar neurons.
7. Multipolar neurons.
8. Ganglion cells.
9. Macroglial cells.
10. Peripheral neuroglia.
11. Neurons in retina.
12. Golgi Type II cells.
13. Linkage cells between Purkinje cells.
14. Smallest type of neuroglial cells.
15. Predominant neurons in cerebellar cortex.
16. Neurons lacking in axons.
17. Linkage cells between pyramidal cells.
18. Linkage cells between sensory and motor neurons.
19. Predominant type of neuron in cerebral cortex.
20. Small cells covering the perikaryon of a sensory neuron.

amacrine cells—1
astrocytes—2
basket cells—3
interneurons—4
microglea—5
motor neuron—6
oligodendrocytes—7
Purkinje cells—8
pyramidal cells—9
satellite cells—10
Schwann cells—11
sensory neuron—12
stellate cells—13

D. True-False
Indicate whether the following statements concerning nerve cells are True or False.

1. Most neurons have many dendrites with only a single axon.
2. Part of the axon of a sensory neuron functions like a dendrite.
3. Unmyelinated fibers lack Schwann cells.
4. Perikarya of afferent neurons are located in the spinal ganglia.
5. Neuroglial cells do not carry nerve impulses.
6. Myelinated fibers carry nerve impulses at a slower speed than unmyelinated ones.
7. Perikarya of motor neurons are located in the white matter of the spinal cord.
8. Some neurons lack axons.
9. Both the myelin sheath and neurilemma are formed by Schwann cells.
10. Motor neurons carry nerve impulses away from the CNS.

Answers

Nerve Cell Types	Fig. 13.1
1. _____	_____
2. _____	_____
3. _____	_____
4. _____	_____
5. _____	_____
6. _____	_____
7. _____	_____
8. _____	_____
9. _____	_____
10. _____	_____
11. _____	_____
12. _____	_____
13. _____	_____
14. _____	_____
15. _____	_____
16. _____	_____
17. _____	_____
18. _____	_____
19. _____	_____
20. _____	

True-False
1. _____
2. _____
3. _____
4. _____
5. _____
6. _____
7. _____
8. _____
9. _____
10. _____

E. Neuron Structure

From the list of neuron structures select those that are described by the following statements. More than one structure may apply in some cases.

1. Long process on motor neuron.
2. Short process on motor neuron.
3. Process that receives nerve impulses.
4. Side branch of an axon.
5. Slender filaments in perikaryon.
6. Synonym for axon.
7. Nerve cell body.
8. Formed by Schwann cells.
9. Synonym for sheath of Schwann.
10. Interruptions in sheath of Schwann.

axis cylinder—1
axon—2
collateral—3
dendrite—4
myelin sheath—5
neurilemma—6
neurofibrils—7
Nissl bodies—8
nodes of Ranvier—9
perikaryon—10
Schwann cells—11
sheath of Schwann—12

11. Bodies on the ER that synthesize protein.
12. Outer membrane covering myelin sheath.
13. Promotes regeneration of damaged nerve fiber.
14. Process that carries nerve impulses away from perikaryon.
15. Enables nerve fiber to carry impulses by saltatory conduction.
16. Bodies in perikaryon that stain well with basic aniline dyes.

F. Muscle Cell Characteristics

Identify the types of muscle fibers that possess these characteristics:

1. Syncytial structure.
2. Non-syncytial.
3. Spindle-shaped cells.
4. Attached to bones.
5. In walls of large major veins that enter heart.
6. In walls of other veins and arteries.
7. Long multinucleated cells.
8. Lack cross-striae.
9. Voluntarily controlled (generally).
10. Involuntarily controlled (generally).
11. Nuclei near surface of cell.
12. Bifurcated cells.

skeletal—1
smooth—2
cardiac—3

G. Muscle Cell Structures

Identify the muscle cell structures that are described by the following statements.

endomysium—1   perimysium—4
fascicle—2   sarcolemma—5
intercalated disks—3   sarcoplasm—6

1. Cell membrane of skeletal muscle fibers.
2. Cell membranes between individual cardiac muscle fibers.
3. Connective tissue between individual skeletal muscle fibers.
4. Cytoplasmic material of skeletal muscle fibers.
5. Bundle of skeletal muscle fibers.

Answers
Structure
1. _____
2. _____
3. _____
4. _____
5. _____
6. _____
7. _____
8. _____
9. _____
10. _____
11. _____
12. _____
13. _____
14. _____
15. _____
16. _____
Chars.
1. _____
2. _____
3. _____
4. _____
5. _____
6. _____
7. _____
8. _____
9. _____
10. _____
11. _____
12. _____
Structures
1. _____
2. _____
3. _____
4. _____
5. _____

Student: _____

Desk No: _____ Section: _____

The Neuromuscular Junction

A. Figure 15.1

Record the labels for this illustration in the answer column.

B. Summary

Supply the correct terms to complete the following statement concerning the neuromuscular junction.

The unmyelinated ending of a motor neuron consists of several knob-like structures called ___1___ . These knobs lie in depressions of the sarcolemma that are called ___2___ gutters. The modified sarcolemma that lines these gutters contains receptors that are activated by the hormone, ___3___ , which is exocytosed by ___4___ . The exocytosis of this hormone is initiated when an ___5___ potential in the motor neuron reaches these vesicles. This hormone bridges the gap between the ___6___ and ___7___ membranes and initiates an ___8___ potential which spreads through the T-tubule system and the SR.

___9___ ions are released by the SR to bring about contraction of the ___10___ of the muscle fiber. Acetylcholine is inactivated by ___11___ within ___12___ of a second.

Answers	
Summary	
1. _____	
2. _____	
3. _____	
4. _____	
5. _____	
6. _____	
7. _____	
8. _____	
9. _____	
10. _____	
11. _____	
12. _____	

Figure 15.1	
_____	_____
_____	_____
_____	_____
_____	_____
_____	_____

Electronic Instrumentation

A. Questions

1. What determines whether one uses an electrode or a transducer for monitoring a particular biological phenomenon?

   _____

   _____

2. Differentiate between:

   Input Transducer _____

   _____

   Output Transducer _____

   _____

3. What is the function of the following controls on an amplifier?

   Gain _____

   Centering Control _____

   _____

   Mode _____

4. Why are shielded cables used on electrical components? _____

   _____

   _____

5. What creates the pattern on the screen of an oscilloscope? _____

   _____

   _____

6. Which specific controls would you use to accomplish the following on the Unigraph?

   a. Turn on the unit _____

   b. Change the speed of the paper _____

   c. Increase the intensity of the tracing _____

   d. Record on the chart the exact time that a new event occurs _____

e. Increase the sensitivity of the amplifier _____

f. Balance the transducer _____

g. Start and stop the movement of the chart _____

7. What is the advantage of a polygraph over a unit such as a Unigraph? _____

_____

_____

8. How would you set the controls of the electronic stimulator in figure 23.3 to produce the following?

   a. 4.3 Volts (Voltage Controls):

      Round control knob .... _____

      Decade switch .... _____

   b. 140 milliseconds (Duration Controls):

      Round control knob .... _____

      Decade switch .... _____

   c. 120 impulses per second:

      Frequency Control

         Round knob .... _____

         Decade switch .... _____

      Delay Controls .... _____

      Regular/Twin/Mode Switch .... _____

      Mode Switch .... _____

Student: _____

Desk No: _____ Section: _____

The Physicochemical Properties of Muscle Contraction

A. Length Changes

Record muscle fiber lengths before and after adding activants.

Original length _____ mm.

Length after adding ATP, K^+, and Mg^{++} _____ mm.

B. Drawings

Sketch the appearance of the muscle fibers as seen under high-dry or oil immersion before and during contraction.

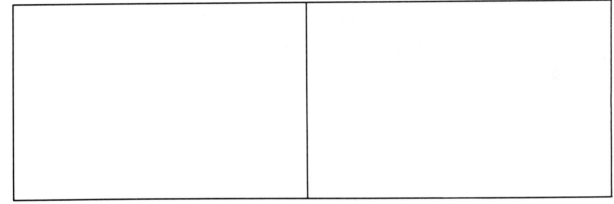

Before Contraction   During Contraction

C. Structure

Select the muscle fiber components that are described by the following statements.

actin—1   myosin—5
F-actin protein—2   sarcomere—6
heavy meromyosin—3   tropomyosin—7
light meromyosin—4   troponin—8

1. Thick myofilament with cross-bridges.

2. Principal constituent of A band.

3. Two types of myofilaments that make up a myofibril.

4. Three components of actin.

5. Type of myofilament that makes up I band.

6. Protein of which cross-bridges are made.

7. Component of a myofibril that extends between two Z lines.

Answers
Structure
1. _____
2. _____
3. _____
4. _____
5. _____
6. _____
7. _____

D. *Physiology*

Supply the terms that are missing in the following explanation of muscle contraction.

actin—1   F-actin protein—5   sarcoplasm reticulum—9
ATP—2   magnesium—6   troponin—10
calcium—3   myosin—7   tropomyosin—11
cross-bridges—4 potassium—8

The interaction of actin and myosin during muscle contraction depends on the presence of ample amounts of ATP, calcium, and ____1____ ions. The calcium ions are released by the ____2____ during depolarization. When these ions are released, ____3____ myofilaments are pulled toward each other by the ratcheting action of ____4____ that are present on the ____5____ myofilaments. The function of the calcium ions is to combine with ____6____ , causing the ____7____ molecule to be drawn deeper which, in turn, exposes active sites on the ____8____ filament. When active sites are exposed, the ____9____ pull the ____10____ filaments together, causing contraction. The energy for this ratcheting effect is derived from ____11____ .

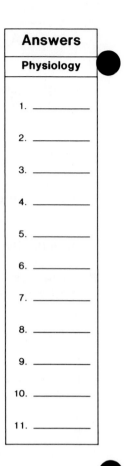

Answers

Physiology

1. _____
2. _____
3. _____
4. _____
5. _____
6. _____
7. _____
8. _____
9. _____
10. _____
11. _____

The Physiology of Muscle Contraction

A. Visible Twitches (Unrecorded)

Record here the **threshold stimuli** that cause visible twitches without using the Unigraph. Note that the muscle is stimulated directly and through the nerve.

A visible twitch with probe on **muscle** .... _____ volts

_____ msec.

Extension of foot with probe on **muscle** .... _____ volts

_____ msec.

A visible twitch with probe on **nerve** .... _____ volts

_____ msec.

Extension of foot with probe on **nerve** .... _____ volts

_____ msec.

Comparisons

Determine how much greater the stimulus requirement is for direct muscle stimulation by dividing one voltage into the other as follows:

$$\frac{\text{Voltage by Direct Muscle Stimulation}}{\text{Voltage by Nerve Stimulation}} =$$

for visible twitch .... _____

for foot extension .... _____

B. Recorded Twitches (On Unigraph)

Attach a segment of the chart in the space provided below. Use Scotch tape, glue, or staples.

C. Evaluation of Recorded Twitches

From the chart record on the previous page, determine the following:

Latent Period (A) .... _____ msec.

Contraction Phase (B) .... _____ msec.

Relaxation Phase (C) .... _____ msec.

Entire Duration (A + B + C) .... _____ msec.

D. *Summation Results*

Attach chart segments which illustrate *multiple motor unit summation* and *wave summation* in the proper spaces below. Be sure to indicate the voltages that were used for each stimulation.

Multiple Motor Unit Summation   **Wave Summation**

E. *Definitions*

Define the following terms:

Motor Unit _____

Threshold Stimulus _____

Recruitment _____

_____

Tetany _____

Electromyography (Ex. 19)

A. *Recordings*

Attach samples of myograms in the space below.

B. *Applications*

List several applications in which electromyography can be very useful: _____

_____

_____

_____

_____

Student: _____

Desk No: _____ Section: _____

Muscle Structure (Ex. 20)

	Answers

A. Figure 20.1

Record the labels for figure 20.1 in the answer column.

B. Muscle Terminology

Match the terms of the right hand column to the following statements.

1. Movable end of a muscle.

2. Immovable end of muscle.

3. Bundle of muscle fibers.

4. Outer connective tissue that covers an entire muscle.

5. Layer of connective tissue that surrounds a group of fasciculi.

6. Flat sheet of connective tissue that attaches a muscle to the skeleton or another muscle.

7. Sheath of fibrous connective tissue which surrounds each fasciculus.

8. Band of fibrous connective tissue that connects muscle to bone or to another muscle.

9. Thin layer of connective tissue surrounding each muscle fiber.

10. Layer of connective tissue between the skin and deep fascia.

aponeurosis—1
deep fascia—2
endomysium—3
epimysium—4
fasciculus—5
insertion—6
ligament—7
origin—8
perimysium—9
superficial fascia—10
tendon—11

Answers

Terms

1. _____
2. _____
3. _____
4. _____
5. _____
6. _____
7. _____
8. _____
9. _____
10. _____

Types

1. _____
2. _____
3. _____
4. _____

Fig 20.1

_____
_____
_____
_____
_____
_____
_____
_____
_____
_____
_____
_____

Body Movements (Ex. 21)

A. Figure 21.1

Record the letters for the various types of movement in this illustration in the answer column.

B. Muscle Types

Determine the types of muscles described by the following statements.

1. Opposing muscles.

2. Prime movers.

3. Muscles that assist prime movers.

4. Muscles that hold structures steady to allow agonists to act smoothly.

agonists—1
antagonists—2
fixation muscles—3
synergists—4

C. Movements

Identify the types of movements described by the following statements.

1. Turning of sole of foot inward.
2. Turning of sole of foot outward.
3. Flexion of toes.
4. Backward movement of head.
5. Movement around central axis.
6. Movement of limb toward median line of body.
7. Movement of limb away from median line of body.
8. Movement of hand from palm-up to palm-down position.
9. Cone-like rotational movement.
10. Movement of hand from palm-down to palm-up position.
11. Angle between two bones is decreased.
12. Angle between two bones is increased.
13. Extension of foot at the ankle.

abduction—1
adduction—2
circumduction—3
dorsiflexion—4
eversion—5
extension—6
flexion—7
hyperextension—8
inversion—9
plantar flexion—10
pronation—11
rotation—12
supination—13

Answers	
Movements	Fig. 21.1
1. _____	_____
2. _____	_____
3. _____	_____
4. _____	_____
5. _____	_____
6. _____	_____
7. _____	_____
8. _____	_____
9. _____	_____
10. _____	_____
11. _____	_____
12. _____	
13. _____	

Head and Trunk Muscles

A. Labels

Record the labels for all the illustrations in Exercise 22 in the answer columns.

B. Head Muscles

Consult figure 22.1 to select the muscles that apply to the following statements. Verify your selections by referring to the text material.

Facial Expressions

1. Sadness.
2. Irony.
3. Contempt or disdain.
4. Blinking or squinting.
5. Smiling.
6. Pouting or pursing lips.
7. Horror.
8. Horizontal wrinkling of forehead.

Other Actions

9. Depresses corners of mouth.
10. Raises lower jaw.
11. Compresses cheeks.

Origins

12. On frontal bone.
13. On zygomatic bone.

Insertions

14. On eyebrows.
15. On lips.
16. On eyelids.

buccinator—1
frontalis—2
masseter—3
orbicularis oculi—4
orbicularis oris—5
platysma—6
quadratus labii
  inferioris—7
quadratus labii
  superioris—8
temporalis—9
triangularis—10
zygomaticus—11

C. Trunk Muscles

Consult figure 22.2 to select the muscles that apply to the following statements. Verify from text.

1. Abducts humerus.
2. Adducts humerus.
3. Pulls scapula forward, downward, and inward toward chest.
4. Pulls scapula toward median line.
5. Compress abdominal organs.
6. Draws head backward.
7. Elevation of ribs in breathing.

Origins

8. On sacrum, iliac crest and lower thoracic vertebrae.
9. On occipital bone, ligamentum nuchae and thoracic vertebrae.
10. On sternum, clavicle and ribs.
11. On lower margin of upper rib of adjacent ribs.

Insertions

12. On humerus.
13. Near vertebral border of scapula.
14. On upper border of lower rib of adjacent ribs.

deltoideus—1
intercostales externus—2
intercostales internus—3
latissimus dorsi—4
pectoralis major—5
rectus abdominis—6
serratus anterior—7
trapezius—8

Answers

Head Muscles	Fig. 22.1
1. _____	_____
2. _____	_____
3. _____	_____
4. _____	_____
5. _____	_____
6. _____	_____
7. _____	_____
8. _____	_____
9. _____	_____
10. _____	_____
11. _____	_____
12. _____	_____
13. _____	_____
14. _____	_____
15. _____	_____
16. _____	_____

Trunk Muscles	Fig. 22.2
1. _____	_____
2. _____	_____
3. _____	_____
4. _____	_____
5. _____	_____
6. _____	_____
7. _____	_____
8. _____	_____
9. _____	_____
10. _____	_____
11. _____	_____
12. _____	_____
13. _____	_____
14. _____	_____

23

Student: _____

Desk No: _____ Section: _____

Arm Muscles

A. Labels

Record the labels for all the illustrations of this exercise in the answer column.

B. Arm Movements

Consult figures 23.1 and 23.2 to select the muscles that apply to the following statements. Verify your answers by referring to the text material.

1. Raises the arm.
2. Extends the forearm.
3. Flexes the forearm.
4. Rotates the humerus laterally.
5. Pulls the humerus forward as in flexion.
6. Flexes forearm and rolls the radius outward to supinate hand.
7. Acts as antagonist of triceps brachii.

biceps brachii—1
brachialis—2
brachioradialis—3
coracobrachialis—4
infraspinatus—5
supraspinatus—6
teres major—7
teres minor—8
triceps brachii—9

Origins

8. On humerus.
9. On infraspinous fossa of scapula.
10. Near inferior angle of scapula.
11. On scapula above scapular spine.
12. On coracoid process of scapula.

Insertions

13. On humerus.
14. On distal end of radius.
15. On proximal end of radius.
16. On olecranon process of ulna.

Answers

Arm Movements	Fig. 23.1
1. _____	_____
2. _____	_____
3. _____	_____
4. _____	_____
5. _____	_____
6. _____	**Fig. 23.2**
7. _____	_____
8. _____	_____
9. _____	_____
10. _____	_____
11. _____	_____
12. _____	**Fig. 23.3**
13. _____	_____
14. _____	_____
15. _____	_____
16. _____	_____
	_____
	_____
	Fig. 23.4

C. Hand Movements

Consult figures 23.3 and 23.4 to select the muscles that apply to the following statements. Verify as above.

1. Flexes the thumb.
2. Pronates the hand.
3. Supinates the hand.
4. Flexes and abducts hand.
5. Flexes all fingers except the thumb.
6. Flexes all distal phalanges except the thumb.
7. Extends thumb.
8. Abducts the thumb.
9. Extends all fingers except the thumb.
10. Extends wrist and hand.

abductor pollicis—1
extensor carpi radialis brevis—2
extensor carpi radialis longus—3
extensor carpi ulnaris—4
extensor digitorum—5
extensor pollicis longus—6
flexor carpi radialis—7
flexor carpi ulnaris—8
flexor digitorum profundus—9
flexor digitorum superficialis—10
flexor pollicis longus—11
pronator quadratus—12
pronator teres—13
supinator—14

Origins

11. On medial epicondyle of humerus.
12. On lateral epicondyle of humerus and part of ulna.
13. On lateral epicondyle of humerus.
14. On posterior surfaces of ulna and radius.
15. On distal fourth of ulna.
16. On anterior proximal surface of ulna and interosseous membrane.
17. On portions of radius, ulna and interosseous membrane.
18. On medial epicondyle of humerus and on olecranon process.
19. On interosseous membrane between radius and ulna.
20. On supracondylar ridge of humerus above lateral epicondyle.
21. On three bones: medial epicondyle of humerus, medial surface of ulna and oblique line of radius.

Insertions

22. On upper lateral surface of radius.
23. On base of fifth metacarpal.
24. On proximal portions of second and third metacarpals.
25. On distal lateral portion of radius.
26. On anterior surfaces of middle phalanges of 2nd, 3rd, 4th, and 5th fingers.
27. On posterior surface of fifth metacarpal.
28. On posterior surface of second metacarpal near its base.
29. On anterior surfaces of distal phalanges of 2nd, 3rd, 4th, and 5th fingers.
30. On anterior surface of distal phalanx of thumb.
31. On lateral portion of the first metacarpal and greater Trapezium.
32. On posterior surfaces of distal phalanges of the 2nd, 3rd, 4th, and 5th fingers.
33. On posterior proximal portion of middle metacarpal.
34. On posterior surface of distal phalanx of thumb.

Answers
1. _____
2. _____
3. _____
4. _____
5. _____
6. _____
7. _____
8. _____
9. _____
10. _____
11. _____
12. _____
13. _____
14. _____
15. _____
16. _____
17. _____
18. _____
19. _____
20. _____
21. _____
22. _____
23. _____
24. _____
25. _____
26. _____
27. _____
28. _____
29. _____
30. _____
31. _____
32. _____
33. _____
34. _____

Leg Muscles

Student: _____

Desk No: _____ Section: _____

Leg Muscles

A. Illustrations
Record the labels for this exercise in the answer column.

B. Thigh Movements
Consult figure 24.1 to select the muscles that apply to the following statements. More than one muscle may apply to a statement. Verify your answers by referring to the text material.

1. Flexes the femur.
2. Extends the femur.
3. Abducts the femur.
4. Adducts the femur.
5. Rotates the femur inward.
6. Rotates the femur outward.

adductor brevis—1
adductor longus—2
adductor magnus—3
gluteus maximus—4
gluteus medius—5
gluteus minimus—6
piriformis—7

Origins

7. On the pubis.
8. On anterior surface of sacrum.
9. On external surface of ilium.
10. On inferior surface of ischium and portion of pubis.
11. On ilium, sacrum and coccyx.

Insertions

12. On anterior border of greater trochanter.
13. On upper border of greater trochanter of femur.
14. On linea aspera of femur.
15. On lateral part of greater trochanter of femur.
16. On iliotibial tract and posterior part of femur.

C. Thigh and Lower Leg Movements
Consult figure 24.2 to select the muscles that apply to the following statements. In some cases more than one muscle may apply to a statement. Verify your answers as suggested above.

1. Adducts the thigh.
2. Rotates leg inward (medially).
3. Flexes the thigh.
4. Flexes calf upon thigh.
5. Extends the leg.

biceps femoris—1
gracilis—2
rectus femoris—3
sartorius—4
semimembranosus—5
semitendinosus—6
vastus intermedius—7
vastus lateralis—8
vastus medialis—9

Origins

6. On linea aspera of femur.
7. On lower margin of pubic bone.
8. On ischial tuberosity.
9. On anterior superior spine of ilium.
10. On front and lateral surfaces of shaft of femur.
11. Two heads: one on ischial tuberosity and other on linea aspera of femur.
12. Two tendons: one on anterior inferior iliac spine and other in groove above acetabulum.

Insertions

13. On head of fibula and lateral condyle of tibia.
14. On tuberosity of tibia.
15. On medial surface of body of tibia.
16. On medial condyle of tibia.
17. On upper end of shaft of tibia.

Answers

Thigh Movements	Fig. 24.1	Fig. 24.7
1. _____	_____	_____
2. _____	_____	_____
3. _____	_____	_____
4. _____	_____	_____
5. _____	_____	_____
6. _____	_____	_____
7. _____	_____	_____
8. _____	_____	_____
9. _____	**Fig. 24.2**	_____
10. _____	_____	_____
11. _____	_____	_____
12. _____	_____	_____
13. _____	_____	_____
14. _____	_____	_____
15. _____	_____	_____
16. _____	_____	_____

Thigh and Lower Leg		
1. _____	_____	_____
2. _____	_____	**Fig. 24.8**
3. _____	_____	_____
4. _____	**Fig. 24.3**	_____
5. _____	_____	_____
6. _____	_____	_____
7. _____	_____	_____
8. _____	_____	_____
9. _____	_____	_____
10. _____	_____	_____
11. _____	**Fig. 24.4**	_____
12. _____	_____	_____
13. _____	_____	_____
14. _____	_____	_____
15. _____	_____	_____
16. _____	_____	_____
17. _____	_____	_____

D. Lower Leg and Foot Movements

Consult figures 24.3 and 24.4 to select the muscles that apply to the following statements. More than one muscle may apply to a statement. Verify answers as above.

1. Extends the foot.
2. Flexes the foot.
3. Inversion of foot.
4. Eversion of foot.
5. Flexes distal phalanges of four smaller toes.
6. Extends proximal phalanges of four smaller toes.
7. Supports foot arches.
8. Flexes the great toe.
9. Flexes the leg.

extensor digitorum longus—1
extensor hallucis longus—2
flexor digitorum longus—3
flexor hallucis longus—4
gastrocnemius—5
peroneus brevis—6
peroneus longus—7
peroneus tertius—8
soleus—9
tibialis anterior—10
tibialis posterior—11

Origins

10. On anterior surface of fibula and interosseous membrane.
11. On distal two-thirds of fibula and intermuscular septa.
12. On anterior surface of the distal end of the fibula.
13. On posterior surfaces of head of fibula and tibia.
14. On upper posterior surfaces of tibia and fibula and on the interosseous membrane.
15. On posterior surfaces of femoral condyles.
16. On the head and upper two-thirds of fibula and the deep fascia.
17. On posterior surface of tibia and fascia of tibialis posterior.
18. On lateral condyle and upper two-thirds of body of tibia.
19. On lateral condyle of tibia, anterior surface of fibula and interosseous membrane.

Insertions

20. On the base of the distal phalanx of the great toe.
21. On the proximal end of the fifth metatarsal.
22. On the superior surfaces of the second and third phalanges of the four smaller toes.
23. On the base of the proximal portion of the fifth metatarsal.
24. On the inferior surfaces of the navicular, cuneiforms, cuboid and metatarsals.
25. On the calcaneous.
26. On the proximal portions of the first metatarsal and second cuneiform.
27. On the inferior surfaces of the first cuneiform and first metatarsal bones.
28. On the bases of the last phalanges of the second, third, fourth and fifth toes.

E. General Questions

Record the answers to the following questions in the answer column.

1. List the four muscles that are collectively referred to as the quadriceps femoris.
2. What muscle is associated with the iliotibial tract?
3. List three muscles that constitute the hamstrings.
4. What two muscles are associated with Achilles' tendon?
5. What two muscles constitute the triceps surae?

F. Figure 21.1

Refer back to figure 21.1 and identify the muscles that cause each type of movement. Record the answers in the answer column.

Answers

Lower Leg and Foot

1. _____	15. _____
2. _____	16. _____
3. _____	17. _____
4. _____	18. _____
5. _____	19. _____
6. _____	20. _____
7. _____	21. _____
8. _____	22. _____
9. _____	23. _____
10. _____	24. _____
11. _____	25. _____
12. _____	26. _____
13. _____	27. _____
14. _____	28. _____

General Questions

1a. _____
b. _____
c. _____
d. _____
2. _____
3a. _____
b. _____
c. _____
4a. _____
b. _____
5a. _____
b. _____

Figure 21.1

A. _____
B. _____
C. _____
D. _____
E. _____
F. _____
G. _____
H. _____
I. _____
J. _____
K. _____
L. _____
M. _____

Student: _____

Desk No: _____ Section: _____

The Spinal Cord and Reflex Arcs

A. Illustrations

Record the labels for figures 25.1, 25.2, and 25.4 in the answer columns.

B. Completion Questions

Supply the information that is necessary to complete the following statements.

1. The caudal end of the spinal cord terminates at the lower border of the _____ lumbar vertebra.
2. The caudal end of the spinal cord is called the _____ .
3. The cluster of nerves that extend downward from the end of the spinal cord is called the _____ .
4. The dural sac of the spinal cord is continuous with the _____ mater that surrounds the brain.
5. The iliohypogastric and ilioinguinal nerves are branches of the first _____ nerve.
6. The femoral nerve is formed by the union of the following three lumbar nerves: _____ , _____ , and _____ .
7. The largest nerve emerging from the sacral plexus is the _____ nerve.
8. The coccygeal nerve emerges from the _____ plexus.
9. The spinal ganglion is an enlargement of the _____ root.
10. The central canal of the spinal cord is lined with _____ ependymal cells.
11. The inner portion of the spinal cord, forming the pattern of a butterfly, consists of _____ matter.
12. The innermost meninx surrounding the spinal cord is the _____ mater.
13. The space between the dura mater and vertebral bone is called the _____ space.
14. The meninx that lies just within the dura mater is the _____ .
15. The cavity within the spinal cord that is continuous with the ventricles of the brain is called the _____ canal.
16. List the two divisions of the autonomic nervous system.
17. If a somatic reflex lacks an interneuron it is said to be _____ .
18. Cell bodies of afferent neurons of visceral reflexes are located in _____ ganglia.
19. A somatic reflex arc with one or more interneurons is said to be _____ .
20. _____ (*vertebral* or *collateral*) ganglia are usually located close to the organ that is innervated.

Answers

Completion

1. _____
2. _____
3. _____
4. _____
5. _____
6. _____
7. _____
8. _____
9. _____
10. _____
11. _____
12. _____
13. _____
14. _____
15. _____
16. _____
17. _____
18. _____
19. _____
20. _____

Figure 25.1

_____ _____
_____ _____
_____ _____
_____ _____
_____ _____
_____ _____
_____ _____
_____ _____
_____ _____
_____ _____
_____ _____
_____ _____
_____ _____

C. True-False

Indicate the validity of the following statements with a T or F.

1. The spinal cord during fetal life fills the entire length of the spinal cavity.
2. The filum terminale externa is a downward extension of the filum terminale interna.
3. The cervical plexus is formed by the union of the first three pairs of cervical nerves.
4. The brachial plexus is formed by the union of the lower four cervical nerves and the first thoracic nerve.
5. The lumbar plexus is formed by the union of L_1, L_2, L_3, and most of L_4 nerves.
6. The sacral plexus is formed by the union of the femoral and sciatic nerves.
7. The posterior cutaneous nerve emerges from the lumbar plexus.
8. The white matter of the spinal cord consists of myelinated fibers of nerve cells.
9. Cerebrospinal fluid fills the subarachnoid space.
10. The innermost fiber of the cauda equina (on the median line) is the filum terminale interna.
11. The central canal of the spinal cord is continuous with the ventricles of the brain.
12. The dura mater of the spinal cord extends peripherally to form a covering for spinal nerves.
13. The central canal of the spinal cord is lined with ciliated epithelium.
14. All somatic reflex arcs have at least one interneuron.
15. All visceral reflex arcs have two efferent neurons.
16. Spinal ganglia are located in the anterior roots of spinal nerves.
17. Collateral ganglia of the autonomic nervous system are united to form a chain along the vertebral column.
18. All visceral reflex arcs have two afferent neurons.
19. Most organs of the body are innervated by fibers of the sympathetic and parasympathetic divisions of the autonomic nervous system.
20. The effector in a somatic reflex is always skeletal muscle.

D. Numbers

Indicate the number of pairs of the following that are normally present.

1. Cervical nerves.
2. Thoracic nerves.
3. Lumbar nerves.
4. Sacral nerves.
5. Coccygeal nerves.

Answers

Terms	Fig. 25.2
1. _____	_____
2. _____	_____
3. _____	_____
4. _____	_____
5. _____	_____
6. _____	_____
7. _____	_____
8. _____	_____
9. _____	_____
10. _____	_____
11. _____	_____
12. _____	_____
13. _____	_____
14. _____	_____
15. _____	_____
16. _____	_____
17. _____	**Fig. 25.4**
18. _____	_____
19. _____	_____
20. _____	_____

Numbers	
1. _____	_____
2. _____	_____
3. _____	
4. _____	
5. _____	

26

Student: _____

Desk No: _____ Section: _____

Somatic Reflexes

A. Functional Nature

Unsuspended Frog: Does the spinal frog attempt to control the position of its hind legs? _____1_____ . Do leg muscles of the spinal frog seem *flaccid* or *firm*? _____2_____ . Does tonus exist in these muscles? _____3_____ . Does the animal jump when prodded? _____4_____ . Can the animal swim? _____5_____ . Does the animal float or sink? _____6_____ .

Suspended Frog: Does the frog attempt to remove the acid? _____7_____ . Do you think that the frog experiences a burning sensation? _____8_____ . Explain: _____

_____

Does the addition of acid to the abdomen or other leg cause a response in the animal? _____9_____ . If you try to restrain the animal's attempt to remove the acid, does it attempt some other maneuver? _____10_____ .

Make a statement as to the apparent functional nature of reflexes. ____

_____

_____

_____

_____

Answers
Functional Nature
1. _____
2. _____
3. _____
4. _____
5. _____
6. _____
7. _____
8. _____
9. _____
10. _____
Inhibition
1. _____
2. _____

B. Reaction Time

Record the reaction times (seconds) for each concentration in the table below. Plot these times in the graph.

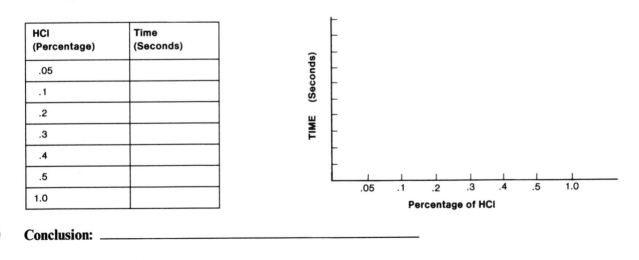

HCl (Percentage)	Time (Seconds)
.05	
.1	
.2	
.3	
.4	
.5	
1.0	

Conclusion: _____

_____

C. Reflex Radiation

Describe the response of the spinal frog to increased electrical stimulation of the foot. _____

_____

_____

_____

_____

D. Reflex Inhibition

Was it possible to inhibit right foot withdrawal from the acid by electrically stimulating the left foot?

_____ . If so, how much voltage was required? _____ .

E. Synaptic Fatigue

Where do you believe fatigue occurred in this experiment? _____

_____

F. Diagnostic Reflex Tests

Record the responses for the first five reflexes with a 0 for *no response,* a + for *diminished response,* and + + for *good response.* For Hoffmann's reflex place a check in the proper space. List the nerves in the last column that are affected in abnormal responses.

REFLEX	RESPONSES		Nerves Affected When
	Right Side	Left Side	Response Is Abnormal
1. Biceps			
2. Triceps			
3. Brachioradialis			
4. Patellar			
5. Achilles			
6. Hoffmann's	Absent (Normal)	Present (Abnormal)	

Brain Anatomy: External

A. Illustrations

Record the labels for all the illustrations of this exercise in the answer column.

B. Brain Dissections

Answer the following questions that pertain to your observations of the sheep brain dissections.

1. Does the arachnoid mater appear to be attached to the dura

   mater? _____ to the brain? _____

2. Describe the appearance of the dura mater.

   _____

   _____

   _____

3. Within what structure is the sagittal sinus enclosed?

   _____

4. Differentiate:

   Gyrus: _____

   Sulcus: _____

5. Of what value are the sulci and gyri of the cerebrum?

   _____

   _____

6. Do you find the tissue of fresh brain tissue soft or firm?

   _____

7. Do you find the tissue of formaldehyde preserved brains soft

   or firm by comparison with fresh brains? _____

   _____

8. Is the cerebellum of the sheep brain divided on the median line

   as is the case of the human brain? _____

   _____

9. What significance is there to the size differences of the olfactory bulbs on the sheep and human brains?

   _____

   _____

   _____

10. Which cranial nerve is the largest in diameter?

   _____

Answers	
Fig. 27.1	**Fig. 27.5**
_____	_____
_____	_____
_____	_____
_____	_____
_____	_____
_____	_____
_____	_____
_____	_____
_____	_____
Fig. 27.3	_____
_____	_____
_____	_____
_____	_____
_____	_____
_____	**Fig. 27.6**
_____	_____
_____	_____
_____	_____
_____	_____
_____	_____

C. Meninges

Select the meninx or meningeal space described by the following statements.

1. The middle meninx.
2. Fibrous outer meninx.
3. Meninx attached to the brain surface.
4. Meninx that surrounds the sagittal sinus.
5. Meninx that forms the falx cerebri.
6. Space that contains cerebrospinal fluid.

arachnoid mater—1
dura mater—2
epidural space—3
pia mater—4
subarachnoid space—5
subdural space—6

D. Fissures and Sulci

Select the fissure or sulcus that is described by the following statements.

1. Between the frontal and parietal lobes.
2. On the median line between the two cerebral hemispheres.
3. Between the occipital and parietal lobes.
4. Between the temporal and parietal lobes.

anterior central sulcus—1
central sulcus—2
lateral cerebral fissure—3
longitudinal cerebral fissure—4
parieto-occipital fissure—5

E. Brain Functions

Select the part of the brain that is responsible for the following activities.

1. Cardiac control.
2. Respiratory control.
3. Maintenance of posture.
4. Consciousness.
5. Vasomotor control.
6. Voluntary muscular movements.
7. Brightness and sound discrimination (animals only).
8. Contains fibers that connect the two cerebral hemispheres.
9. Contains nuclei of fifth, sixth, seventh and eighth cranial nerves.
10. Coordination of complex muscular movements.
11. Function in man is unknown.

cerebellum—1
cerebrum—2
corpora quadrigemina—3
corpus callosum—4
medulla oblongata—5
midbrain—6
pineal body—7
pons Varolii—8

F. Cerebral Functional Localization

Select those areas of the cerebrum that are described by the following statements.

Location

1. On parietal lobe.
2. On superior temporal gyrus.
3. On posterior central gyrus.
4. On frontal lobe.
5. On occipital lobe.
6. On temporal lobe.
7. On angular gyrus.
8. In area anterior to anterior central gyrus.
9. On anterior central gyrus.

association area—1
auditory area—2
common integrative area—3
motor speech area—4
olfactory area—5
somatomotor area—6
somatosensory area—7
premotor area—8
visual area—9

Answers

Meninges	Fissures and Sulci
1. ___	1. ___
2. ___	2. ___
3. ___	3. ___
4. ___	4. ___
5. ___	
6. ___	

Fig. 27.7

Brain Functions: 1–11

Localization: 1–9

F. Cerebral Functional Localization (continued)

Function

10. Cutaneous sensibility.
11. Sight.
12. Visual interpretation.
13. Voluntary muscular movement.
14. Speech production.
15. Speech understanding.
16. Influences motor area function.
17. Sense of smell.
18. Integration of sensory association areas.

association area—1
auditory area—2
common integrative area—3
motor speech area—4
olfactory area—5
somatomotor area—6
somatosensory area—7
premotor area—8
visual area—9

G. Cranial Nerves

Record the number of the cranial nerve that applies to the following statements. More than one nerve may apply to some statements.

1. Sensory nerves.
2. Mixed nerves.
3. Emerges from midbrain.
4. Emerges from pons Varolii.
5. Emerges from medulla.

Structures Innervated

6. Cochlea of ear.
7. Heart.
8. Salivary glands.
9. Abdominal viscera.
10. Retina of eye.
11. Receptors in nasal membranes.
12. Taste buds at back of tongue.
13. Lateral rectus muscle of eye.
14. Taste buds of anterior 2/3 of tongue.
15. Semicircular canals of ear.
16. Thoracic viscera.
17. Three extrinsic muscles (superior rectus, medial rectus, inferior oblique) and levator palpebrae.
18. Superior oblique muscle of eye.
19. Pharynx, upper larynx, uvula and palate.

1. Olfactory
2. Optic
3. Oculomotor
4. Trochlear
5. Trigeminal
6. Abducens
7. Facial
8. Statoacoustic
9. Glossopharyngeal
10. Vagus
11. Accessory
12. Hypoglossal

H. Trigeminal Nerve

Select the branch of the trigeminal nerve that innervates the following structures.

1. All upper teeth.
2. All lower teeth.
3. Lacrimal gland.
4. Tongue.
5. Outer surface of nose.
6. Buccal gum tissues of mandible.
7. Lower teeth, tongue, muscles of mastication and gum surfaces.

incisal—1
infraorbital—2
inferior alveolar—3
lingual—4
long buccal—5
mandibular—6
maxillary—7
ophthalmic—8

Answers
Localization
10. _____
11. _____
12. _____
13. _____
14. _____
15. _____
16. _____
17. _____
18. _____
Cranial Nerves
1. _____
2. _____
3. _____
4. _____
5. _____
6. _____
7. _____
8. _____
9. _____
10. _____
11. _____
12. _____
13. _____
14. _____
15. _____
16. _____
17. _____
18. _____
19. _____
Trigeminal Nerve
1. _____
2. _____
3. _____
4. _____
5. _____
6. _____
7. _____

I. General Questions

Select the correct answers that complete the following statements.

1. The infundibulum supports the
   (1) mammillary body (2) hypophysis (3) hypothalamus.

2. Deep furrows on the surface of the cerebrum are called
   (1) sulci (2) fissures (3) gyri.

3. Shallow furrows on the surface of the cerebrum are called
   (1) sulci (2) fissures (3) gyri.

4. The corpora quadrigemina are located on the
   (1) cerebrum (2) medulla (3) midbrain.

5. The major ganglion of the trigeminal nerve is the
   (1) Gasserian (2) sphenopalatine (3) ciliary.

6. The mammillary bodies are a part of the
   (1) medulla (2) midbrain (3) hypothalamus.

7. The part of the brain known as the hippocampus is associated with
   (1) taste (2) smell (3) hearing.

8. The spinal bulb is the
   (1) medulla oblongata (2) pons Varolii (3) midbrain.

9. The hypophysis is located on the inferior surface of the
   (1) medulla (2) midbrain (3) hypothalamus.

10. The pineal gland is located on the
    (1) medulla (2) pons Varolii (3) midbrain.

11. Convolutions on the surface of the cerebrum are called
    (1) sulci (2) fissures (3) gyri.

Answers
General Questions
1. _____
2. _____
3. _____
4. _____
5. _____
6. _____
7. _____
8. _____
9. _____
10. _____
11. _____

28

Brain Anatomy: Internal

A. Illustrations

Record the labels for all the illustrations of this exercise in the answer columns.

B. Location

Identify that part of the brain in which the following structures are located.

cerebellum—1
cerebrum—2
diencephalon—3
medulla oblongata—4
midbrain—5
pons Varolii—6

1. Corpus callosum.
2. Caudate nuclei.
3. Globus pallidus.
4. Aqueduct of Sylvius.
5. Hypothalamus.
6. Fornix.
7. Cerebral peduncles.
8. Third ventricle.
9. Rhinencephalon.
10. Thalamus.
11. Mammillary bodies.
12. Lateral ventricles.
13. Putamen.

C. Functions

Identify the structures that perform the following functions.

aqueduct of Sylvius—1
arachnoid granulations—2
caudate nuclei—3
cerebral peduncles—4
choroid plexus—5
corpus callosum—6
fornix—7
foramen of Luschka—8
foramen of Magendie—9
foramen of Monro—10
hypothalamus—11
infundibulum—12
intermediate mass—13
lentiform nucleus—14
rhinencephalon—15
thalamus—16

1. Entire olfactory mechanism of cerebrum.
2. Assists in muscular coordination.
3. Relay station for all messages to cerebrum.
4. Fiber tracts of olfactory mechanism.
5. Temperature regulation.
6. A commissure which unites the cerebral hemispheres.
7. Secretes cerebrospinal fluid into ventricle.
8. Connects the two halves of the thalamus.
9. Consists of ascending and descending tracts.
10. Allows cerebrospinal fluid to pass from third to fourth ventricle.
11. Exerts steadying effect on voluntary movements.
12. Allows cerebrospinal fluid to return to blood.
13. Supporting stalk of hypophysis.
14. Coordinates autonomic nervous system.

D. General Questions

Select the best answer that completes the following statements. Supplementary reading material should be consulted for some of these questions.

1. The lateral ventricles are separated by the
   (1) thalamus (2) fornix (3) septum pellucidum.
2. The brain stem consists of the
   (1) cerebrum, pons, midbrain and medulla
   (2) cerebellum, medulla and pons
   (3) pons, medulla and midbrain.

Answers

Location	Fig. 28.2
1. _____	_____
2. _____	_____
3. _____	_____
4. _____	_____
5. _____	_____
6. _____	_____
7. _____	_____
8. _____	_____
9. _____	_____
10. _____	_____
11. _____	_____
12. _____	_____
13. _____	_____

Functions	
1. _____	_____
2. _____	_____
3. _____	_____
4. _____	_____
5. _____	_____
6. _____	Fig. 28.3
7. _____	_____
8. _____	_____
9. _____	_____
10. _____	_____
11. _____	_____
12. _____	_____
13. _____	_____
14. _____	_____

General	
1. _____	_____
2. _____	_____
	_____
	_____
	_____

3. The hypothalamus regulates
   (1) the hypophysis, appetite and wakefulness
   (2) body temperature, hypophysis and vision
   (3) reproductive functions, body temperature and voluntary movements.
4. Pain is perceived in the
   (1) somatomotor area   (2) thalamus   (3) pons.
5. The reticular formation consists of
   (1) gray matter   (2) white matter
   (3) an interlacement of white and gray matter.
6. Fiber tracts of the spinal cord cross from one side to the other (decussate) in the pyramids of the
   (1) medulla   (2) pons   (3) midbrain   (4) thalamus.
7. Feelings of pleasantness and unpleasantness appear to be associated with the
   (1) somatomotor area   (2) hypothalamus   (3) thalamus   (4) medulla oblongata.
8. The reticular formation is seen in the
   (1) spinal cord   (2) cerebrum   (3) pons and cerebellum   (4) spinal cord, brain stem and diencephalon.
9. Centers for vomiting, coughing, swallowing and sneezing are located in the
   (1) pons   (2) medulla   (3) midbrain   (4) thalamus.
10. The intermediate mass passes through the
    (1) midbrain   (2) third ventricle   (3) hypothalamus

E. Cerebrospinal Fluid

You should be able to trace the path of the cerebrospinal fluid from its point of origin to where it is reabsorbed into the blood. If you can complete the following paragraph without referring to the text, you know the sequence fairly well. If you can state the sequence from memory, better yet.

Ventricles	Structures
lateral—1	arachnoid granulations—10
third—2	cerebellum—11
fourth—3	cerebrum—12
Passageways	choroid plexus—13
acoustic meatus—4	cisterna cerebello-medullaris—14
aqueduct of Sylvius—5	cisterna superior—15
foramen magnum—6	dura mater—16
foramen of Magendie—7	sagittal sinus—17
foramen of Monro—8	septum pellucidum—18
foramina of Luschka—9	subarachnoid space—19

Cerebrospinal fluid is secreted into each ventricle by a ___1___. From the ___2___ ventricles, which are located in the cerebral hemispheres, the fluid passes to the ___3___ ventricle through an opening called the ___4___. A canal, called the ___5___, allows the fluid to pass from the latter ventricle to the ___6___ ventricle. From this last ventricle the cerebrospinal fluid passes into a subarachnoid space called the ___7___ through three foramina: one ___8___ and two ___9___. From this cavity the fluid passes over the cerebellum into another subarachnoid space called the ___10___. It also passes ___11___ *(up, down)* the posterior side of the spinal cord and ___12___ *(up, down)* the anterior side of the spinal cord. From the cisterna superior the cerebrospinal fluid passes to the subarachnoid space around the ___13___. This fluid is reabsorbed back into the blood through delicate structures called the ___14___. The blood vessel that receives the cerebrospinal fluid is the ___15___.

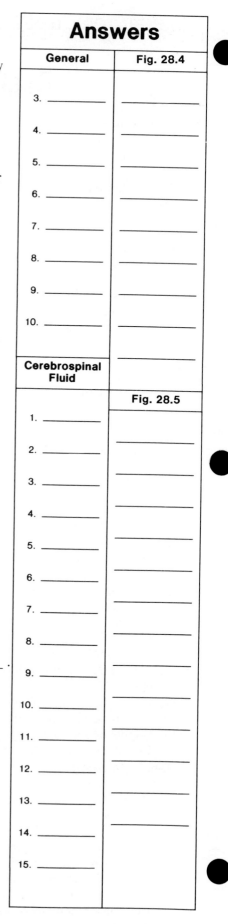

Answers

General	Fig. 28.4
3. _____	_____
4. _____	_____
5. _____	_____
6. _____	_____
7. _____	_____
8. _____	_____
9. _____	_____
10. _____	_____

Cerebrospinal Fluid

	Fig. 28.5
1. _____	_____
2. _____	_____
3. _____	_____
4. _____	_____
5. _____	_____
6. _____	_____
7. _____	_____
8. _____	_____
9. _____	_____
10. _____	_____
11. _____	_____
12. _____	_____
13. _____	_____
14. _____	_____
15. _____	_____

Electroencephalography

A. Records

Attach samples of the EEG in spaces provided below. Use tape, glue, or staples for attachment. It will be necessary to trim excess paper from the chart.

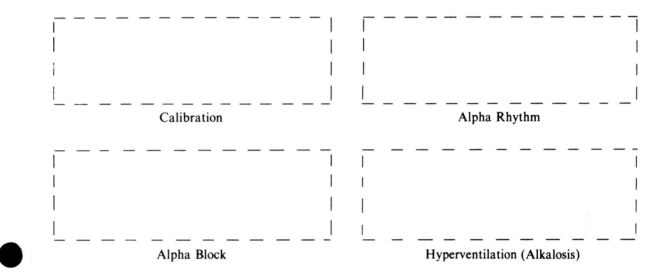

Calibration   Alpha Rhythm

Alpha Block   Hyperventilation (Alkalosis)

B. Results

1. Were you able to demonstrate, successfully, changes in the EEG pattern? _____

   If not, provide a possible explanation: _____

   _____

   _____

2. In terms of millivolts, what was the maximum amplitude seen in a brain wave? _____

   _____

3. Give the significance of the following in a normal person:

   Alpha Rhythm _____

   _____

   Beta Rhythm _____

   _____

   Delta Waves _____

   _____

Anatomy of the Eye (Ex. 30)

A. Illustrations
Record the labels from figures 30.1, 30.2, and 30.3 in the answer columns.

B. Structures
Identify the structures described by the following statements.

1. Outer layer of wall of eye.
2. Fluid between lens and cornea.
3. Fluid between lens and retina.
4. Inner light sensitive layer.
5. Small pit in retina of eye.
6. Round yellow spot on retina.
7. Small non-photosensitive area on retina.
8. Delicate membrane that lines eyelids and covers cornea and sclera.
9. Middle vascular layer of wall of eyeball.
10. Drainage tubes for tears in eyelids.
11. Tube that drains tears from lacrimal sac.
12. Clear transparent portion of front of eyeball.
13. Circular color band between lens and cornea.
14. Circular band of smooth muscle tissue surrounding lens.
15. Cartilaginous loop through which superior oblique muscle acts.
16. Connective tissue between lens perimeter and surrounding muscle.
17. Chamber between iris and cornea.
18. Conical body in medial corner of the eye.
19. Chamber between the iris and lens.
20. A semicircular fold of conjunctiva in the medial canthus of the eye.

anterior chamber—1
aqueous humor—2
blind spot—3
caruncula—4
choroid coat—5
ciliary body—6
cornea—7
conjunctiva—8
fovea centralis—9
iris—10
lacrimal ducts—11
lacrimal sac—12
macula lutea—13
medial canthus—14
nasolacrimal duct—15
plica semilunaris—16
posterior chamber—17
pupil—18
retina—19
scleroid coat—20
suspensory ligament—21
trochlea—22
vitreous body—23

C. Extrinsic Muscles
Select the muscles of the eye that are described by the following statements.

1. Inserted on side of eyeball.
2. Inserted on bottom of eyeball.
3. Inserted on top of eyeball.
4. Inserted on medial surface.
5. Inserted on eyelid.
6. Raises the eyelid.
7. Rotates eyeball downward.
8. Rotates eyeball inward.
9. Rotates eyeball outward.
10. Rotates eyeball upward.

inferior oblique—1
inferior rectus—2
lateral rectus—3
medial rectus—4
superior levator palpebrae—5
superior oblique—6
superior rectus—7

Answers

Structures	Fig. 30.1
1. _____	_____
2. _____	_____
3. _____	_____
4. _____	_____
5. _____	_____
6. _____	_____
7. _____	_____
8. _____	_____
9. _____	_____
10. _____	_____
11. _____	_____
12. _____	Fig. 30.2
13. _____	_____
14. _____	_____
15. _____	_____
16. _____	_____
17. _____	_____
18. _____	_____
19. _____	_____
20. _____	_____

Muscles	
1. _____	
2. _____	
3. _____	
4. _____	
5. _____	
6. _____	
7. _____	
8. _____	
9. _____	
10. _____	

D. Functions

Select the part of the eye that performs the following functions. More than one answer may apply in some cases.

aqueous humor—1
blind spot—2
canal of Schlemm—3
choroid coat—4
ciliary body—5
conjunctiva—6
iris—7
lacrimal ducts—8
lacrimal puncta—9
lacrimal sac—10

macula lutea—11
nasolacrimal duct—12
pupil—13
retina—14
scleroid coat—15
suspensory ligament—16
trabeculae—17
trochlea—18
vitreous body—19

1. Exerts force on the lens, changing its contour.
2. Controls the amount of light that enters the eye.
3. Place where nerve fibers of retina leave eyeball.
4. Maintains firmness and roundness of eyeball.
5. Furnishes blood supply for retina and sclera.
6. Part of retina where critical vision occurs.
7. Provides most of the strength to the wall of the eyeball.
8. Large vessel in wall of the eye that collects aqueous humor from trabeculae.

Answers	
Functions	Fig. 30.3
1. _____	_____
2. _____	_____
3. _____	_____
4. _____	_____
5. _____	_____
6. _____	_____
7. _____	_____
8. _____	_____

E. Beef Eye Dissection

Answer the following questions that pertain to the dissection of the beef eye.

1. What is the shape of the pupil? _____

2. Why do you suppose it is so difficult to penetrate the sclera with a sharp scalpel?
   _____

3. What is the function of the black pigment in the eye?
   _____
   _____

4. Compare the consistency of the two fluids in the eye.
   aqueous humor: _____
   vitreous humor: _____

5. When you hold the lens up and look through it what is unusual about the image?
   _____

6. Does the lens magnify printed matter when placed directly on it? _____

7. Compare the consistencies of the following portions of the lens.
   center: _____
   edge: _____

8. What is the reflective portion of the choroid coat called? _____

9. Is there a macula lutea on the retina of the beef eye? _____

F. Ophthalmoscopy

Answer the following questions that pertain to your study of the eye interior with an ophthalmoscope.

1. What lens diopter, if any, did you have to use to examine the fundus of the eye of your laboratory

partner? _____

2. If you were able to examine your laboratory partner's eye with "O" in the diopter window of the ophthalmoscope, what would this indicate: about the curvature of the **lens of your eye?**

_____

about your laboratory **partner's eye?** _____

3. List any abnormalities that were observed: _____

_____

_____

Visual Tests (Ex. 31)

A. The Purkinje Tree

Describe in a few sentences the image you observed. _____

_____

_____

_____

B. Tabulations

Record the results that were obtained in the following five eye tests:

Test	Right Eye	Left Eye
Blind Spot (inches)		
Near Point (inches)		
Scheiner's Experiment (inches)		
Visual Acuity (X / 20)		
Astigmatism (present or absent)		

C. Peripheral Vision Testing

Sketch in an approximate pattern of the field of vision for each eye:

Left Eye

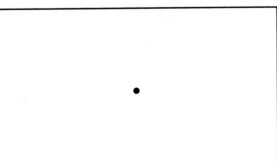

Right Eye

D. Color Blindness Test

Have your laboratory partner record your responses to each test plate. The mark *x* indicates inability to read correctly.

Plate Number	Subject's Response	Normal Response	Response if Red-Green Deficiency				Response if Totally Colorblind
1		12	12				12
2		8	3				x
3		5	2				x
4		29	70				x
5		74	21				x
6		7	x				x
7		45	x				x
8		2	x				x
9		x	2				x
10		16	x				x
11		traceable	x				x
			Protan		Deutan		
			Strong	Mild	Strong	Mild	
12		35	5	(3)5	3	3(5)	
13		96	6	(9)6	9	9(6)	
14	can trace two lines	can trace two lines	purple	purple (red)	red	red purple	x

x—Indicates inability of subject to respond in any way to test.

E. Pupillary Reflexes

Answer the following questions related to your observations in the two experiments on pupillary reflexes.

1. Light Intensity and Pupil Size

   a. Did the pupil of the unexposed eye become smaller when the right eye was exposed to light?

   _____

   b. Trace the pathways of the nerve impulses in effecting the response. _____

   _____

   _____

   c. Can you suggest a possible benefit that might result from this phenomenon? _____

   _____

2. Accommodation Pupillary Reflex

   a. What pupillary size change occurred in this experiment? _____

   _____

   b. Can you suggest a benefit that might result from this happening? _____

   _____

Student: _____

Desk No: _____ Section: _____

The Ear: Its Role in Hearing

A. Figure 32.2

Record the labels for this figure in the answer column.

B. Structure

Identify the structures described by the following statements.

basilar membrane—1	malleus—6	scala vestibuli—12
endolymph—2	perilymph—7	stapes—13
hair cells—3	reticular lamina—8	tectorial membrane—14
helicotrema—4	rods of Corti—9	tympanic membrane—15
incus—5	scala media—10	vestibular membrane—16
	scala tympani—11	

1. Membrane set in vibration by sound waves in the air.
2. Ossicle activated by tympanic membrane.
3. Ossicle that fits in oval window.
4. Chamber of cochlea into which oval window opens.
5. Chamber of cochlea into which round window opens.
6. Two fluids found in cochlea.
7. Fluid within cochlear duct.
8. Fluid within scala tympani.
9. Fluid within scala vestibuli.
10. Parts of organ of Corti.
11. Membrane that contains 20,000 fibers of varying lengths.
12. Membrane in which hair tips of hair cells are embedded.
13. Transfers vibrations from basilar membrane to reticular lamina.
14. Receptors that initiate action potentials in cochlear nerve.
15. Opening between scala vestibuli and scala tympani.

C. Physiology of Hearing (True-False)

Indicate the validity of each of the following statements.

1. The pitch of a sound is determined by the amplitude of the sound wave.
2. Quality and pitch are two different characteristics of sound.
3. The loudness of sound is directly proportional to the square of the amplitude.
4. While there is no cure for nerve deafness, conduction deafness can often be corrected.
5. Some hearing takes place through the bones of the skull rather than through the auditory ossicles.
6. The bending of hairs of hair cells causes an alternating electrical charge known as the endocochlear potential.
7. Nerve deafness can be detected by placing a tuning fork on the mastoid process.
8. The frequency range used in speech is from 30 to 20,000 c.p.s.
9. Endolymph in the scala media flows into the scala vestibuli through the helicotrema.
10. Before using an audiometer, it is necessary for the operator to calibrate it.

Answers

Structure	Fig. 32.2
1. _____	_____
2. _____	_____
3. _____	_____
4. _____	_____
5. _____	_____
6. _____	_____
7. _____	_____
8. _____	_____
9. _____	_____
10. _____	_____
11. _____	_____
12. _____	_____
13. _____	_____
14. _____	_____
15. _____	_____

Physiology	
1. _____	_____
2. _____	_____
3. _____	_____
4. _____	_____
5. _____	_____
6. _____	_____
7. _____	_____
8. _____	_____
9. _____	_____
10. _____	_____

D. Screening Test

Record the distances at which the watch tick could be heard with your ears, as well as in three other persons.

Subject	Inches From Ear			
	Right Ear		Left Ear	
	Approaching	Receding	Approaching	Receding

E. Tuning Fork Methods

Record the results of the two tuning fork methods.

Rinne Test

Right Ear _____

Left Ear _____

Weber Test

Right Ear _____

Left Ear _____

F. Audiometry

Record the threshold levels in the right ear with a red "O" and in the left ear with a blue "X". Connect the points with appropriately colored lines.

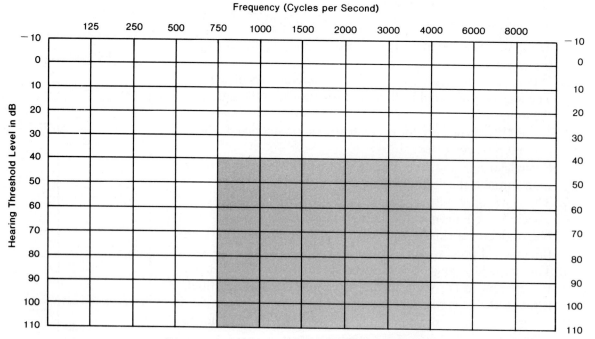

(Shaded area indicates critical area of speech interpretation)

33

The Ear: Its Role in Equilibrium

A. Anatomy

Answer the following questions that pertain to the anatomy of the vestibular apparatus.

1. What are the sensory receptors for static equilibrium called?

2. Where are the static equilibrium sensory receptors located? (two places)

3. What are the small calcium carbonate crystals of the macula called?

4. Where are the sensory receptors of dynamic equilibrium located?

5. What nerve branch of the eighth cranial nerve supplies the vestibular apparatus?

6. What part of the vestibular apparatus connects directly to the cochlear duct?

7. What is the name of the fluid within the vestibular apparatus?

8. What structures in joints prevent a sense of malequilibrium when one tilts the head to one side?

9. What fluid lies between the semicircular ducts and the surrounding bone of the semicircular canals?

10. Within how many planes do the semicircular ducts lie?

Answers
Anatomy
1. _____
2. _____
_____
3. _____
4. _____
5. _____
6. _____
7. _____
8. _____
9. _____
10. _____

B. Nystagmus

After having observed nystagmus, answer these questions:

1. What direction do the eyes move in relation to the direction of rotation? _____

2. What seems to be the function of slow movement of the eye? _____

_____

_____

3. What seems to be the function of fast movement of the eye? _____

_____

_____

4. What part of the vestibular apparatus triggers the reflexes that controls the eye muscles? _____

_____

C. Proprioceptive Influences

Record your observations here:

1. At Rest Reactions

a. Was the subject able to place the heel of his foot on the toes of the other foot with eyes closed?

_____

b. Was subject able to touch nose with eyes closed? _____

c. Explain what mechanisms in the body make this possible: _____

_____

d. Describe the subject's ability to touch the pencil eraser with eyes closed after practicing

with eyes open. _____

_____

2. Effects of Rotation

a. Was the subject, with eyes open, able to point directly to the pencil eraser with finger after

rotation? _____

_____

b. When blindfolded subject tried to locate pencil eraser after rotation, what was the result?

_____

_____

Hematological Tests

TEST RESULTS

Except for blood typing, record all test results in Table II on page 380. Calculations for blood cell counts should be performed as stated below.

A. Differential White Blood Cell Count

As you move the slide in the pattern indicated in figure 34.3, record all the different types of cells in Table I. Refer to figure 34.1 for cell identification. Use this method of tabulation: ̶H̶H̶T̶ ̶H̶H̶T̶ 11. Identify and tabulate 100 leukocytes. Divide the total of each kind of cell by 100 to determine percentages.

Table I. Leukocyte Tabulation

Neutrophils	Lymphocytes	Monocytes	Eosinophils	Basophils
Totals				
Percent				

B. Calculations for Blood Cell Counts

1. **Total White Blood Cell Count**
   Total WBC's counted in 4 W areas $\times$ 50 = **WBC's per cu. mm.**

2. **Red Blood Cell Count**
   Total RBC's counted in 5 R areas $\times$ 10,000 = **RBC's per cu. mm.**

C. Blood Typing

Record your blood type here: _____

1. If you needed a blood transfusion, what types of blood could you be given?

   _____

2. If your blood were to be given to someone in need, what type should that person have?

   _____

QUESTIONS

Although most of the answers to the following questions can be derived from this manual, it will be necessary to consult your lecture text or medical dictionary for some of the answers.

A. Blood Components

Identify the blood components described by the following statements. More than one answer may apply.

antibodies—1
basophils—2
eosinophils—3
erythrocytes—4
fibrin—5
fibrinogen—6
lymphocytes—7

lysozyme—8
monocytes—9
neutrophils—10
platelet factor—11
platelets—12
prothrombin—13
thromboplastin—14

1. Phagocytic cell concerned, primarily, with local infections.
2. Phagocytic cell concerned, primarily, with generalized infections.
3. Cells that transport O_2 and CO_2.
4. Gelatinous-like material formed in blood clotting.
5. Substances necessary for blood clotting.
6. Non-cellular formed elements essential for blood clotting.
7. Protein substances in blood that inactivate foreign protein.
8. Enzyme in blood plasma that destroys some kinds of bacteria.
9. Substance released by injured cells that initiates blood clotting.
10. Most numerous type of blood cell.
11. Contains hemoglobin.
12. Most numerous type of leukocyte.
13. Leukocyte that probably produces heparin.
14. Leukocyte distinguished by having distinctive red stained granules.

B. Blood Diseases

Identify the pathological conditions characterized by the following statements. More than one condition may apply to some statements.

anemia—1
eosinophilia—2
erythroblastosis fetalis—3
hemophelia—4
leukemia—5
leukocytosis—6
leukopenia—7

lymphocytosis—8
neutropenia—9
neutrophilia—10
pernicious anemia—11
polycythemia—12
sickle-cell anemia—13

1. Too few leukocytes.
2. Too many neutrophils.
3. Too many red blood cells.
4. Too few red blood cells.
5. Too little hemoglobin.
6. Too many lymphocytes.
7. Cancerous-like condition in which there are too many leukocytes.
8. Heritable bleeder's disease.
9. Hereditary blood disease of blacks; RBCs of abnormal shape.
10. Rh factor incompatibility present at birth.
11. Defective RBCs due to stomach enzyme deficiency.

Answers

Components

1. _____
2. _____
3. _____
4. _____
5. _____
6. _____
7. _____
8. _____
9. _____
10. _____
11. _____
12. _____
13. _____
14. _____

Diseases

1. _____
2. _____
3. _____
4. _____
5. _____
6. _____
7. _____
8. _____
9. _____
10. _____
11. _____

C. *Terminology*
Differentiate between the following pairs of related terms:

Plasma _____

Lymph _____

Coagulation _____

Agglutination _____

Antibody _____

Antigen _____

D. *General Questions*
Indicate in the answer column whether the following statements are true or false.

	Answers
1. Hemoglobin combines with both oxygen and carbon dioxide.	1. _____
2. The maximum life span of an erythrocyte is about 30 days.	2. _____
3. Anemia is usually due to an iodine deficiency.	3. _____
4. Diapedesis is the ability of leukocytes to move between the cells of the capillary walls.	4. _____
5. White blood cells live much longer than red blood cells.	5. _____
6. The most common types of blood are types O and A.	6. _____
7. A universal donor has O-Rh negative blood.	7. _____
8. The vitamin which is essential to blood clotting is vitamin D.	8. _____
9. An infant born with erythroblastosis would best be treated with a transfusion of Rh negative instead of Rh positive blood.	9. _____
10. Fibrin is formed when prothrombin reacts with fibrinogen.	10. _____
11. A deficiency of calcium will result in the failure of blood to clot.	11. _____
12. Approximately 87% of the population is Rh negative.	12. _____
13. Blood typing may be used to prove that an individual is the father in a paternity suit.	13. _____
14. Blood typing may be used to prove that an individual could not be the father in a paternity suit.	14. _____
15. Heparin is essential to fibrin formation.	15. _____
16. The rarest type of blood is AB negative.	16. _____
17. Bloods that are matched for ABO and Rh factor can be mixed with complete assurance of no incompatibility.	17. _____
18. Megakaryocytes of bone marrow produce blood platelets.	18. _____
19. Blood clotting can be enhanced with dicumarol.	19. _____
20. Viral infections may cause an increase in the number of lymphocytes or monocytes.	20. _____
21. Worm infestations may cause eosinophilia.	21. _____

E. Summarization of Results

Record blood cell counts and other test results in the following table:

Table II.

Test	Normal Values		Test Results	Evaluation (over, under, normal)
Differential WBC Count	Neutrophils: 50%-70%			
	Lymphocytes: 20%-30%			
	Monocytes: 2%-6%			
	Eosinophils: 1%-5%			
	Basophils: 0.5%-1%			
Total WBC Count	5000-9000 per cu. mm.			
RBC Count	Males: 4.8-6.0 million/cu. mm.			
	Females: 4.1-5.1 million/cu. mm.			
Hemoglobin Percentage	Males: 13.4-16.4 gms./100 ml.			
	Females: 12.2-15.2 gms./100 ml.			
Hematocrit (VPRC)	Males: 40%-54% (Av. 47%)			
	Females: 37%-47% (Av. 42%)			
Coagulation Time	2 to 6 minutes			

F. Materials

Identify the various blood tests in which the following supplies are used.

1. Capillary tubes.
2. Centrifuge.
3. Hemacytometer.
4. Hemoglobinometer.
5. Wright's stain.
6. Microscope slides.
7. Hemolysis applicators.
8. Seal-Ease.
9. Diluting fluid.
10. Typing antisera.

hematocrit (VPRC)—1
clotting time—2
RBC and WBC counts—3
differential WBC count—4
hemoglobin determination—5
none of these—6

Materials

1. _____
2. _____
3. _____
4. _____
5. _____
6. _____
7. _____
8. _____
9. _____
10. _____

Anatomy of the Heart

A. Illustrations

Record the labels for figures 39.1 and 39.2 in the answer columns.

B. Structures

Identify the structures of the heart described by the following statements.

aorta—1
aortic semilunar valve—2
bicuspid valve—3
chordae tendineae—4
endocardium—5
epicardium—6
inferior vena cava—7
left atrium—8
left pulmonary artery—9
left ventricle—10
ligamentum arteriosum—11
mitral valve—12

myocardium—13
papillary muscles—14
parietal pericardium—15
pulmonary artery—16
pulmonic semilunar valve—17
right atrium—18
right pulmonary artery—19
right ventricle—20
superior vena cava—21
tricuspid valve—22
ventricular septum—23
visceral pericardium—24

1. Partition between right and left ventricles.
2. Muscular portion of cardiac wall.
3. Lining of the heart.
4. Thin covering on surface of heart. (2 names)
5. Fibroserous sac-like structure surrounding the heart.
6. Large vein that empties blood into top of right atrium.
7. Two chambers of the heart that contain deoxygenated blood.
8. Two chambers of the heart that contain oxygenated blood.
9. Large artery that carries blood from the right ventricle.
10. Chamber of the heart that receives blood from the lungs.
11. Blood vessel that returns blood to heart from head and arms.
12. Large artery that carries blood out from left ventricle.
13. Remnant of a functional prenatal vessel between the pulmonary artery and the aorta.
14. Large vein that empties blood into the lower part of the right atrium.
15. Structure formed from ductus arteriosum.
16. Atrioventricular valves.
17. Valve at the base of the pulmonary artery.
18. Valve at the base of the aorta.
19. Two arteries that are branches of the pulmonary artery.
20. Blood vessel in which openings to the coronary arteries are located.
21. Synonym for bicuspid valve.
22. Atrium into which the coronary sinus empties.
23. Structures on the cardiac wall to which the chordae tendineae are attached.
24. Valvular restraints that prevent the atrioventricular valve cusps from being forced back into the atria.

Answers

Structures	Fig. 39.1
1. _____	_____
2. _____	_____
3. _____	_____
4. _____	_____
5. _____	_____
6. _____	_____
7. _____	_____
8. _____	_____
9. _____	_____
10. _____	_____
11. _____	_____
12. _____	_____
13. _____	_____
14. _____	_____
15. _____	_____
16. _____	_____
17. _____	_____
18. _____	_____
19. _____	_____
20. _____	_____
21. _____	_____
22. _____	_____
23. _____	_____
24. _____	_____

C. Coronary Circulation

Identify the arteries and veins of the coronary circulatory system that are described by the following statements.

anterior interventricular artery—1
circumflex artery—2
coronary sinus—3
great cardiac vein—4
left coronary artery—5
middle cardiac vein—6

posterior descending right coronary artery—7
posterior interventricular vein—8
right coronary artery—9
small cardiac vein—10

1. Two major coronary arteries that take their origins in the wall of the aorta.
2. Two principal branches of the left coronary artery.
3. Large coronary vessel that empties into the right atrium.
4. Coronary vein that lies in the anterior interventricular sulcus.
5. Coronary vein that parallels the right coronary artery in the right atrioventricular sulcus.
6. Coronary artery that lies in the right atrioventricular sulcus.
7. Branch of the right coronary artery on the posterior surface of the heart.
8. Vein that receives blood from the great cardiac vein.
9. Vein that lies alongside of the posterior descending right coronary artery.
10. Vein that lies in the posterior interventricular sulcus.

D. Sheep Heart Dissection

After completing the sheep heart dissection answer the following questions. Some of these questions were encountered during the dissection; others pertain to structures not shown in figures 39.1 and 39.2.

1. Identify the following structures:

   Pectinate muscle: _____

   _____

   Moderator band: _____

   _____

   Ligamentum arteriosum: _____

   _____

2. How many papillary muscles did you find in the

   right ventricle: _____ left ventricle: _____

3. How many pouches are present in each of the following?

   pulmonary semilunar valve: _____

   aortic semilunar valve: _____

4. Where does blood enter the myocardium? _____

   _____

5. Where does blood leave the myocardium and return to the circulatory system? _____

Answers	
Coronary Circulation	Fig. 39.2 Anterior
1. _____	_____
2. _____	_____
3. _____	_____
4. _____	_____
5. _____	_____
6. _____	_____
7. _____	_____
8. _____	_____
9. _____	_____
10. _____	_____
Fig. 39.2 Posterior	_____

Cardiovascular Sounds

A. *Phonocardiogram*

Attach below a segment of a phonocardiogram that you made on a Unigraph. Identify on the chart those segments that represent first and second sounds.

B. *Causes*

Explain what causes:

1. First sound _____

_____

_____

2. Second Sound _____

_____

_____

3. Third Sound _____

_____

4. Murmur _____

_____

_____

C. Questions

1. Why does the aortic valve close sooner than the pulmonic valve during inspiration? _____

_____

_____

2. What effect would complete right bundle branch block have on the splitting of the second sound?

_____

_____

_____

3. Explain why pulmonary stenosis is associated with greater splitting of the second heart sound.

_____

_____

_____

Student: _____

Desk No: _____ Section: _____

The Electrocardiogram

A. Tracings

Attach three tracings of your experiment in the spaces provided below. It will be necessary to trim excess paper from the tracings to fit them into the allotted spaces. Attach with Scotch tape.

Tracings	Evaluation
Calibration	
	'Heart Rate: _____ per min. QR Potential: _____ millivolts "Duration of cycle: _____ msec.
Subject At Rest	
	'Heart Rate: _____ per min. QR Potential: _____ millivolts "Duration of cycle: _____ msec.
After Exercise	

'Space between two margin lines is 3 seconds.
''Time from beginning of P wave to end of T wave.

B. Questions

1. During what part of ECG wave pattern does atrial depolarization and contraction occur? ____

   _____

2. During what part of ECG wave pattern does ventricular depolarization and contraction occur?

   _____

3. What infectious disease causes the highest incidence of heart disease? _____

   _____

4. Would a heart murmur necessarily show up on an ECG? _____

   Explain. _____

   _____

Pulse Monitoring

A. Tracings

Attach samples of tracings made in this experiment in the spaces provided below. Use the right hand column for relevant comments.

Tracings	Evaluation
	Pulse Rate: _____
Slow Speed, At Rest	
	Pulse Rate: _____
Fast Speed, At Rest	
	Pulse Rate: _____
Slow Speed, Showing Change of Hand Position	

B. Optional Tracings

Attach here any tracings related to exercise, smoking, or other activities. Provide captions.

Tracings	Evaluation
	Pulse Rate: _____
	Pulse Rate: _____

C. Pulse Tabulations

Record your pulse (while at rest) on the chalkboard. Once the pulse rates of all students are recorded, record the highest, lowest, and median pulses:

Highest _____ Lowest _____ Median _____

Significance _____

_____

D. Questions

1. What conceivably might be an explanation for changes occurring in the dicrotic notch with aging?

_____

_____

2. What might be an explanation for an alteration in the pulse amplitude during smoking? _____

_____

3. The finger pulse is a manifestation of blood volume, velocity, direction, and pressure. Which one of these factors is directly responsible for the dicrotic notch and why? _____

_____

Student: _____

Desk No: _____ Section: _____

Peripheral Circulation Control

A. Results

1. What observable change occurred in the frog's foot when histamine was applied to the foot?

   _____

   _____

2. What was the effect of histamine on the arterioles to produce this effect (vasodilation or vaso-

   constriction)? _____

3. What observable change occurred in the frog's foot when epinephrine was applied to the foot?

   _____

   _____

4. What was the effect of epinephrine on the arterioles to produce this effect (vasodilation or va-

   soconstriction)? _____

B. Questions

1. Does epinephrine have the same effect on arterioles of skeletal muscles? _____

   _____

2. Provide an explanation (theoretical) of how epinephrine can cause vasodilation in one part of

   the body and vasoconstriction in another region. _____

   _____

   _____

   _____

3. What are baroreceptors and what role do they play in regulating blood flow? _____

   _____

   _____

4. Why are arteries called "resistance" vessels and veins "capacitance" vessels? _____

   _____

   _____

Student: _____

Desk No: _____   Section: _____

The Arteries and Veins

A. Illustrations

Record the labels for figures 44.2 and 44.3 in the answer columns. Note that the labels for figure 44.2 are on both sides of this sheet.

B. Arteries

Refer to the left hand illustration in figure 44.2 to identify the arteries described by the following statements.

anterior tibial—1
aorta—2
aortic arch—3
axillary—4
brachial—5
celiac—6
common iliac—7
deep femoral—8
external iliac—9
femoral—10
inferior mesenteric—11
innominate—12

internal iliac—13
left common carotid—14
left subclavian—15
popliteal—16
posterior tibial—17
radial—18
renal—19
right common carotid—20
subclavian—21
superior mesenteric—22
ulnar—23

1. Artery of upper arm.
2. Artery of the armpit.
3. Artery of the shoulder.
4. Medial artery of the forearm.
5. Lateral artery of the forearm.
6. Gives rise to right common carotid and right subclavian.
7. Major artery of the thigh.
8. Three branches of the aortic arch.
9. Supplies the stomach, spleen and liver.
10. Major artery of chest and abdomen.
11. Artery of calf region.
12. Artery of knee region.
13. Artery in interior portion of lower leg.
14. Supplies the kidney.
15. Also known as the hypogastric artery.
16. Supplies the large intestine and rectum.
17. Large branch of common iliac.
18. Supplies blood to most of small intestines and part of colon.
19. Curved vessel that receives blood from left ventricle.
20. Branch of femoral that parallels medial surface of femur.
21. Gives rise to femoral artery.
22. Small branch of common iliac.

Answers	
Arteries	**Fig. 44.2 Arteries**
1. _____	_____
2. _____	_____
3. _____	_____
4. _____	_____
5. _____	_____
6. _____	_____
7. _____	_____
8. _____	_____
9. _____	_____
10. _____	_____
11. _____	_____
12. _____	_____
13. _____	_____
14. _____	_____
15. _____	_____
16. _____	_____
17. _____	_____
18. _____	_____
19. _____	_____
20. _____	_____
21. _____	_____
22. _____	_____

C. Veins

Refer to the right hand illustration in figure 44.2 and to figure 44.3 to identify the veins described by the following statements.

accessory cephalic—1
axillary—2
basilic—3
brachial—4
cephalic—5
common iliac—6
coronary—7
dorsal venous arch—8
external iliac—9
external jugular—10
femoral—11
great saphenous—12
hepatic—13
inferior vena cava—14
innominate—15
inferior mesenteric—16
internal iliac—17
internal jugular—18
median cubital—19
popliteal—20
portal—21
posterior tibial—22
pyloric—23
subclavian—24
superior mesenteric—25
superior vena cava—26

1. Vein of armpit.

2. Largest vein in neck.

3. Small vein in neck.

4. Collects blood from veins of head and arms.

5. Short vein between basilic and cephalic veins.

6. Empty into innominate veins.

7. Vein that great saphenous empties into.

8. Collects blood from veins of chest, abdomen and legs.

9. Empties into popliteal.

10. Collects blood from two innominate veins.

11. Collects blood from top of foot.

12. Vein from liver to inferior vena cava.

13. Receives blood from descending colon and rectum.

14. Large veins that unite to form the inferior vena cava.

15. On posterior surface of humerus.

16. Superficial vein on medial surface of leg.

17. Vein of knee region.

18. On medial surface of upper arm.

19. On lateral portion of forearm.

20. On lateral portion of upper arm.

21. Vein that receives blood from posterior tibial.

22. Receives blood from ascending colon and part of ileum.

23. Empties into great saphenous.

24. Collects blood from intestines, stomach and colon.

25. Two veins that drain blood from stomach into portal vein.

26. Small vein which empties into common iliac at juncture of external iliac.

Answers	
Veins	**Fig. 44.2 Veins**
1. _____	_____
2. _____	_____
3. _____	_____
4. _____	_____
5. _____	_____
6. _____	_____
7. _____	_____
8. _____	_____
9. _____	_____
10. _____	_____
11. _____	_____
12. _____	_____
13. _____	_____
14. _____	_____
15. _____	_____
16. _____	_____
17. _____	_____
18. _____	_____
19. _____	_____
20. _____	_____
21. _____	_____
22. _____	_____
23. _____	_____
24. _____	_____
25. _____	**Fig. 44.3**
26. _____	_____
	_____
	_____
	_____
	_____
	_____
	_____
	_____
	_____
	_____
	_____

Fetal Circulation (Ex. 45)

A. Figure 45.1

Record the labels for this illustration in the answer column.

B. Questions

Record the answers for the following questions in the answer column.

1. What blood vessel in the umbilical cord supplies the fetus with nutrients?
2. What blood vessels in the umbilical cord returns blood to the placenta from the fetus?
3. What blood vessel shunts blood from the pulmonary artery to the aorta?
4. What vessel in the liver carries blood from the umbilical vein to the inferior vena cava?
5. What is the name of the blood vessel that forms from the umbilical artery?
6. What structure forms from the ductus arteriosus?
7. What structure forms from the umbilical vein?
8. What structure enables blood to flow from the right atrium to the left atrium before birth?
9. Give 2 occurrences at childbirth that prevent excessive blood loss by the infant through cut umbilical cord.

The Lymphatic System and the Immune Response (Ex. 46)

A. Figure 46.1

Record the labels for this illustration in the answer column.

B. Components

By referring to figures 46.1 and 46.2 identify the following structures in the lymphatic system.

1. Large vessel in thorax and abdomen that collects lymph from lower extremities.
2. Short vessel which collects lymph from right arm and right side of head.
3. Microscopic lymphatic vessels situated among cells of tissues.
4. Sac-like structure that receives chyle from intestine.
5. Blood vessel that receives fluid from thoracic duct.
6. Filters of the lymphatic system.
7. Vessels of arms and legs that convey lymph to collecting ducts.
8. Depression in lymph node through which blood vessels enter node.
9. Outer covering of lymph node.
10. Structural units of lymph node that are packed with lymphocytes.
11. Cells involved in organ transplant rejections.
12. Cells that produce antibodies.

capsule—1
cisterna chyli—2
germinal centers—3
hilum—4
left subclavian vein—5
lymphatics—6
lymph capillaries—7
lymphoblasts—8
lymph nodes—9
plasma cells—10
right lymphatic duct—11
thoracic duct—12
none of these—13

Answers

Questions

1. _____
2. _____
3. _____
4. _____
5. _____
6. _____
7. _____
8. _____
9a. _____
b. _____

Components	Fig. 45.1
1. _____	_____
2. _____	_____
3. _____	_____
4. _____	_____
5. _____	_____
6. _____	_____
7. _____	_____
8. _____	_____
9. _____	_____
10. _____	_____
11. _____	_____
12. _____	_____

C. Questions

1. List three forces that move lymph through the lymphatic vessels:

   (a) _____

   (b) _____ (c) _____ .

2. How does lymph in the lymphatics of the legs differ in composition from the lymph in the thoracic duct?

   _____

   _____

3. Where do stem cells for all lymphocytes originate? _____

   _____

4. Give the cell type that accounts for each type of immunity:

   Humoral Immunity: _____

   Cellular Immunity: _____

5. What tissues program stem cells for

   Humoral Immunity: _____

   Cellular Immunity: _____

6. What type of tissue is common to the thymus gland and bursal tissues?

   _____

Answers
Fig. 46.1
_____
_____
_____
_____
_____
_____
_____
_____
_____
_____
_____
_____
_____
_____
_____
_____
_____
_____

Blood Pressure Monitoring (Ex. 47)

A. Tabulation

Record here the blood pressures of the test subject.

	BEFORE EXERCISE	AFTER EXERCISE
Systolic		
Diastolic		
*Mean Arterial Pressure		
**Pulse Pressure		

*Mean Arterial Pressure $\simeq$ diastolic pressure $+ \frac{1}{3}$ pulse pressure.

**Pulse Pressure = systolic pressure minus diastolic pressure.

B. Questions

1. Give the systolic and diastolic pressures for normal blood pressure: _____

2. What systolic pressure indicates hypertension? _____

3. Why is a low diastolic pressure considered harmful? _____

4. Why are diuretics of value in treating hypertension? _____

   _____

5. Why is reduced salt intake of value in regulating blood pressure? _____

   _____

   _____

Student: _____

Desk No: _____ Section: _____

The Respiratory Organs

A. Illustrations

Record the labels for the illustrations of this exercise in the answer column.

B. Sheep Pluck Dissection

After completing the sheep pluck dissection, answer the following questions.

1. Are the cartilaginous rings of the trachea continuous all the way around the organ? _____

2. Describe the texture of the surface of the lung. _____
_____
_____

3. How many lobes exist on the right lung? _____
. . . on the left lung? _____

4. Why does lung tissue collapse so readily when you quit blowing into it with a straw? _____
_____
_____

5. What does the "pulmonary membrane" consist of? _____
_____
_____
_____

C. Histological Study

On a separate sheet of plain paper make drawings, as required by your instructor, of tracheal and lung tissue.

Answers	
Fig. 48.1	Fig. 48.2
_____	_____
_____	_____
_____	_____
_____	_____
_____	_____
_____	_____
_____	_____
_____	_____
_____	_____
_____	_____
_____	
_____	
_____	
_____	
_____	
_____	
_____	
_____	

D. Organ Identification

Identify the respiratory structures according to the following statements.

alveoli—1
bronchioles—2
bronchi—3
cricoid cartilage—4
epiglottis—5
hard palate—6
larynx—7
lingual tonsils—8
nasal cavity—9
nasal conchae—10

nasopharynx—11
oral cavity proper—12
oral vestibule—13
oropharynx—14
palatine tonsils—15
pharyngeal tonsils—16
pleural cavity—17
soft palate—18
thyroid cartilage—19
trachea—20

1. Fleshy lobes in nasal cavity.
2. Partition between nasal and oral cavities.
3. Cavity between lips and teeth.
4. Tonsils located on sides of pharynx.
5. Tonsils attached to base of tongue.
6. Flexible flap-like cartilage over larynx.
7. Cartilage of Adam's apple.
8. Most inferior cartilaginous ring of larynx.
9. Tubes formed by bifurcation of trachea.
10. Small air sacs of lung tissue.
11. Potential cavity between lung and thoracic wall.
12. Cavity above soft palate.
13. Cavity above hard palate.
14. Cavity near palatine tonsils.
15. Cavity that contains the tongue.
16. Voice box.
17. Tube between larynx and bronchi.
18. Small tubes leading into alveoli.
19. Another name for adenoids.

E. Organ Function

Select the structures in the above list that perform the following functions.

1. Assists in destruction of harmful bacteria in the oral region.
2. Prevents food from entering the larynx when swallowing.
3. Initiates swallowing.
4. Provides supporting walls for vocal folds.
5. Prevents food from entering nasal cavity during chewing and swallowing.
6. Warms the air as it passes through the nasal cavity.
7. Essential for speech.
8. Provides surface for gas exchange in the lungs.

Answers
Organ Identification
1. _____
2. _____
3. _____
4. _____
5. _____
6. _____
7. _____
8. _____
9. _____
10. _____
11. _____
12. _____
13. _____
14. _____
15. _____
16. _____
17. _____
18. _____
19. _____
Organ Function
1. _____
2. _____
3. _____
4. _____
5. _____
6. _____
7. _____
8. _____

Regulation of Respiration

A. Deep Breathing

1. Why does breathing into a bag after hyperventilating restore normal breathing more rapidly than breathing into the air? _____

_____

_____

2. How long were you able to hold your breath with only one deep inspiration? _____

3. How long were you able to hold your breath after hyperventilating?

_____

4. After exercising, how long were you able to hold your breath? _____

_____

Generalizations on above. _____

_____

_____

_____

_____

_____

_____

B. Deglutition Apnea

1. Did the desire to inhale disappear as you sipped water? _____

2. Postulate in a few words your interpretation of the nerve pathway in this reflex. _____

_____

_____

_____

_____

_____

C. The Valsalva Maneuver
Piece together matching sections of blood pressure and finger pulse tracings to fit the space provided below:

D. The Diving Reflex
Attach matching sections of the tracings made while performing this experiment.

E. Questions
1. Why does one get dizzy while hyperventilating? _____

_____

2. Rank oxygen, pH, and carbon dioxide according to importance in controlling respiration. _____

_____

_____

Student: _____

Desk No: _____ Section: _____

Spirometry: Lung Capacities (Ex. 50)

A. Tabulation

Record in the following table the results of your spirometer readings and calculations.

Lung Capacities	Normal (ml.)	Your Capacities
Tidal Volume (TV)	500	
Minute Respiratory Volume (MRV) (MRV × TV × Resp. Rate)	6,000	
Expiratory Reserve Volume (ERV)	1,100	
Vital Capacity (VC) (VC = TV + ERV + IRV)	See Appendix A (Tables IV and V)	
Inspiratory Capacity (IC) (IC = VC − ERV)	3,000	
Inspiratory Reserve Volume (IRV) (IRV = IC − TV)	2,500	

B. Evaluation

List below any of your lung capacities that are significantly low (22% or more). _____

_____

_____

_____

Spirometry: The FEV$_T$ Test (Ex. 51)

A. Expirogram

Trim the tracing of your expirogram to as small a piece as you can without removing volume values and tape it in space shown. Do not tape sides and bottom.

Tape Here

B. Calculations

Record your computations for determining the various percentages here.

1. **Vital Capacity** From the expirogram determine the total volume of air expirated (Vital Capacity) .... _____

2. **Corrected Vital Capacity** By consulting Table III, Appendix A, determine the corrected vital capacity (Step 4).

   VC × Conversion Factor = ... _____

   Math:

3. **One-Second Timed Vital Capacity (FEV$_1$)** After determining the FEV$_1$ (Steps 1, 2, and 3) record your results here .... _____

4. **Corrected FEV$_1$** Correct the FEV$_1$ for the temperature of the spirometer (Step 5).

   FEV$_1$ × Conversion Factor = ... _____

5. **FEV$_1$ Percentage** Divide the corrected FEV$_1$ by the corrected vital capacity to determine the percentage expired in the first second (Step 6).

   $$\frac{\text{Corrected FEV}_1}{\text{Corrected Vital Capacity}} = \text{...}$$ _____

   Math:

6. **Percent of Predicted VC** Determine what percentage your vital capacity is of the predicted vital capacity (Step 7).

   $$\frac{\text{Your Corrected VC}}{\text{Predicted VC}} = \text{...}$$ _____

   Math:

C. Questions

1. Why isn't a measure of one's vital capacity as significant as the FEV$_T$? _____

   _____

   _____

   _____

2. What respiratory diseases are most readily detected with this test? _____

   _____

   _____

Student: _____

Desk No: _____ Section: _____

Anatomy of the Digestive System

A. Illustrations

Record the labels for the illustrations of this exercise in the answer column.

B. Microscopic Studies

Record here the drawings of the microscopic examinations that are required for this exercise. If there is insufficient space for all required drawings, utilize a separate sheet of paper for all of them.

Taste Buds	Salivary Glands
Stomach Wall	Intestinal Wall

Answers

Fig. 52.1	Fig. 52.2
_____	_____
_____	_____
_____	_____
_____	_____
_____	_____
_____	_____
_____	_____
_____	_____
_____	**Fig. 52.3**
_____	_____
_____	_____
_____	_____
_____	_____
_____	_____
_____	_____
_____	_____
_____	_____
_____	_____
_____	_____
_____	_____
_____	_____
_____	_____
_____	_____

C. Alimentary Canal

Identify the parts of the digestive system described by the following statements.

anus—1
cecum—2
colon—3
colon, sigmoid—4
duct, common bile—5
duct, cystic—6
duct, hepatic—7
duct, pancreatic—8
duodenum—9

esophagus—10
ileum—11
pharynx—12
rectum—13
stomach, fundus—14
stomach, pyloric portion—15
valve, cardiac—16
valve, ileocecal—17
valve, pyloric—18

1. Entrance opening of stomach.
2. Exit opening of stomach.
3. Place where swallowing (peristalsis) begins.
4. First twelve inches of small intestine.
5. Tube between mouth and stomach.
6. Distal coiled portion of small intestine.
7. Section of large intestine between descending colon and rectum.
8. Most active portion of stomach.
9. Proximal pouch or compartment of large intestine.
10. Duct that drains the gall bladder.
11. Part of small intestine where most digestion occurs.
12. Valve between small and large intestines.
13. Part of small intestine where most absorption occurs.
14. Duct that drains the liver.
15. Duct that conveys bile to intestine.
16. Exit of alimentary canal.
17. Part of tract where most water absorption (conservation) occurs.
18. Last six inches of alimentary canal.
19. Structure on which appendix is located.
20. Duct that joins the common bile duct before entering the intestine.

Answers	
Alimentary Canal	**Fig. 52.4**
1. _____	_____
2. _____	_____
3. _____	_____
4. _____	_____
5. _____	_____
6. _____	
7. _____	
8. _____	
9. _____	
10. _____	
11. _____	
12. _____	_____
13. _____	
14. _____	
15. _____	
16. _____	
17. _____	
18. _____	
19. _____	
20. _____	

D. Oral Cavity

Identify the structures of the mouth that are described by the following statements.

arch, glossopalatine—1
arch, pharyngopalatine—2
buccae—3
duct, Stenson's—4
duct, Wharton's—5
ducts of Rivinus—6
frenulum, labial—7
frenulum, lingual—8
gingiva—9
glands, parotid—10
glands, sublingual—11
glands, submandibular—12
mucosa—13
papillae, filiform—14
papillae, foliate—15
papillae, fungiform—16
papillae vallate—17
tonsil, lingual—18
tonsil, palatine—19
tonsil, pharyngeal—20
uvula—21

1. Lining of the mouth.
2. Duct that drains the parotid gland.
3. Name for the cheeks.
4. Duct that drains the submandibular gland.
5. Fold of skin between the lip and gums.
6. Fold of skin between the tongue and floor of mouth.
7. Portion of mucosa around the teeth.
8. Tonsils seen at back of mouth.
9. Tonsils located at root of tongue.
10. Finger-like projection at end of soft palate.
11. Ducts that drain sublingual gland.
12. Large papillae at back of tongue.
13. Small tactile papillae on surface of tongue.
14. Rounded papillae on dorsum of tongue.
15. Vertical ridges on side of tongue.
16. Membrane located in front of palatine tonsil.
17. Membrane located in back of palatine tonsil.
18. Salivary glands located under the tongue.
19. Salivary glands located in cheeks.
20. Salivary glands located inside and below the mandible.
21. Place where taste buds are located.

E. The Teeth

Select the correct answer that completes each of the following statements.

1. A complete set of primary teeth consists of
   (1) ten teeth   (2) twenty teeth (3) thirty-two teeth.
2. A complete set of permanent teeth consists of
   (1) twenty-four teeth   (2) twenty-eight teeth
   (3) thirty-two teeth.
3. The smallest permanent molars are the
   (1) first molars   (2) second molars   (3) third molars.
4. The permanent teeth with the longest roots are the
   (1) incisors   (2) cuspids   (3) molars.
5. The primary dentition lacks
   (1) molars   (2) incisors   (3) bicuspids.
6. All deciduous teeth usually erupt by the time a child is
   (1) one year   (2) two years   (3) four years old.

Answers

Oral Cavity	Fig. 52.6
1. _____	_____
2. _____	_____
3. _____	_____
4. _____	_____
5. _____	_____
6. _____	_____
7. _____	_____
8. _____	_____
9. _____	_____
10. _____	_____
11. _____	_____
12. _____	_____
13. _____	_____
14. _____	_____
15. _____	_____
16. _____	_____
17. _____	_____
18. _____	
19. _____	
20. _____	
21. _____	

Teeth
1. _____
2. _____
3. _____
4. _____
5. _____
6. _____

7. Trifurcated roots exist on
   (1) upper cuspids   (2) upper molars   (3) lower molars.
8. Bifurcated roots exist on
   (1) upper first bicuspids and lower molars
   (2) cuspids and lower bicuspids   (3) all molars.
9. The following teeth are said to be succedaneous
   (1) incisors, cuspids, and bicuspids
   (2) cuspids, bicuspids, and molars
   (3) all permanent teeth.
10. The wisdom tooth is
    (1) a supernumerary tooth   (2) a succedaneous tooth
    (3) the third molar.
11. Cementum is secreted by cells called
    (1) odontoblasts   (2) ameloblasts   (3) cementocytes.
12. Dentin is produced by cells called
    (1) odontoblasts   (2) ameloblasts   (3) cementocytes.
13. A tooth is considered "dead" or devitalized if
    (1) the enamel is destroyed   (2) the pulp is destroyed
    (3) The peridental membrane is infected.
14. Caries (cavities) are caused primarily by
    (1) using the wrong toothpaste   (2) acid production by bacteria (3) fluorides in water.
15. Most teeth that have to be extracted have become useless because of
    (1) caries   (2) gingivitis   (3) periodontitis (pyorrhea).
16. The tooth is held in the alveolus by
    (1) bone   (2) dentin   (3) peridental membrane.
17. The tooth receives nourishment through
    (1) the apical foramen   (2) the peridental membrane
    (3) both the apical foramen and the peridental membrane.
18. Enamel covers the
    (1) entire tooth   (2) clinical crown only
    (3) anatomical crown only.

Answers
Teeth
7. _____
8. _____
9. _____
10. _____
11. _____
12. _____
13. _____
14. _____
15. _____
16. _____
17. _____
18. _____

LABORATORY REPORT **53**

Student: ______________

Desk No: ________ Section: ______________

The Chemistry of Hydrolysis

A. Carbohydrate Digestion

1. **IKI Test** Record here the presence (+) or absence (−) of starch as revealed by IKI test on spot plates.

1A	2A	3A	4A	5A

a. Did the pancreatic extract hydrolyze starch? ______________

b. Which tube is your evidence for this conclusion? ______________

c. How do tubes 1A and 5A compare in color? ______________

d. What do you conclude from question *c?* ______________

e. What is the value of tube 4A? ______________

____

f. What is the value of tube 2A? ______________

____

2. **Barfoed's and Benedict's Tests** After performing Barfoed's and Benedict's tests on tubes 1A and 5A, what are your conclusions about the degree of digestion of starch by amylase? ______________

____

____

B. Protein Digestion

Record the color and Optical Densities (OD) of each tube in the following chart.

	1P	2P	3P	4P	5P
Color					
O.D.					

1. Did pancreatic juice hydrolyze the BAPNA? ______________
2. Which tube is your evidence for this conclusion? ______________

3. What other tube shows protein hydrolysis? _____

4. What can you conclude from the previous observation in question 3? _____

_____

5. Why is there no hydrolysis in tube 4P? _____

6. Why is there no hydrolysis in tube 3P? _____

7. Why didn't tube 2P show hydrolysis? _____

_____

C. Fat Digestion
Determine the pH of each tube by comparing with a bromthymol blue color standard. Record these values in the table below.

1L	2L	3L	4L	5L	6L

1. Did the pancreatic extract hydrolyze fat in the presence of bile? _____

2. Did the pancreatic extract hydrolyze fat in the absence of bile? _____

3. How do tubes 1L and 6L compare as to degree of hydrolysis? _____

_____

4. What can you conclude from the previous observation in question 3? _____

_____

5. What is the value of tube 2L? _____

_____

D. Summarization
Make an itemized statement of all the facts you learned in this experiment.

Enzyme Review

Consult your text and lecture notes to answer the following questions concerning digestive enzymes.

A. Digestive Juices

Select the enzymes that are present in the following digestive juices. Use the list of enzymes of the next group of questions.

1. Saliva
2. Gastric juice
3. Pancreatic fluid
4. Intestinal juice (succus entericus)

B. Substrates

Select the enzymes that act on the following substrates.

1. Starch
2. Dextrins
3. Maltose
4. Sucrose
5. Lactose
6. Fats
7. Proteins
8. Proteoses and Peptones
9. Casein
10. Peptides

Carbohydrases
amylase, pancreatic—1
amylase, salivary—2
lactase—3
maltase—4
sucrase—5
Proteases
carboxypeptidase—6
chymotrypsin—7
erepsin (peptidase)—8
pepsin—9
rennin—10
trypsin—11
Lipases
lipase, pancreatic—12

C. End Products

Select the enzymes from the above list that produce the following end-products in digestion.

1. Maltose
2. Glucose
3. Fructose
4. Galactose
5. Fatty acids
6. Glycerol
7. Polypeptides
8. Amino acids

Answers

Digestive Juices

1. _____
2. _____
3. _____
4. _____

Substrates

1. _____
2. _____
3. _____
4. _____
5. _____
6. _____
7. _____
8. _____
9. _____
10. _____

End Products

1. _____
2. _____
3. _____
4. _____
5. _____
6. _____
7. _____
8. _____

Student: _____

Desk No: _____ Section: _____

Anatomy of the Urinary System

A. Illustrations

Record the labels for the three illustrations of this exercise in the answer column. Note that figure 54.2 is on the reverse side of this sheet.

B. Microscopy

In the space provided below sketch a renal corpuscle as observed under the high-dry objective.

C. Anatomy

Identify the structures of the urinary system that are described by the following statements.

calyces—1	renal capsule—8
collecting tubule—2	renal column—9
cortex—3	renal papilla—10
glomerular capsule—4	renal pelvis—11
glomerulus—5	renal pyramids—12
medulla—6	ureters—13
nephron—7	urethra—14

1. Tube that drains the urinary bladder.
2. Portion of kidney that contains renal corpuscles.
3. Tubes that drain the kidneys.
4. Portion of kidney that consists primarily of collecting tubules.
5. Basic functioning unit of the kidney.
6. Cone-shaped areas of medulla.
7. Two portions of renal corpuscle.
8. Distal tip of renal pyramid.
9. Short tubes that receive urine from renal papillae.
10. Funnel-like structure that collects urine from calyces of each kidney.
11. Thin fibrous outer covering of kidney.
12. Cortical tissue between renal pyramids.
13. Tuft of capillaries that produces dilute urine.
14. Structure that receives urine from several nephrons.
15. Cup-shaped membranous structure that surrounds glomerulus.

Answers

Anatomy	Fig. 54.1
1. _____	_____
2. _____	_____
3. _____	_____
4. _____	_____
5. _____	_____
6. _____	_____
7. _____	_____
8. _____	_____
9. _____	_____
10. _____	_____
11. _____	**Fig. 54.3**
12. _____	_____
13. _____	_____
14. _____	_____
15. _____	_____

D. Physiology

Select the best answer that completes the following statements concerning the physiology of urine production.

1. Water reabsorption from the glomerular filtrate into the peritubular blood is facilitated by
   (1) antidiuretic hormone   (2) renin
   (3) aldosterone   (4) both 1 and 3

2. Blood enters the glomerulus through the
   (1) efferent arteriole   (2) arcuate artery
   (3) afferent arteriole   (4) none of these

3. The following substances are reabsorbed through the walls of the nephron into the peritubular blood:
   (1) glucose and water   (2) urea and water
   (3) glucose, amino acids, salts and water
   (4) glucose, amino acids, urea, salts and water

4. The amount of urine produced is affected by
   (1) blood pressure
   (2) environmental temperature
   (3) amount of solute in glomerular filtrate
   (4) 1, 2, 3 and additional factors

5. The amount of urine normally produced in 24 hours is about
   (1) 100 ml.   (2) 500 ml.   (3) 1.5 liters   (4) 4.5 liters

6. The reabsorption of sodium ions from the glomerular filtrate into the peritubular blood draws the following back into the blood
   (1) potassium ions   (2) chloride ions
   (3) water   (4) chloride ions and water

7. Most glucose is reabsorbed in the
   (1) proximal convoluted tubule
   (2) Henle's loop
   (3) distal convoluted tubule
   (4) none of these

8. The desire to micturate normally occurs when the following amount of urine is present in the bladder:
   (1) 100 ml.   (2) 200 ml.   (3) 300 ml.   (4) 400 ml.

9. The presence of glucose in the urine and a low level of insulin in the blood would be diagnosed as
   (1) diabetes insipidis   (2) diabetes mellitus
   (3) renal diabetes   (4) none of these

10. The reabsorption of sodium ions into the blood from the glomerular filtrate may bring about the return of the following substances from the blood to the urine:
    (1) hydrogen ions   (2) ammonia
    (3) potassium ions   (4) hydrogen or potassium ions

11. Renal diabetes is due to
    (1) a lack of ADH   (2) a lack of insulin
    (3) faulty reabsorption of glucose in the nephron
    (4) both 1 and 2

12. The mechanism of electrolyte reabsorption in the nephron is
    (1) diffusion   (2) osmosis
    (3) active transport   (4) both 1 and 3

13. Surgical removal of a kidney is called
    (1) nephrectomy   (2) nephrotomy   (3) nephrolithotomy

Answers
Physiology
1. _____
2. _____
3. _____
4. _____
5. _____
6. _____
7. _____
8. _____
9. _____
10. _____
11. _____
12. _____
13. _____

Fig. 54.2

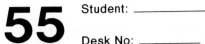
Urine: Composition and Tests

A. Test Results

Record on the chart below the results of any urine test performed. If the tests are done as a demonstration, the results will be tabulated in columns A and B. If students perform the tests on their own urine, the last column will be used for their results.

TEST	Normal Values	Abnormal Values	A — Positive Test Control	B — Unknown Sample	C — Student's Urine
COLOR	Colorless Pale straw Straw Amber	Milky Reddish amber Brownish yellow Green Smoky brown			
CLOUDINESS	Clear	+ slight + + moderate + + + cloudy + + + + very cloudy			
SP. GRAVITY	1.001–1.060	Above 1.060			
pH	4.8–7.5	Below 4.8 Above 7.5			
ALBUMIN (Protein)	None	+ barely visible + + granular + + + flocculent + + + + large flocculent			
MUCIN	None	Visible amounts			
GLUCOSE	None	+ yellow green + + greenish yellow + + + yellow + + + + orange			
KETONES	None	+ pink-purple ring			
HEMOGLOBIN	None	See color chart			

B. Microscopy

Record here by sketch and word any structures seen on micro-scopic examination.

C. Interpretation

Indicate the probable significance of each of the following in urine. More than one condition may apply in some instances.

bladder infection—1
cirrhosis of liver—2
diabetes insipidis—3
diabetes mellitus—4
exercise (extreme)—5

kidney tumor—6
glomerulonephritis—7
gonorrhea—8
hepatitis—9
normally present—10

1. Blood.
2. Glucose.
3. Urea.
4. Albumin (constantly).
5. Albumin (periodic).
6. Bacteria.
7. Mucin.
8. Creatinine.
9. Bile pigments.
10. Uric acid.
11. Pus cells (neutrophils).
12. Porphyrin.
13. Sodium chloride.
14. Specific gravity (low).
15. Specific gravity (high).

D. Terminology

Select the terms described by the following statements.

albuminurea—1
anuria—2
catheter—3
cystitis—4
diuretic—5

enuresis—6
glycosurea—7
nephritis—8
pyelitis—9
uremia—10

1. Absence of urine production.
2. Device for draining bladder.
3. Inflammation of pelvis of kidney.
4. High urea level in blood.
5. Sugar in urine.
6. Inflammation of nephrons.
7. Albumin in urine.
8. Involuntary bed-wetting during sleep.
9. Bladder infection.
10. Chemical that stimulates urine production.

Answers

Interpretation

1. _____
2. _____
3. _____
4. _____
5. _____
6. _____
7. _____
8. _____
9. _____
10. _____
11. _____
12. _____
13. _____
14. _____
15. _____

Terminology

1. _____
2. _____
3. _____
4. _____
5. _____
6. _____
7. _____
8. _____
9. _____
10. _____

56

Student: _____

Desk No: _____ Section: _____

The Endocrine Glands

A. Illustrations
Record the labels for figures 56.1 and 56.2 in the answer columns.

B. Microscopy
Make the drawings of the various microscopic studies on separate drawing paper. Include these drawings with the Laboratory Report.

C. Sources
Select the glandular tissues that produce the following hormones.

1. ACTH.
2. ADH.
3. Aldosterone.
4. Calcitonin.
5. Chorionic gonadotropin.
6. Cortisol.
7. Epinephrine.
8. Estrogens.
9. FSH.
10. Glucagon.
11. ICSH.
12. Insulin.
13. LH.
14. Melatonin.
15. Norepinephrine.
16. Oxytocin.
17. Parathyroid hormone.
18. PIF.
19. Prolactin.
20. Progesterone.
21. Somatostatin.
22. Testosterone.
23. Thymosin.
24. Thyrotropin.
25. Thyroxine.
26. Triiodothyronine.
27. Vasopressin.

adrenal cortex
  zona fasciculata—1
  zona glomerulosa—2
  zona reticularis—3
adrenal medulla—4
hypothalamus—5
ovary
  corpus luteum—6
  Graafian follicle—7
pancreas
  A cells—8
  B cells—9
  D cells—10
  F cells—11
parathyroid gland—12
pineal gland—13
pituitary gland
  adenohypophysis—14
  neurohypophysis—15
placenta—16
testis—17
thymus gland—18
thyroid gland
  follicular cells—19
  parafollicular cells—20

D. Physiology
Select the hormones that produce the following physiological effects. Note that a separate list of hormones is provided for each group.

GROUP I
1. Promotes gluconeogenesis.
2. Increases metabolism.
3. Increases blood pressure.
4. Causes vasoconstriction in all organs.
5. Increases strength of heartbeat.
6. Promotes sodium absorption in nephron.
7. Causes vasodilation in skeletal muscles.
8. Causes vasoconstriction in all organs except muscles and liver.
9. Inhibits the estrus cycle in lower animals.

aldosterone—1
epinephrine—2
glucocorticoids—3
melatonin—4
norepinephrine—5
none of these—6

Answers

Sources	Fig. 56.1
1. _____	_____
2. _____	_____
3. _____	_____
4. _____	_____
5. _____	_____
6. _____	_____
7. _____	_____
8. _____	_____
9. _____	_____
10. _____	_____
11. _____	_____
12. _____	_____
13. _____	_____
14. _____	_____
15. _____	_____
16. _____	_____
17. _____	_____
18. _____	_____
19. _____	_____
20. _____	_____
21. _____	_____
22. _____	_____
23. _____	_____
24. _____	_____
25. _____	_____
26. _____	
27. _____	

Physiology
1. _____
2. _____
3. _____
4. _____
5. _____
6. _____
7. _____
8. _____
9. _____

D. Physiology (continued)

Select the hormones that produce the following physiological effects.

GROUP II

1. Stimulates osteoclasts.
2. Inhibits glucagon production.
3. Mobilizes fatty acids.
4. Inhibits insulin production.
5. Raises calcium level in blood.
6. Increases numbers of mitochondria.
7. Inhibits production of SH.
8. Stimulates RNA production.
9. Promotes storage of glucose, fatty acids, and amino acids.
10. Controls bone remodeling during growing years.
11. Stimulates metabolism of all cells in body.
12. Inhibits conversion of osteoclasts to osteoblasts.
13. Mobilizes glucose, fatty acids, and amino acids.

calcitonin—1
glucagon—2
insulin—3
parathyroid hormone—4
somatostatin—5
thyroid hormone—6
none of these—7

GROUP III

1. Promotes rapid bone growth.
2. Stimulates testicular descent.
3. Promotes breast enlargement.
4. Promotes maturation of spermatozoa.
5. Causes constriction of arterioles.
6. Causes milk ejection from breasts.
7. Stimulates glucocorticoid production.
8. Stimulates production of thyroxine.
9. Causes thickening and vascularization of endometrium.
10. Causes interstitial cells to produce testosterone.
11. Inhibits prolactin production.
12. Suppresses development of female genitalia in male.
13. Promotes stratification of vaginal epithelium.
14. Prepares the endometrium for egg cell implantation.
15. Promotes milk production after childbirth.
16. Promotes growth of all cells in the body.
17. Causes myometrium contraction during childbirth.

ACTH—1
ADH—2
chorionic gonadotropin—3
estrogens—4
FSH—5
LH—6
oxytocin—7
PIF—8
progesterone—9
somatotropin—10
testosterone—11
thyrotropin—12
none of these—13

E. Hormonal Imbalance

Select the hormones that are responsible for the following disorders. After each hormone number indicate with a (+) or (−) whether the condition is due to excess (+) or deficiency (−). More than one hormone may apply in some cases.

1. Acromegaly.
2. Addison's disease.
3. Adrenal diabetes.
4. Cretinism.
5. Cushing's syndrome.
6. Diabetes insipidis.
7. Diabetes mellitus.
8. Dwarfism.
9. Exophthalmos.
10. Gigantism.
11. Hyperglycemia.
12. Hypothyroidism.
13. Myxedema.
14. Tetany.

ADH—1
glucagon—2
glucocorticoids—3
insulin—4
mineralocorticoids—5
parathyroid hormone—6
somatotropin—7
thyroid hormone—8
none of these—9

Answers

Physiology Group II	Fig. 56.2
1. _____	_____
2. _____	_____
3. _____	_____
4. _____	_____
5. _____	_____
6. _____	_____
7. _____	_____
8. _____	_____
9. _____	_____
10. _____	_____
11. _____	_____
12. _____	_____
13. _____	_____

Group III		Imbalance
1. _____		
2. _____		
3. _____		
4. _____		
5. _____		1. _____
6. _____		2. _____
7. _____		3. _____
8. _____		4. _____
9. _____		5. _____
10. _____		6. _____
11. _____		7. _____
12. _____		8. _____
13. _____		9. _____
14. _____		10. _____
15. _____		11. _____
16. _____		12. _____
17. _____		13. _____
		14. _____

The Reproductive Organs

A. Labels
Record the labels for the illustrations of this exercise in the answer columns.

B. Microscopy
Draw the various stages of spermatogenesis on a separate piece of paper to be included with this Laboratory Report.

C. Male Organs
Identify the structures described by the following statements.

common ejaculatory duct—1
corpora cavernosa—2
corpus spongiosum—3
Cowper's gland—4
epididymis—5
glans penis—6
penis—7
prepuce—8
prostate gland—9
prostatic urethra—10
seminal vesicle—11
seminiferous tubule—12
testes—13
vas deferens—14

1. Copulatory organ of male.
2. Erectile tissue of penis (3).
3. Source of spermatozoa.
4. Fold of skin over end of penis.
5. Coiled up structures in testes where spermatozoa originate.
6. Duct that passes from urinary bladder through the prostate gland.
7. Structure that stores spermatozoa.
8. Contribute fluids to semen (3).
9. Distal end of penis.
10. Tube that carries sperms from epididymis to ejaculatory duct.

D. Female Organs
Identify the structures described by the following statements.

cervix—1
clitoris—2
external os—3
fimbriae—4
greater vestibular glands—5
hymen—6
infundibulum—7
internal os—8
labia majora—9
labia minora—10
mons pubis—11
myometrium—-12
paraurethral glands—13
ovaries—14
uterine tubes—15
vagina—16

1. Copulatory organ of female.
2. Muscular wall of uterus.
3. Neck of uterus.
4. Source of ova in female.
5. Protuberance of erectile tissue sensitive to sexual excitation.
6. Finger-like projections around edge of infundibulum.
7. Entrance to uterus at vaginal end.
8. Provides vaginal lubrication during coitus.
9. Ducts that convey ovum to uterus.
10. Open funnel-like end of uterine tube.
11. Membranous fold of tissue surrounding entrance to vagina.
12. Homologous to prostate gland of male.
13. Homologous to scrotum of male.
14. Homologous to penis of male.
15. Homologous to Cowper's glands of male.

Answers	
Male Organs	**Fig. 57.1**
1. _____	_____
2. _____	_____
3. _____	_____
4. _____	_____
5. _____	_____
6. _____	_____
7. _____	_____
8. _____	_____
9. _____	_____
10. _____	_____
Female Organs	
1. _____	_____
2. _____	_____
3. _____	_____
4. _____	_____
5. _____	_____
6. _____	_____
7. _____	_____
8. _____	_____
9. _____	_____
10. _____	_____
11. _____	_____
12. _____	_____
13. _____	_____
14. _____	_____
15. _____	_____

E. Ligaments

Identify the following supporting structures that hold the ovaries, uterus, and uterine tubes in place.

broad ligament—1
infundibulopelvic ligament—2
mesosalpinx—3
ovarian ligament—4

round ligament—5
sacro-uterine ligament—6
suspensory ligament—7
none of these—8

1. Ligament between the cervix and sacral part of the pelvic wall.
2. A mesentery between the ovary and uterine tubes that contains blood vessels leading to the uterine tube.
3. A ligament that extends from the uterus to the ovary.
4. A large flat ligament of peritoneal tissue that extends from the uterine tube to the cervical part of the uterus.
5. A fold of peritoneum that attaches the ovary to the uterine tube.
6. A ligament that extends from the infundibulum to the wall of the pelvic cavity.
7. A ligament that extends from the corpus of the uterus to the body wall.

F. Germ Cells

Identify the types of cells of oogenesis and spermatogenesis described by the following statements. More than one answer may apply.

Spermatogenesis
primary spermatocyte—1
secondary spermatocyte—2
spermatids—3
spermatogonia—4
spermatozoa—5

Oogenesis
oogonia—6
polar bodies—7
primary oocyte—8
secondary oocyte—9

1. Cells at periphery of ovary that produce all ova.
2. Cells that undergo mitosis.
3. Cells in which first meiotic division occurs.
4. Cells at periphery of seminiferous tubule that give rise to all spermatozoa.
5. Cells in which second meiotic division occurs.
6. Cells that are haploid.
7. Cells that contain tetrads.
8. Cells that contain monads.
9. Cells that contain dyads.
10. Cells that are diploid.
11. Cells that develop directly into mature spermatozoa.

G. Physiology of Reproduction and Development

Select the best answer that completes the following statements.

1. The birth canal consists of the
   (1) vagina   (2) uterus   (3) the vagina and uterus.
2. The vaginal lining is normally
   (1) alkaline   (2) acid   (3) neutral.
3. The prostatic secretion is
   (1) alkaline   (2) acid   (3) neutral.
4. The ovum gets from the ovary to the uterus by
   (1) amoeboid movement   (2) ciliary action
   (3) peristalsis and ciliary action of uterine tube.
5. The seminal vesicle secretion is
   (1) alkaline   (2) acid   (3) neutral.
6. Spermatozoan viability is enhanced by a temperature that is
   (1) 98.6°F.   (2) above 98.6°F.   (3) below 98.6°F.

Answers	
Ligaments	**Fig. 57.3**
1. _____	_____
2. _____	_____
3. _____	_____
4. _____	_____
5. _____	_____
6. _____	_____
7. _____	_____
Germ Cells	_____
1. _____	_____
2. _____	_____
3. _____	_____
4. _____	**Fig. 57.4**
5. _____	_____
6. _____	_____
7. _____	_____
8. _____	_____
9. _____	_____
10. _____	_____
11. _____	_____
Physiology	_____
1. _____	_____
2. _____	_____
3. _____	_____
4. _____	_____
5. _____	_____
6. _____	_____

7. Circumcision involves excisement of the
   (1) prepuce   (2) perineum   (3) glans penis.

8. Fertilization of the human ovum usually occurs in the
   (1) uterus   (2) vagina   (3) uterine tube.

9. The fetal stage of the human is
   (1) the first 8 weeks   (2) from the 9th week till birth
   (3) the entire prenatal term.

10. The innermost fetal membrane surrounding the embryo is the
    (1) amnion   (2) chorion   (3) decidua.

11. The life span of an ovum without fertilization is
    (1) six hours   (2) 24–28 hours   (3) 7 days.

12. Implantation of the fertilized ovum occurs usually
    (1) within 2 days after fertilization
    (2) within 6 to 8 days after fertilization
    (3) after 10 days.

13. The outermost fetal membrane surrounding the embryo is the
    (1) amnion   (2) chorion   (3) decidua.

14. The lining of the uterus is the
    (1) myometrium   (2) endometrium   (3) epimetrium

15. Uterine muscle consists of
    (1) smooth muscle   (2) striated muscle
    (3) both smooth and striated muscle.

16. Abortion is correctly known as
    (1) criminal emptying of the uterus
    (2) interruption of pregnancy during fetal life
    (3) interruption of pregnancy during embryonic life.

17. Parturition is the
    (1) process of giving birth   (2) period of pregnancy
    (3) first few days of postnatal life.

18. The following substances pass from the mother to the fetus
    through the placenta:
    (1) nutrients, gases and blood cells
    (2) nutrients and all blood cells except red blood cells
    (3) nutrients, gases, hormones and antibodies.

19. The afterbirth refers to the
    (1) placenta   (2) damaged uterus
    (3) the placenta, umbilical vessels and fetal membranes.

20. Milk production by the breasts usually occurs
    (1) immediately after birth   (2) on the second day
    (3) on the third or fourth day.

H. Menstrual Cycle

Select the correct answer to the following statements.

1. The first menstrual flow is known as
   (1) menarche   (2) climacteric   (3) menopause.

2. The cessation of menstrual flow in a woman in her late forties
   is known as
   (1) menarche   (2) menopause   (3) amenorrhea.

3. A distinct rise in the basal body temperature usually occurs
   (1) at ovulation   (2) during the proliferative phase
   (3) during the quiescent period.

Answers

Fig. 57.5

7. ____   ____

8. ____   ____

9. ____   ____

10. ____   ____

11. ____   ____

12. ____   ____

13. ____   ____

14. ____   ____

15. ____   ____

16. ____   ____

17. ____   ____

18. ____   ____

19. ____   ____

20. ____   ____

Menstrual Cycle

1. ____

2. ____

3. ____

4. Ovulation usually (not always) occurs the following number of days before the next menstrual period:
(1) three   (2) seven   (3) fourteen   (4) eighteen.

5. Pain during ovulation, as experienced by some women is known as
(1) dysmenorrhea   (2) oligomenorrhea
(3) mittleschmerz.

6. Menstrual flow is the result of
(1) a deficiency of progesterone and estrogen
(2) a deficiency of estrogen
(3) an excess of progesterone and estrogen.

7. Progesterone secretion ceases entirely within the following number of days after the onset of menses:
(1) ten   (2) fourteen   (3) twenty-six   (4) twenty-eight.

8. Repair of the endometrium after menstruation is due to
(1) estrogen   (2) progesterone   (3) luteinizing hormone.

9. The proliferative phase in the menstrual cycle begins at about the
(1) second day   (2) fifth day   (3) seventh day of menstrual cycle.

10. Absence of menstruation is called
(1) dysmenorrhea   (2) oligomenorrhea   (3) amenorrhea.

11. Occasional or irregular menses is known as
(1) amenorrhea   (2) oligomenorrhea   (3) dysmenorrhea.

12. Excessive discomfort and pain during menstruation is known as
(1) dysmenorrhea   (2) oligomenorrhea   (3) menorrhagia.

13. Excessive blood flow during menstruation is known as
(1) oligomenorrhea   (2) dysmenorrhea   (3) menorrhagia.

I. Medical

Select the type of surgery or condition that is described by the following statements.

anteflexion—1
cryptorchidism—2
endometritis—3
episiotomy—4
gonorrhea—5
hysterectomy—6
mastectomy—7
oophorectomy—8
oophorhysterectomy—9
oophoroma—10
retroflexion—11
salpingitis—12
salpingectomy—13
syphilis—14
vasectomy—15

1. Failure of testes to descend into scrotum.
2. Surgical removal of uterus.
3. Surgical removal of breast.
4. Sterilization procedure in males.
5. Spirochaetal venereal disease.
6. Surgical removal of one or both ovaries.
7. Type of incision made in perineum at childbirth to prevent excessive damage to anal sphincter.
8. Venereal disease that affects the mucous membranes rather than the blood.
9. Ovarian malignancy.
10. Most common venereal disease.
11. Surgical removal or sectioning of uterine tubes.
12. Malpositioned uterus (2 types).
13. Inflammation of uterine wall.
14. Venereal disease caused by a coccoidal (spherical) organism.
15. Surgical removal of ovaries and uterus.

Answers

4. _____
5. _____
6. _____
7. _____
8. _____
9. _____
10. _____
11. _____
12. _____
13. _____

Medical

1. _____
2. _____
3. _____
4. _____
5. _____
6. _____
7. _____
8. _____
9. _____
10. _____
11. _____
12. _____
13. _____
14. _____
15. _____

Appendix
Tables

Table I. International Atomic Weights

Element	Symbol	Atomic Number	Atomic Weight
Aluminum	Al	13	26.97
Antimony	Sb	51	121.76
Arsenic	As	33	74.91
Barium	Ba	56	137.36
Beryllium	Be	4	9.013
Bismuth	Bi	83	209.00
Boron	B	5	10.82
Bromine	Br	35	79.916
Cadmium	Cd	48	112.41
Calcium	Ca	20	40.08
Carbon	C	6	12.010
Chlorine	Cl	17	35.457
Chromium	Cr	24	52.01
Cobalt	Co	27	58.94
Copper	Cu	29	63.54
Fluorine	F	9	19.00
Gold	Au	79	197.2
Hydrogen	H	1	1.0080
Iodine	I	53	126.92
Iron	Fe	26	55.85
Lead	Pb	82	207.21
Magnesium	Mg	12	24.32
Manganese	Mn	25	54.93
Mercury	Hg	80	200.61
Nickel	Ni	28	58.69
Nitrogen	N	7	14.008
Oxygen	O	8	16.0000
Palladium	Pd	46	106.7
Phosphorus	P	15	30.98
Platinum	Pt	78	195.23
Potassium	K	19	39.096
Radium	Ra	88	226.05
Selenium	Se	34	78.96
Silicon	Si	14	28.06
Silver	Ag	47	107.880
Sodium	Na	11	22.997
Strontium	Sr	38	87.63
Sulfur	S	16	32.066
Tin	Sn	50	118.70
Titanium	Ti	22	47.90
Tungsten	W	74	183.92
Uranium	U	92	238.07
Vanadium	V	23	50.95
Zinc	Zn	30	65.38
Zirconium	Zr	40	91.22

Table II.   Temperature conversion table Centigrade to Fahrenheit.

°C.	0	1	2	3	4	5	6	7	8	9
−50	**−58.0**	**−59.8**	**−61.6**	**−63.4**	**−65.2**	**−67.0**	**−68.8**	**−70.6**	**−72.4**	**−74.2**
−40	−40.0	−41.8	−43.6	−45.4	−47.2	−49.0	−50.8	−52.6	−54.4	−56.2
−30	−22.0	−23.8	−25.6	−27.4	−29.2	−31.0	−32.8	−34.6	−36.4	−38.2
−20	− 4.0	− 5.8	− 7.6	− 9.4	−11.2.	−13.0	−14.8	−16.6	−18.4	−20.2
−10	+14.0	+12.2	+10.4	+ 8.6	+ 6.8	+ 5.0	+ 3.2	+ 1.4	− 0.4	− 2.2
− 0	+32.0	+30.2	+28.4	+26.6	+24.8	+23.0	+21.2	+19.4	+17.6	+15.8
0	**32.0**	**33.8**	**35.6**	**37.4**	**39.2**	**41.0**	**42.8**	**44.6**	**46.4**	**48.2**
10	50.0	51.8	53.6	55.4	57.2	59.0	60.8	62.6	64.4	66.2
20	68.0	69.8	71.6	73.4	75.2	77.0	78.8	80.6	82.4	84.2
30	86.0	87.8	89.6	91.4	93.2	95.0	96.8	98.6	100.4	102.2
40	104.0	105.8	107.6	109.4	111.2	113.0	114.8	116.6	118.4	120.2
50	122.0	123.8	125.6	127.4	129.2	131.0	132.8	134.6	136.4	138.2
60	**140.0**	**141.8**	**143.6**	**145.4**	**147.2**	**149.0**	**150.8**	**152.6**	**154.4**	**156.2**
70	158.0	159.8	161.6	163.4	165.2	167.0	168.8	170.6	172.4	174.2
80	176.0	177.8	179.6	181.4	183.2	185.0	186.8	188.6	190.4	192.2
90	194.0	195.8	197.6	199.4	201.2	203.0	204.8	206.6	208.4	210.2
100	212.0	213.8	215.6	217.4	219.2	221.0	222.8	224.6	226.4	228.2
110	**230.0**	**231.8**	**233.6**	**235.4**	**237.2**	**239.0**	**240.8**	**242.6**	**244.4**	**246.2**
120	248.0	249.8	251.6	253.4	255.2	257.0	258.8	260.6	262.4	264.2
130	266.0	267.8	269.6	271.4	273.2	275.0	276.8	278.6	280.4	282.2
140	284.0	285.8	287.6	289.4	291.2	293.0	294.8	296.6	298.4	300.2
150	302.0	303.8	305.6	307.4	309.2	311.0	312.8	314.6	316.4	318.2
160	**320.0**	**321.8**	**323.6**	**325.4**	**327.2**	**329.0**	**330.8**	**332.6**	**334.4**	**336.2**
170	338.0	339.8	341.6	343.4	345.2	347.0	348.8	350.6	352.4	354.2
180	356.0	357.8	359.6	361.4	363.2	365.0	366.8	368.6	370.4	372.2
190	374.0	375.8	377.6	379.4	381.2	383.0	384.8	386.6	388.4	390.2
200	392.0	393.8	395.6	397.4	399.2	401.0	402.8	404.6	406.4	408.2
210	**410.0**	**411.8**	**413.6**	**415.4**	**417.2**	**419.0**	**420.8**	**422.6**	**424.4**	**426.2**
220	428.0	429.8	431.6	433.4	435.2	437.0	438.8	440.6	442.4	444.2
230	446.0	447.8	449.6	451.4	453.2	455.0	456.8	458.6	460.4	462.2
240	464.0	465.8	467.6	469.4	471.2	473.0	474.8	476.6	478.4	480.2
250	482.0	483.8	485.6	487.4	489.2	491.0	492.8	494.6	496.4	498.2

$$°F. = °C. \times 9/5 + 32 \qquad °C. = °F. - 32 \times 5/9$$

Table III. Conversion factors for temperature differentials (spirometry).

°C	°F	Conversion Factor
20	68.0	1.102
21	69.8	1.096
22	71.6	1.091
23	73.4	1.085
24	75.2	1.080
25	77.0	1.075
26	78.8	1.068
27	80.6	1.063
28	82.4	1.057
29	84.2	1.051
30	86.0	1.045
31	87.8	1.039
32	89.6	1.032
33	91.4	1.026
34	93.2	1.020
35	95.0	1.014
36	96.8	1.007
37	98.6	1.000

Table IV. Predicted vital capacities for males.

HEIGHT IN CENTIMETERS AND INCHES

AGE	CM. 152 / IN. 59.8	154 / 60.6	156 / 61.4	158 / 62.2	160 / 63.0	162 / 63.7	164 / 64.6	166 / 65.4	168 / 66.1	170 / 66.9	172 / 67.7	174 / 68.5	176 / 69.3	178 / 70.1	180 / 70.9	182 / 71.7	184 / 72.4	186 / 73.2	188 / 74.0
16	3,920	3,975	4,025	4,075	4,130	4,180	4,230	4,285	4,335	4,385	4,440	4,490	4,540	4,590	4,645	4,695	4,745	4,800	4,850
18	3,890	3,940	3,995	4,045	4,095	4,145	4,200	4,250	4,300	4,350	4,405	4,455	4,505	4,555	4,610	4,660	4,710	4,760	4,815
20	3,860	3,910	3,960	4,015	4,065	4,115	4,165	4,215	4,265	4,320	4,370	4,420	4,470	4,520	4,570	4,625	4,675	4,725	4,775
22	3,830	3,880	3,930	3,980	4,030	4,080	4,135	4,185	4,235	4,285	4,335	4,385	4,435	4,485	4,535	4,585	4,635	4,685	4,735
24	3,785	3,835	3,885	3,935	3,985	4,035	4,085	4,135	4,185	4,235	4,285	4,330	4,380	4,430	4,480	4,530	4,580	4,630	4,680
26	3,755	3,805	3,855	3,905	3,955	4,000	4,050	4,100	4,150	4,200	4,250	4,300	4,350	4,395	4,445	4,495	4,545	4,595	4,645
28	3,725	3,775	3,820	3,870	3,920	3,970	4,020	4,070	4,115	4,165	4,215	4,265	4,310	4,360	4,410	4,460	4,510	4,555	4,605
30	3,695	3,740	3,790	3,840	3,890	3,935	3,985	4,035	4,080	4,130	4,180	4,230	4,275	4,325	4,375	4,425	4,470	4,520	4,570
32	3,665	3,710	3,760	3,810	3,855	3,905	3,950	4,000	4,050	4,095	4,145	4,195	4,240	4,290	4,340	4,385	4,435	4,485	4,530
34	3,620	3,665	3,715	3,760	3,810	3,855	3,905	3,950	4,000	4,045	4,095	4,140	4,190	4,235	4,285	4,330	4,380	4,425	4,475
36	3,585	3,635	3,680	3,730	3,775	3,825	3,870	3,920	3,965	4,010	4,060	4,105	4,155	4,200	4,250	4,295	4,340	4,390	4,435
38	3,555	3,605	3,650	3,695	3,745	3,790	3,840	3,885	3,930	3,980	4,025	4,070	4,120	4,165	4,210	4,260	4,305	4,350	4,400
40	3,525	3,575	3,620	3,665	3,710	3,760	3,805	3,850	3,900	3,945	3,990	4,035	4,085	4,130	4,175	4,220	4,270	4,315	4,360
42	3,495	3,540	3,590	3,635	3,680	3,725	3,770	3,820	3,865	3,910	3,955	4,000	4,050	4,095	4,140	4,185	4,230	4,280	4,325
44	3,450	3,495	3,540	3,585	3,630	3,675	3,725	3,770	3,815	3,860	3,905	3,950	3,995	4,040	4,085	4,130	4,175	4,220	4,270
46	3,420	3,465	3,510	3,555	3,600	3,645	3,690	3,735	3,780	3,825	3,870	3,915	3,960	4,005	4,050	4,095	4,140	4,185	4,230
48	3,390	3,435	3,480	3,525	3,570	3,615	3,655	3,700	3,745	3,790	3,835	3,880	3,925	3,970	4,015	4,060	4,105	4,150	4,190
50	3,345	3,390	3,430	3,475	3,520	3,565	3,610	3,650	3,695	3,740	3,785	3,830	3,870	3,915	3,960	4,005	4,050	4,090	4,135
52	3,315	3,353	3,400	3,445	3,490	3,530	3,575	3,620	3,660	3,705	3,750	3,795	3,835	3,880	3,925	3,970	4,010	4,055	4,100
54	3,285	3,325	3,370	3,415	3,455	3,500	3,540	3,585	3,630	3,670	3,715	3,760	3,800	3,845	3,890	3,930	3,975	4,020	4,060
56	3,255	3,295	3,340	3,380	3,425	3,465	3,510	3,550	3,595	3,640	3,680	3,725	3,765	3,810	3,850	3,895	3,940	3,980	4,025
58	3,210	3,250	3,290	3,335	3,375	3,420	3,460	3,500	3,545	3,585	3,630	3,670	3,715	3,755	3,800	3,840	3,880	3,925	3,965
60	3,175	3,220	3,260	3,300	3,345	3,385	3,430	3,470	3,510	3,555	3,595	3,635	3,680	3,720	3,760	3,805	3,845	3,885	3,930
62	3,150	3,190	3,230	3,270	3,310	3,350	3,390	3,440	3,480	3,520	3,560	3,600	3,640	3,680	3,730	3,770	3,810	3,850	3,890
64	3,120	3,160	3,200	3,240	3,280	3,320	3,360	3,400	3,440	3,490	3,530	3,570	3,610	3,650	3,690	3,730	3,770	3,810	3,850
66	3,070	3,110	3,150	3,190	3,230	3,270	3,310	3,350	3,390	3,430	3,470	3,510	3,550	3,600	3,640	3,680	3,720	3,760	3,800
68	3,040	3,080	3,120	3,160	3,200	3,240	3,280	3,320	3,360	3,400	3,440	3,480	3,520	3,560	3,600	3,640	3,680	3,720	3,760
70	3,010	3,050	3,090	3,130	3,170	3,210	3,250	3,290	3,330	3,370	3,410	3,450	3,480	3,520	3,560	3,600	3,640	3,680	3,720
72	2,980	3,020	3,060	3,100	3,140	3,180	3,210	3,250	3,290	3,330	3,370	3,410	3,450	3,490	3,530	3,570	3,610	3,650	3,680
74	2,930	2,970	3,010	3,050	3,090	3,130	3,170	3,200	3,240	3,280	3,320	3,360	3,400	3,440	3,470	3,510	3,550	3,590	3,630

From: Archives of Environmental Health
February 1966, Vol. 12, pp. 146-189
E. A. Gaensler, MD and G. W. Wright, MD

Table V. Predicted vital capacities for females.

HEIGHT IN CENTIMETERS AND INCHES

AGE	CM. 152 / IN. 59.8	154 / 60.6	156 / 61.4	158 / 62.2	160 / 63.0	162 / 63.7	164 / 64.6	166 / 65.4	168 / 66.1	170 / 66.9	172 / 67.7	174 / 68.5	176 / 69.3	178 / 70.1	180 / 70.9	182 / 71.7	184 / 72.4	186 / 73.2	188 / 74.0
16	3,070	3,110	3,150	3,190	3,230	3,270	3,310	3,350	3,390	3,430	3,470	3,510	3,550	3,590	3,630	3,670	3,715	3,755	3,800
17	3,055	3,095	3,135	3,175	3,215	3,255	3,295	3,335	3,375	3,415	3,455	3,495	3,535	3,575	3,615	3,655	3,695	3,740	3,780
18	3,040	3,080	3,120	3,160	3,200	3,240	3,280	3,320	3,360	3,400	3,440	3,480	3,520	3,560	3,600	3,640	3,680	3,720	3,760
20	3,010	3,050	3,090	3,130	3,170	3,210	3,250	3,290	3,330	3,370	3,410	3,450	3,490	3,525	3,565	3,605	3,645	3,695	3,720
22	2,980	3,020	3,060	3,095	3,135	3,175	3,215	3,255	3,290	3,330	3,370	3,410	3,450	3,490	3,530	3,570	3,610	3,650	3,685
24	2,950	2,985	3,025	3,065	3,100	3,140	3,180	3,220	3,260	3,300	3,335	3,375	3,415	3,455	3,490	3,530	3,570	3,610	3,650
26	2,920	2,960	3,000	3,035	3,070	3,110	3,150	3,190	3,230	3,265	3,300	3,340	3,380	3,420	3,455	3,495	3,530	3,570	3,610
28	2,890	2,930	2,965	3,000	3,040	3,070	3,115	3,155	3,190	3,230	3,270	3,305	3,345	3,380	3,420	3,460	3,495	3,535	3,570
30	2,860	2,895	2,935	2,970	3,010	3,045	3,085	3,120	3,160	3,195	3,235	3,270	3,310	3,345	3,385	3,420	3,460	3,495	3,535
32	2,825	2,865	2,900	2,940	2,975	3,015	3,050	3,090	3,125	3,160	3,200	3,235	3,275	3,310	3,350	3,385	3,425	3,460	3,495
34	2,795	2,835	2,870	2,910	2,945	2,980	3,020	3,055	3,090	3,130	3,165	3,200	3,240	3,275	3,310	3,350	3,385	3,425	3,460
36	2,765	2,805	2,840	2,875	2,910	2,950	2,985	3,020	3,060	3,095	3,130	3,165	3,205	3,240	3,275	3,310	3,350	3,385	3,420
38	2,735	2,770	2,810	2,845	2,880	2,915	2,950	2,990	3,025	3,060	3,095	3,130	3,170	3,205	3,240	3,275	3,310	3,350	3,385
40	2,705	2,740	2,775	2,810	2,850	2,885	2,920	2,955	2,990	3,025	3,060	3,095	3,135	3,170	3,205	3,240	3,275	3,310	3,345
42	2,675	2,710	2,745	2,780	2,815	2,850	2,885	2,920	2,955	2,990	3,025	3,060	3,100	3,135	3,170	3,205	3,240	3,275	3,310
44	2,645	2,680	2,715	2,750	2,785	2,820	2,855	2,890	2,925	2,960	2,995	3,030	3,060	3,095	3,130	3,165	3,200	3,235	3,270
46	2,615	2,650	2,685	2,715	2,750	2,785	2,820	2,855	2,890	2,925	2,960	2,995	3,030	3,060	3,095	3,130	3,165	3,200	3,235
48	2,585	2,620	2,650	2,685	2,715	2,750	2,785	2,820	2,855	2,890	2,925	2,960	2,995	3,030	3,060	3,095	3,130	3,160	3,195
50	2,555	2,590	2,625	2,655	2,690	2,720	2,755	2,785	2,820	2,855	2,890	2,925	2,955	2,990	3,025	3,060	3,090	3,125	3,155
52	2,525	2,555	2,590	2,625	2,655	2,690	2,720	2,755	2,790	2,820	2,855	2,890	2,925	2,955	2,990	3,020	3,055	3,090	3,125
54	2,495	2,530	2,560	2,590	2,625	2,655	2,690	2,720	2,755	2,790	2,820	2,855	2,885	2,920	2,950	2,985	3,020	3,050	3,085
56	2,460	2,495	2,525	2,560	2,590	2,625	2,655	2,690	2,720	2,755	2,790	2,820	2,855	2,885	2,920	2,950	2,980	3,015	3,045
58	2,430	2,460	2,495	2,525	2,560	2,590	2,625	2,655	2,690	2,720	2,750	2,785	2,815	2,850	2,880	2,920	2,945	2,975	3,010
60	2,400	2,430	2,460	2,495	2,525	2,560	2,590	2,625	2,655	2,685	2,720	2,750	2,780	2,810	2,845	2,875	2,915	2,940	2,970
62	2,370	2,405	2,435	2,465	2,495	2,525	2,560	2,590	2,620	2,655	2,685	2,715	2,745	2,775	2,810	2,840	2,870	2,900	2,935
64	2,340	2,370	2,400	2,430	2,465	2,495	2,525	3,555	2,585	2,620	2,650	2,680	2,710	2,740	2,770	2,805	2,835	2,865	2,895
66	2,310	2,340	2,370	2,400	2,430	2,460	2,495	2,525	2,555	2,585	2,615	2,645	2,675	2,705	2,735	2,765	2,800	2,825	2,860
68	2,280	2,310	2,340	2,370	2,400	2,430	2,460	2,490	2,520	2,550	2,580	2,610	2,640	2,670	2,700	2,730	2,760	2,795	2,820
70	2,250	2,280	2,310	2,340	2,370	2,400	2,425	2,455	2,485	2,515	2,545	2,575	2,605	2,635	2,665	2,695	2,725	2,755	2,780
72	2,220	2,250	2,280	2,310	2,335	2,365	2,395	2,425	2,455	2,480	2,510	2,540	2,570	2,600	2,630	2,660	2,685	2,715	2,745
74	2,190	2,220	2,245	2,275	2,305	2,335	2,360	2,390	2,420	2,450	2,475	2,505	2,535	2,565	2,590	2,620	2,650	2,680	2,710

From: Archives of Environmental Health
February 1966, Vol. 12, pp. 146–189
E. A. Gaensler, MD and G. W. Wright, MD

B Appendix Solutions and Reagents

Physiological Solutions

Working with freshly dissected or excised vertebrate tissues requires that they be perfused or immersed in an environment that approximates as nearly as possible the ionic, osmotic, and pH qualities of their own tissue fluids. Doing so prevents dysfunction and allows for maintenance and observation of the tissue over longer intervals. A partial list of physiological solutions follows:

Physiological Saline a solution of sodium chloride in water that can allow for moisture maintenance and tonicity of most tissues for short periods. Consists of 0.9% for mammals and 0.7% for amphibians.

Frog Ringer's an all-purpose solution for amphibian tissues (nerve, muscle, skin, etc.)

Turtle Ringer's replaces the tissue fluids of most reptiles.

Mammalian Ringer's an all-purpose solution for general applications in mammalian tissue studies, both short and long term.

Locke's Solution devised for use primarily with isolated smooth muscle and cardiac muscle of mammals.

Tyrode's Solution designed for use with mammalian smooth muscle preparations.

Krebs-Henseleit Solution may be used for work with mammalian nerves and for metabolic measurements of other mammalian tissues.

In this appendix, the individual recipes for some of the above solutions will be given in alphabetical order; however, where large quantities of solutions are required during the semester, it will be more convenient to make up stock solutions from which the various solutions can be quickly made. The longer shelf life of stock solutions enables one to be able to make up fresh physiological solutions daily as needed.

Stock Solutions

Make up five flasks in which the amounts listed are *grams per liter of distilled water.* The molarity of each solution is also given.

Table I. Physiological solutions from stock solutions.

Stock	Frog Ringer's	Turtle Ringer's	Mammalian Ringer's	Tyrode's	Locke's	Krebs Henseleit
A NaCl	103 ml.	117 ml.	155 ml.	137 ml.	155 ml.	111 ml.
B KCl	30 ml.	40 ml.	65 ml.	27 ml.	56 ml.	47 ml.
C CaCl$_2$	20 ml.	20 ml.	22 ml.	18 ml.	22 ml.	25 ml.
NaHCO$_3$ (dry)	.2 gm.	.2 gm.	.2 gm.	.1 gm.	.2 gm.	.2 gm.
D NaH$_2$PO$_4$	---	---	8.3 ml.	2.5 ml.	---	---
E MgSO$_4$	---	---	---	---	---	12.5 ml.
*Glucose (dry)	1.0 gm.	1.0 gm.	1.0 gm.	1.0 gm.	1.0 gm.	1.0 gm.

*Addition of glucose is optional. Keep refrigerated. Lasts only 1 week in refrigerator.

Sol'n	Compound	Amount	Molarity
A	NaCl	58.54 gm.	1M
B	KCl	7.45 gm.	0.1M
C	$CaCl_2$	11.1 gm.	0.1M
D	NaH_2PO_4	12.0 gm.	0.1M
E	$MgSO_4$	12.0 gm.	0.1M

In addition to the above solutions, $NaHCO_3$ in dry form is used. Its instability in solution precludes using a stock solution of this ingredient. Glucose, in dry form, is also added for certain applications. *Refrigeration of all glucose solutions is necessary to inhibit bacterial action.*

Dilutions

To prepare a liter of any of the physiological solutions, fill a volumetric flask half full of distilled water and add the amounts given in Table I. Each compound is added in sequence and mixed thoroughly before adding succeeding compounds. After all ingredients are added, the flask is filled to the one liter mark.

Caution If excess dissolved CO_2 is present in the water, it will tend to precipitate out the calcium in the form of $CaCO_3$. This may be prevented by: (1) pre-aerating the water with an oil-free oxygen or air line for 5–10 minutes or (2) quickly adding solution C ($CaCl_2$) with rapid agitation at the very last. If a precipitate does occur, the solution is not usable.

Other Solutions and Reagents

BAPNA (Ex. 53)

BAPNA is benzoyl DL arginine p-nitroanaline HCl. To prepare this substrate for Ex. 53, dissolve 43 mg. of BAPNA in 1 ml. of dimethyl sulfoxide (caution) and make up to a volume of 100 ml. with TRIS buffer solution.

Barfoed's Reagent

Dissolve 12 gm. copper acetate in 450 ml. of boiling distilled water. Do not filter if precipitate forms. To this hot solution, quickly add 13 ml. of 8.5% lactic acid. Most of precipitate will dissolve. Cool the mixture and dilute to 500 ml. Remove any final precipitate by filtration.

Benedict's Solution (Qualitative)

Sodium citrate	173.0 gm.
Sodium carbonate, anhydrous	100.0 gm.
Copper sulfate, pure crystalline $CuSO_4 \cdot 5H_2O$	17.3 gm.

Dissolve the sodium citrate and sodium carbonate in 700 ml. distilled water with aid of heat and filter.

Then, dissolve the copper sulfate in 100 ml. of distilled water with aid of heat and pour this solution slowly into the first solution, stirring constantly. Make up to 1 liter volume with distilled water.

Exton's Reagent (for albumin test)

Sodium sulfate (anhydrous)	88 gm.
Sulfosalicylic acid	50 gm.
Distilled water to make one liter.	

Dissolve the sodium sulfate in 80 ml. of water with heat. Cool, add the sulfosalicylic acid and make up to volume with water.

Frog Ringer's Solution

This solution is used with most amphibian muscle and nerve preparations. Frog skin and toad bladder preparations can also be handled with this saline solution.

NaCl	6.02 gm.
KCl	0.22 gm.
$CaCl_2$	0.22 gm.
$NaHCO_3$	2.00 gm.
Glucose	1.00 gm.
Distilled water to make 1 liter.	

Iodine (IKI) Solution

Potassium iodide	20 gm.
Iodine crystals	4 gm.
Distilled water	1 liter

Dissolve the potassium iodide in 1 liter of distilled water and add the iodine crystals, stirring to dissolve. Store in dark bottles.

Locke's Solution

NaCl	9.06 gm.
KCl	0.42 gm.
$CaCl_2$	0.24 gm.
$NaHCO_3$	2.00 gm.
Glucose	1.00 gm.

Distilled water to make 1 liter.

Mammalian Ringer's Solution

NaCl	9.00 gm.
KCl	0.48 gm.
$CaCl_2$	0.24 gm.
$NaHCO_3$	0.20 gm.
NaH_2PO_4	0.10 gm.
Glucose	1.00 gm.

Distilled water to make one liter.

Methylene Blue (Loeffler's)

Solution A: Dissolve 0.3 gm. of Methylene blue (90% dye content) in 30.0 ml. ethyl alcohol (95%).

Solution B: Dissolve 0.01 gm. potassium hydroxide in 100.0 ml. distilled water. Mix solutions A and B.

Red Blood Cell (RBC) Diluting Fluid (Hayem's)

Mercuric chloride	1.0 gm.
Sodium sulfate (anhydrous)	4.4 gm.
Sodium chloride	2.0 gm.
Distilled water	400.0 ml.

Rothera's Reagent

The addition of 1 gm. of this reagent to 5 ml. of urine with ammonium hydroxide is used to detect ketones.

Sodium nitroprusside	7.5 gm.
Ammonium sulfate	700.0 gm.

Mix and pulverize in mortar with pestle.

Starch Solution (1%)

Add 10 grams of cornstarch to 1 liter of distilled water. Bring to boil, cool, and filter. Keep refrigerated.

TRIS, pH 8.2 (Ex. 53)

The molecular weight of 2-amino-2(hydroxymethyl)-1,3-propanediol, or TRIS, is 121.14. Its formula is $(HOCH_2)_3CNH_2$. To make up this solution for Ex. 53, make up a .05M solution first by dissolving 6.066 gms. in a liter of distilled water. To 500 ml. of the 0.5M TRIS buffer, add 130 ml. of .05N HCl.

Tyrode's Solution

This solution is used with various smooth muscle preparations of mammals.

NaCl	8.00 gm.
KCl	0.20 gm.
$CaCl_2$	0.20 gm.
$NaHCO_3$	1.00 gm.
NaH_2PO_4	0.03 gm.

Distilled water to make one liter.

White Blood Cell (WBC) Diluting Fluid

Hydrochloric acid	5 ml.
Distilled water	495 ml.

Appendix
Tests and Methods
C

Carbohydrate Differentiation

Several experiments in this book require that sugars and starches be differentiated. An understanding of the application of these tests can be derived from the separation outline below.

Barfoed's Test To 5 ml. of reagent in test tube, add 1 ml. of unknown. Place in boiling water bath for 5 minutes. Interpretation is based on above separation outline.

Benedict's Test This is a semi-quantitative test for amounts of reducing sugar present. To 5 ml. of reagent in test tube, add 8 drops of unknown. Place in boiling water bath for 5 minutes. A green, yellow, or orange-red precipitate determines the amount of reducing sugar present. See Ex. 55.

Electronic Equipment Adjustments

Lengthy instructions pertaining to recorder and transducer manipulations that are used in several experiments are outlined here for reference.

Balancing the Transducer (Ex. 17)

The Unigraph is designed to function in a sensitivity range of .1 MV/CM to 2 MV/CM. When an experiment is begun, the sensitivity control is set at 2 MV/CM, its least sensitive position. As the experiment progresses it may become necessary to shift to a more sensitive setting, such as 1 MV/CM or .5 MV/CM. If the Wheatstone bridge in the transducer is not balanced, the shifting from one sensitivity position to another will cause the stylus to change position. This produces

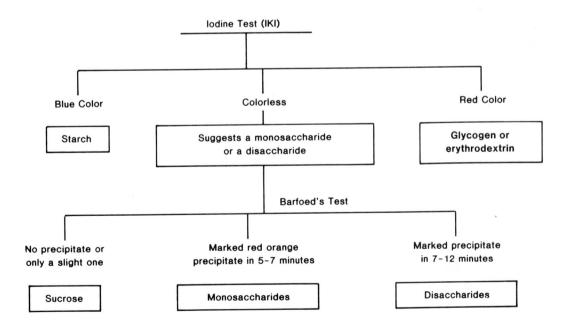

427

an unsatisfactory record. Thus, it is essential that one go through the following steps to balance the transducer.

1. Connect the transducer jack to the transducer socket in the end of the Unigraph. The small upper socket is the one to use.

2. Before plugging in the Unigraph, make sure that the power switch is off, the chart control switch is on STBY, the red gain selector knob is on 2 MV/CM, the blue sensitivity knob is turned completely counterclockwise, the yellow mode selector control is on TRANS, and the Hi Filter and Mean switches are positioned toward Normal. Now plug in the Unigraph.

3. Turn on the Unigraph power switch and set the red heat control knob at the 2:00 position.

4. Place the speed control lever at the slow position and the c.c. lever at Chart On.

5. As the chart moves along, bring the stylus to the center of the paper by turning the centering knob.

6. Turn the sensitivity control completely clockwise to maximum sensitivity. If the bridge is unbalanced, the stylus will move away from the center. To return the stylus to the center, unlock the TRANS-BAL control by pushing the small lever on the control and turning the knob in either direction to return the stylus to the center of the chart.

7. Set the red gain control to 1. If the bridge is unbalanced, the stylus will not be centered. Center with the TRANS-BAL control.

8. Set the gain control to .5 and re-center the stylus with the TRANS-BAL control.
Repeat this same procedure for .2 and .1 settings of the gain control. The bridge is now completely balanced at its highest gain and sensitivity. Lock the TRANS-BAL control with the lock lever.
Note: Although we have gone through this lengthy process to achieve a balanced transducer, we may find that we will still get baseline shifts when going from one setting to another. However, they will be minor compared to one that has not been through this procedure. Any minor deviations should be corrected with the centering control.

9. Return the gain control to 2 and the chart control switch to STBY.

Calibration of Unigraph with Strain Gauge (Ex. 17)

When the transducer has been balanced, it is necessary to calibrate the tracing on the Unigraph to a known force on the transducer. Calibration establishes a linear relationship between the magnitude of deflection and the degree of strain (extent of bending) of the transducer leaf before the muscle is attached to the transducer. Calibrations must be made for each gain at maximum sensitivity that one expects to use in the experiment (your muscle preparations will probably require gain settings of 2, 1, or .5). After calibration, it is possible to directly determine the force of each contraction by simply reading the maximum height of the tracing.

Materials:

small paper clip weighing around .5 gram

1. Weigh a convenient object, such as a small paper clip, to the nearest 0.1 mg.

2. With the chart control switch on STBY, suspend the paper clip on to the two smallest leaves of the transducer.

3. Put the c.c. switch to Chart On and note the extent of deflection on the chart. Record about 1 centimeter on the chart and place the c.c. switch on STBY. Label this deflection in. mm./mg.

4. Set the gain control at 1, cc. to Chart On, and record for another centimeter. Place the c.c. switch on STBY. Label the deflection.

5. Repeat the above procedure for the other two gain settings (.5, .2, and .1). Note how the stylus vibrates as you increase the sensitivity of the Unigraph. This is normal.

6. Return the c.c. switch to STBY, gain to 2, and remove the paper clip. The Unigraph is now calibrated and the muscle can now be attached to it.

Calibration for EMG (Ex. 19)

Calibration of the Unigraph for EMG measurements is necessary to establish that one centimeter deflection of the stylus equals 0.1 millivolt. When calibration is accomplished, the recording is made with the EEG mode since there is no EMG mode. An EMG can be made very well in this mode at lower sensitivities than would be used when making an EEG. Proceed as follows to calibrate:

1. Turn the yellow mode selector control to CC/Cal. (Capacitor Coupled Calibrate).
2. Turn the small main switch to ON.
3. Turn the stylus heat control (red knob) to the 2:00 position.
4. Check the speed selector lever to see that it is in the slow position. The free end of the lever should be pointed toward the styluses.
5. Move the chart-stylus switch to CHART ON and observe the width of the tracing that appears on the paper. Adjust the stylus heat control to produce the desired width of tracing.
6. Set the red gain knob to .1 MV/CM.
7. Push down the .1 MV. button to determine the amount of deflection. Hold the button down for 2 seconds before releasing. If it is released too quickly the tracing will not be perfect. Adjust the deflection so that it deflects exactly 1 centimeter in each direction by turning the blue sensitivity knob. The unit is now calibrated so that you get 1 cm. deflection for .1 millivolt.
8. Reset the red gain knob to .5 MV/CM. This reduces the sensitivity of the instrument.
9. Return the yellow mode selector control to EEG.
10. Return the chart stylus switch to STBY. Place the speed selector in the fast position. The instrument is now ready for use.

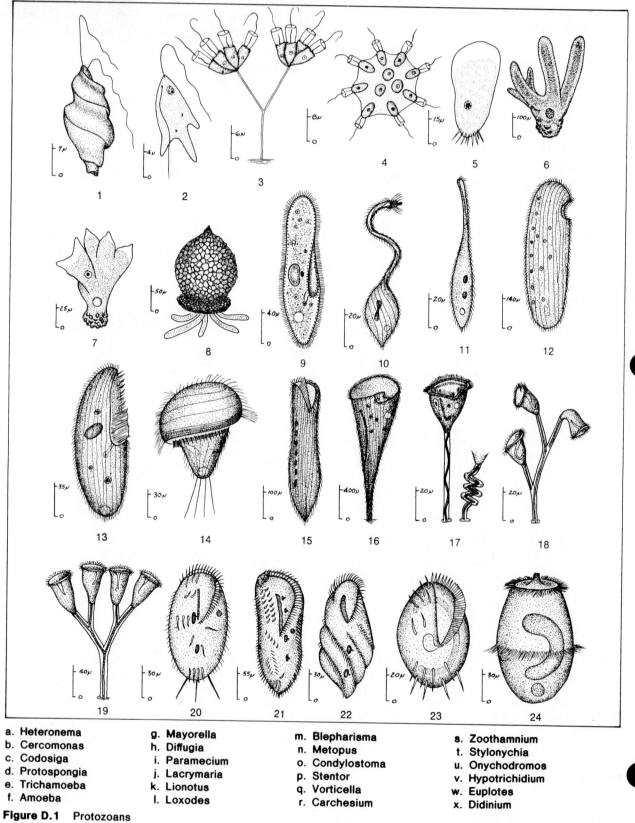

Figure D.1 Protozoans

a. Heteronema
b. Cercomonas
c. Codosiga
d. Protospongia
e. Trichamoeba
f. Amoeba

g. Mayorella
h. Diffugia
i. Paramecium
j. Lacrymaria
k. Lionotus
l. Loxodes

m. Blepharisma
n. Metopus
o. Condylostoma
p. Stentor
q. Vorticella
r. Carchesium

s. Zoothamnium
t. Stylonychia
u. Onychodromos
v. Hypotrichidium
w. Euplotes
x. Didinium

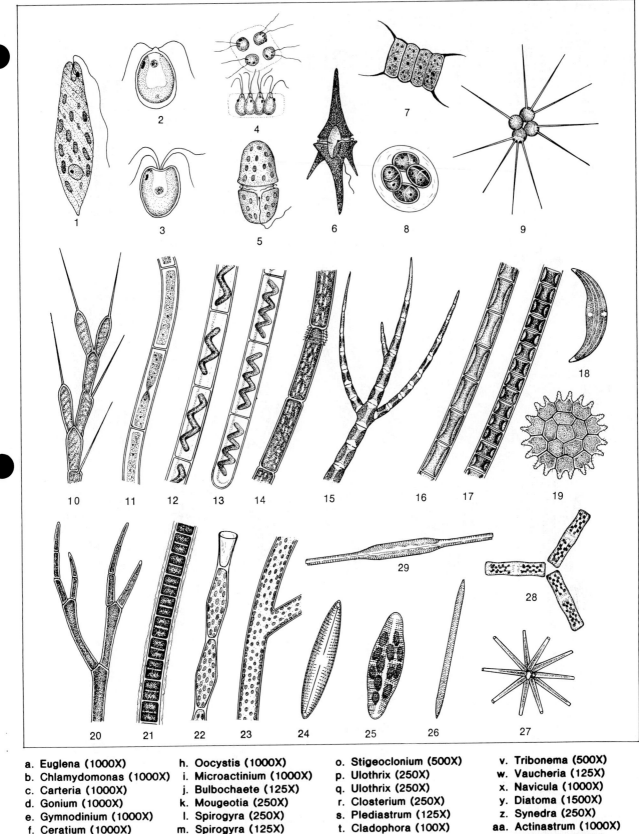

a. Euglena (1000X)
b. Chlamydomonas (1000X)
c. Carteria (1000X)
d. Gonium (1000X)
e. Gymnodinium (1000X)
f. Ceratium (1000X)
g. Scenedesmus (1000X)

h. Oocystis (1000X)
i. Microactinium (1000X)
j. Bulbochaete (125X)
k. Mougeotia (250X)
l. Spirogyra (250X)
m. Spirogyra (125X)
n. Oedogonium (500X)

o. Stigeoclonium (500X)
p. Ulothrix (250X)
q. Ulothrix (250X)
r. Closterium (250X)
s. Plediastrum (125X)
t. Cladophora (100X)
u. Zygnema (250X)

v. Tribonema (500X)
w. Vaucheria (125X)
x. Navicula (1000X)
y. Diatoma (1500X)
z. Synedra (250X)
aa. Actinastrum (1000X)
bb. Tabellaria (250X)
cc. Fragilaria (1000X)

Figure D.2 Eucaryotic Algae

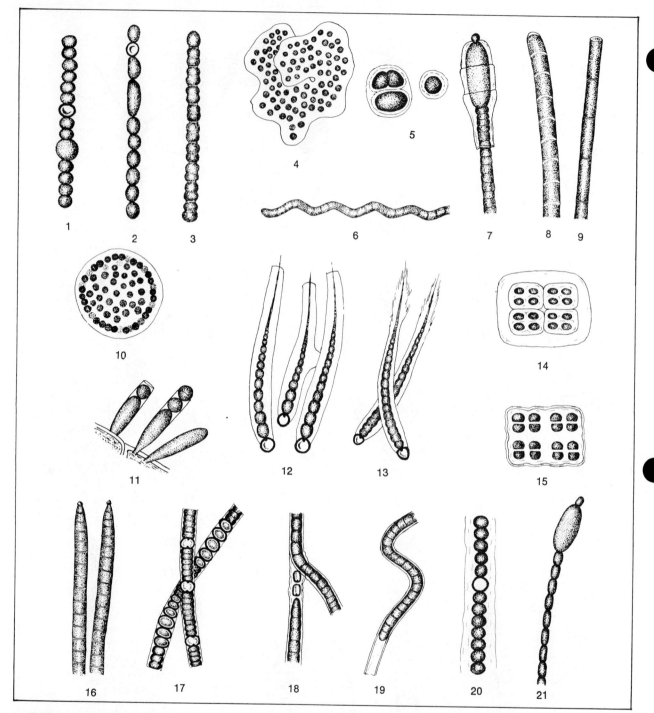

Figure D.3 Blue-Green Algae

a. Anabaena plactonica (250X)
b. Anabaena flos-aquae (500X)
c. Anabaena constricta (500X)
d. Anacystis cyanea (250X)
e. Anacystis dimidata (1000X)
f. Arthrospira jenneri (1000X)
g. Gleotricha natans (250X)
h. Oscillatoria chlorina (1000X)
i. Oscillatoria putrida (1000X)
j. Gomphosphaeria lacustris (500X)
k. Entophysalis lemaniae (1500X)

l. Rivularia dura (250X)
m. Calothrix parietina (500X)
n. Agmenellum quadriduplicatum, tenuissma type (1000X)
o. Agmenellum quadriduplicatum, glauca type (250X)
p. Phormidium autumnale (500X)
q. Nodularia spumigena (500X)
r. Tolypothrix tenuis (500X)
s. Lyngbya lagerheimii (1000X)
t. Nostoc carneum (500X)
u. Cylindrospermum stagnale (250X)

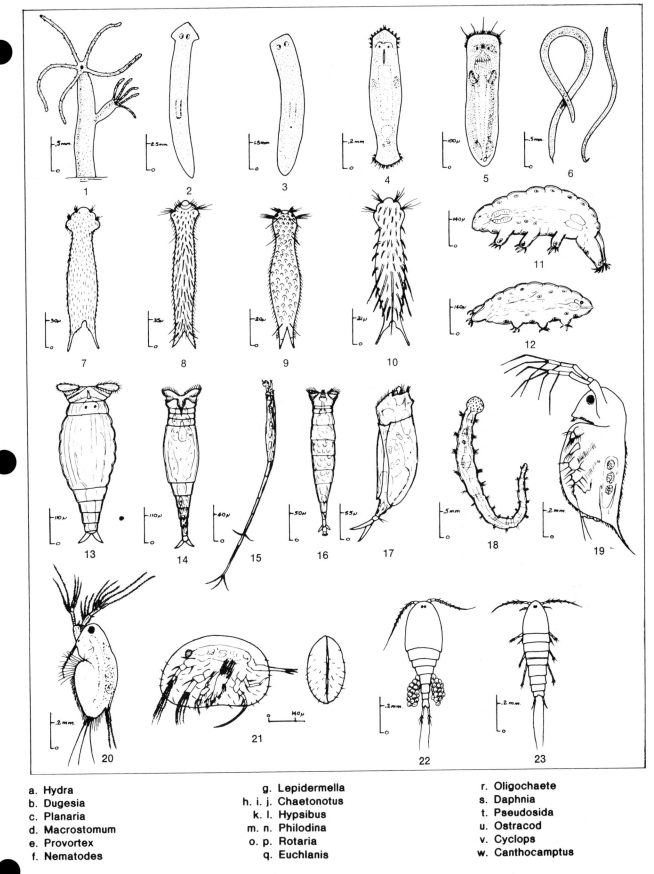

a. Hydra
b. Dugesia
c. Planaria
d. Macrostomum
e. Provortex
f. Nematodes

g. Lepidermella
h. i. j. Chaetonotus
k. l. Hypsibus
m. n. Philodina
o. p. Rotaria
q. Euchlanis

r. Oligochaete
s. Daphnia
t. Pseudosida
u. Ostracod
v. Cyclops
w. Canthocamptus

Figure D.4   Microscopic Invertebrates

Reading References

Anatomy and Physiology

Anthony, C. P., and Kolthoff, N. J. 1978. *Textbook of anatomy and physiology*. 10th ed. St. Louis: C. V. Mosby Co.

Crouch, J. E., and McClintock, J. R. 1976. *Human anatomy and physiology*. 2d ed. New York: John Wiley & Sons.

DeCoursey, R. M. 1974. *The human organism*. 4th ed. New York: McGraw-Hill.

Grollman, S. 1978. *The human body, its structure and physiology*. 4th ed. New York: Macmillan Co.

Hole, John W., Jr. 1978. *Human anatomy and physiology*. 2d ed. Dubuque, Ia.: Wm. C. Brown Co.

Jacob, S. W., and Francone, C. A. 1976. *Elements of anatomy and physiology*. Philadelphia: W. B. Saunders Co.

Langley, L. L., et al. 1980. *Dynamic anatomy and physiology*, 5th ed. New York: McGraw-Hill.

Tortora, G. J., and Anagnostakos, N. P. 1978. *Principles of anatomy and physiology*. 2d ed. San Francisco: Canfield Press.

Anatomy

Crouch, J. 1978. *Functional human anatomy*. 3d ed. Philadelphia: Lea & Febiger.

Goss, C. M. 1973. *Gray's anatomy of the human body*. 29th ed. Philadelphia: Lea & Febiger.

Greenblatt, G. 1981. *Cat musculature*. 2d ed. Chicago: Univ. of Chicago Press.

House, E. L., and Pansky, B. 1979. *A functional approach to neuroanatomy*. 3d ed. New York: McGraw-Hill.

Netter, F. 1957. *The Ciba collection of medical illustrations*. Vols. 1, 2, and 3, Summit. New Jersey: Ciba Pharmaceutical Products, Inc.

Physiology

Ganong, W. F. 1981. *Review of medical physiology*. 10th ed. Los Altos, Ca.: Lange Medical Pubs.

Guyton, A. C. 1977. *Basic human physiology*. 2d ed. Philadelphia: W.B. Saunders Co.

Guyton, A. C. 1981. *Textbook of medical physiology*. 6th ed. Philadelphia: W. B. Saunders Co.

Malasanos, L., Barkauskas, V., and Moss, M. 1981. *Health assessment,* 2d ed. St. Louis: C. V. Mosby Co.

Ruch, T., and Patton, H. D., eds. 1974. *Physiology and biophysics*. 20th ed. Philadelphia: W. B. Saunders Co.

Schottelius, B. A., and Schottelius, D. D. 1978. *Textbook of physiology,* 18th ed. St. Louis: C. V. Mosby Co.

Histology

Arey, L. B. 1974. *Human histology*. 4th ed. Philadelphia: W. B. Saunders Co.

Bevelander, G., and Ramaley, J. A. 1975. *Essentials of histology*. 8th ed. St. Louis: C. V. Mosby Co.

Bloom, W., and Fawcett, D. 1975. *A textbook of histology*. 10th ed. Philadelphia: W. B. Saunders Co.

DiFiore, M. S. H. 1974. *An atlas of human histology*. 5th ed. Philadelphia: Lea & Febiger.

Ham, A. W., and Cormack, D. H. 1979. *Histology*. 8th ed. New York: J. B. Lippincott Co.

Rhodin, J. A. G. 1975. *An atlas of histology*. Oxford Univ. Press.

Chemistry

Arnow, L. E. 1976. *Introduction to physiological and pathological chemistry*. 9th ed. St. Louis: C. V. Mosby Co.

Gilman, A. G., et al. 1980. *The pharmacological basis of therapeutics*. 6th ed. Woodridge: MacMillan Pubs.

Varley, H. 1967. *Practical clinical biochemistry*. 4th ed. London: Wm. Heineman Medical Books, Ltd.

Wilson, C., et al. 1977. *Textbook of organic medicinal and pharmaceutical chemistry*. 7th ed. New York: J. B. Lippincott Co.

Wynter, C. I. 1975. *Chemical analyses for medical technologists*. Springfield: C. C. Thomas Publishers.

Reading References

Instrumentation

Dewhurst, D. J. 1975. *An introduction to biomedical instrumentation.* 2d ed. Elmsford: Pergamon Press.

DuBorg, J. L. 1978. *Introduction to biomedical electronics.* New York: McGraw-Hill.

Geddes, L. A., and Baker, L. E. 1975. *Principles of applied biomedical instrumentation.* 2d ed. Wiley Interscience.

Geddes, L. A. 1972. *Electrodes and the measurement of bioelectric events.* New York: John Wiley & Sons.

Strong, P. 1971. *Biophysical measurements.* Beaverton: Tektronix.

Muscle Physiology

Andersen, K., et al. 1970. *Fundamentals of exercise testing.* Geneva: World Health Organization.

Basmajian, J. V. 1967. *Muscles alive.* 2d ed. Baltimore: The Williams & Wilkins Co.

Cohen, C. 1975. The protein switch of muscle contraction. *Scientific American.* Offprint no. 1329. San Francisco: W. H. Freeman & Co.

Hoyle, G. 1970. How is muscle turned on and off? *Scientific American.* Offprint no. 1175. San Francisco: W. H. Freeman & Co.

Karpovich, P. V., and Sinning, W. E. 1971. *Physiology of muscular activity.* Philadelphia: W. B. Saunders Co.

Lahoda, Frieder, Ross, Arno, and Issel. 1974. *EMG primer, a guide to practical electromyography and electroneurography.* Berlin: Springer-Verlag.

Murray, J. M., and Weber, A. 1974. The cooperative action of muscle proteins. *Scientific American.* Offprint No. 1290. San Francisco: W. H. Freeman & Co.

Eye and Ear

Bellows, J., ed. 1972. *Contemporary ophthalmology.* Huntington: Robert E. Krieger Co.

Corboy, J. M. 1979. *The retinoscopy book: a manual for beginners.* Thorofare: C. B. Slack, Inc.

Glorig, A., ed. 1977. *Audiometry: principles and practices.* Huntington: Robert E. Krieger Pub. Co.

Keeney, A. H. 1976. *Ocular examination: basics and technique.* 2d ed. St. Louis: C. V. Mosby Co.

Lloyd, L. L., and Kaplan, H. 1978. *Audiometric interpretation: a manual of basic audiometry.* Baltimore: University Park Press.

Miscellaneous

Barnes, T. C. 1968. *Synopsis of electroencephalography or guide to brain waves.* New York: Hafner Press.

Basmajian, J. V. 1979. *Principles and practice for clinicians: biofeedback.* Baltimore: The Williams & Wilkins Co.

Blowers, M. G., and Smith, R. J. 1977. *How to read an ECG.* Albany, N.Y.: Delmar Publishers.

Diggs, L. W., et al. *The morphology of human blood cells.* 4th ed. North Chicago: Ill.: Abbott Laboratories.

Kooi, K. A., et al. 1978. *Fundamentals of electroencephalography.* 2d ed. New York: Harper & Row.

Moscovitz, Toni. 1978. *Physiology of diving.* Lorain: Dayton Labs.

Reid, J. E., and Inbau, F. E. 1977. *Truth and deception, the polygraph technique.* Baltimore: Williams & Wilkins Co.

Short, C. 1974. *Clinical veterinary anesthesia: a guide for the practitioner.* St. Louis: C. V. Mosby Co.

Wintrobe, M. M., et al. 1981. *Clinical hematology.* 8th ed. Philadelphia: Lea & Febiger.

Index